THE STORY OF

BALL CLAY, CHINA CLAY
SOAPSTONE & CHINA STONE

EXTRAORDINARY EARTHS

FROM CORNWALL, DEVON & DORSET
1700 TO 1914

RONALD PERRY AND CHARLES THURLOW

The Authors

Charles Thurlow has lived in Cornwall for many years and is a graduate of the School of Mines at Camborne. For most of his career he was employed by English China Clays, where he worked on china clay and china stone production. After retiring he became a publisher and founded Cornish Hillside Publications. Over 20 books have been published, with an emphasis on mineral-related subjects.

Ronald Perry, BSc, MA, PhD, has lived in the West Country since 1965, building up the Faculty of Management, Business and Professional Studies at Cornwall College of Further and Higher Education. During this time he has published numerous socio-economic surveys of the region for the British Social and Economic Society, South West Arts, Exeter and Plymouth Universities and many Cornish institutions.

oOo

First published in 2010 by
Cornish Hillside Publications
St Austell, Cornwall PL25 4DW

ISBN 978 1 900147 491 paperback
ISBN 978 1 900147 507 cloth bound

Designed by SJS Design

Printed by Short Run Press, Exeter, Devon

Contents

Extraordinary Earths
Location Maps

Ball Clay Areas

Petrockstowe Basin

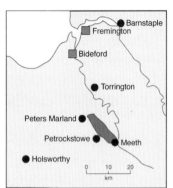

North Devon

Bovey Basin

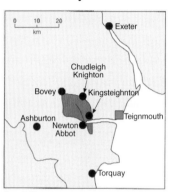

South Devon

Wareham Basin

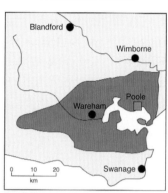

South Dorset

Dark shaded areas show the extent of ball clay. Ports are shown by squares.

China Clay, China Stone and Soapstone Areas

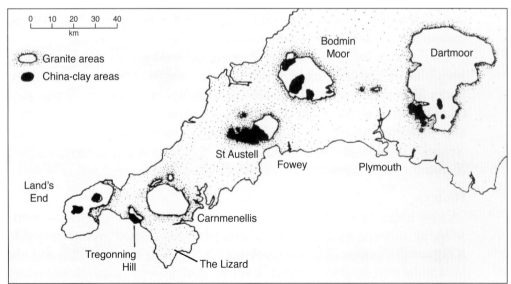

Map showing china-clay areas of Devon and Cornwall. China stone was largely found close to china clay in the St Austell area. Soapstone was found on the western side of the Lizard. The main china clay/stone ports were in St Austell Bay, Fowey and Plymouth.

Acknowledgements

The authors wish to express their gratitude to the large number of people, each an expert in his or her own field, who helped in writing this book. John Tonkin has read early drafts of several chapters, as well as a final draft of the whole book, and has made numerous valuable comments, giving us the benefit of his unequalled knowledge of china clay history. John Pike, Chairman of the Ball clay Heritage Society, read the chapters related to ball clay and made useful corrections and additions, especially on Devon ball clays. Derek Giles, Chairman of the China Clay History Society, read an earlier draft and made many useful suggestions. Dr Alan Bennett, the railway historian, led us around the ball clay area of Dorset. Others who have improved our understanding of particular sections include Professor Colin Bristow on geology, Myles Varcoe on family history, Colin Hanley on ceramics, Jim Lewis and Penny Smith on aspects of metal mining, John Probert on Methodism, Dr Helen Doe on shipping, Lynne Mayers on female workers, Dr Bernard Deacon, Dr Garry Tregidga and Tony Noonan on industrial relations, and Dr Alan Kent on clay country literature.

Thanks are also due to Dr Brian Strathen of the China Clay Museum, Wheal Martyn; Terry Knight, Joanne Laing and Kim Cooper of the Cornwall Centre, Redruth; Angela Broome and Anne Knight of the Courtney Library of the Royal Institution of Cornwall; Christine North and Colin Edwards of Cornwall Record Office and Jacqui Halewood and staff of the Dorset History Centre.

Last, but certainly not least, we wish to record our heartfelt thanks to Ann Perry for typing and retyping innumerable drafts, for correcting errors and improving grammar and for sustaining the authors over more than five years with her delicious cakes and biscuits.

Introduction

White Alps, Black Holes

S ome years ago the authors took part in an Exeter University seminar entitled 'White alps, black holes'[1] . It contrasted the importance of the china clay and china stone industry, symbolised by the dazzling white 'sky tips' of debris that once dominated the landscape, with what were identified as considerable gaps in research into its history. While the shelves of libraries and book shops groaned under the weight of volumes about metal mining, you would be lucky to find, tucked away in a corner, something about china clay and stone, but little if anything on ball clay. The unique historical and cultural significance of West Country metal mining has now been recognised by the designation of Cornwall and West Devon as a world industrial heritage site, but we believe that the story of non-metalliferous extraction in Dorset, Devon and Cornwall is also well worth the telling.

Changes in fashion, the Industrial Revolution and the clay industry

This book is the first to encompass the history of the entire china clay, china stone, soapstone and ball clay industries of the West Country from the seventeenth century to the 1914 War. It contains much new research by the authors and other historians, and offers a wider perspective than previous works on the subject. We cover a period when the growth of the industry was associated with style and taste, and also closely integrated with some of the propellant forces that drove the British Industrial Revolution. The first major commercial applications of clay can be seen in any portrait of the beau monde: gentlemen in wigs, powdered with clay, smoking clay pipes, accompanied by ladies sipping tea from elegant chinaware, and sweetening it with white sugar refined by 'sugar bakers' clay'. Cotton cloth manufacturers, other important customers for china clay, led the Industrial Revolution at that time, their business expanding faster even than that of the iron producers.[2] The next, and eventually the largest, increase in demand for china clay came from paper makers, fuelled by mass literacy in the second half of the nineteenth century. Today, few people realise that, at the time, china clay accounted for about half the weight of some cotton cloth and paper.

Product and process innovation

An inner dynamic of the British Industrial Revolution was the expansion of markets for new products at home and abroad, and our study is as much concerned with the demand as with the supply of clay. While dealing, in more detail than in most books, with innovation in extraction and refining processes, it differs from previous works in attaching equal weight to product innovation — the identification of new applications. It treats the clay industry as a multi-product, not a single-product sector. It is the first work to outline the way that clay became involved in the technical evolution of ceramics, cotton manufacture, paper-making and many other industries. Within this framework we recognise a dramatic change in the technical role of the clay merchants. In around 1800 it had been the customers, mainly potters, who had raised industrial processes from guesswork and rule of thumb to scientific measurement, and who told the clay producers what their requirements were in terms of quality. A century later the clay merchants were at least as knowledgeable as their clients on technical matters involving the use of clay in a whole host of trades, from the production of ultramarine and alum to the manufacture of linoleum and miners' safety fuse.

Marketing and distribution
Since West Country clay and stone were materials of low value for their bulk, and the main users for them were located far away, a significant part of the final cost of clay to the customer consisted of handling, storage and transport charges. Our book places more emphasis than other works on improvements in marketing and distribution, which made possible the enormous expansion in sales during the period under review. It celebrates the flair of the clay merchants who, by the Edwardian era, dominated world trade with their products, so that even such clay producing countries as Germany and the USA were net importers of West Country clay.

Finally, this book examines some under-researched or unexplored issues: the importance of internally generated finance compared with external investment; the reason for significant changes in ownership and control; the social and administrative contribution of the clay merchants to community life; the health and safety of workers; the existence of parallel industrial disputes in Cornish china clay and Devon ball clay with completely different outcomes; and misleading interpretations of the clay industry by well-known contemporary writers who lived and worked in the district.

The scope of this book
In a work that covers such a wide range of topics, a brief summary of the contents may be useful.

In Chapters One and Two we trace the development, from the seventeenth century until the mid-1800s, of the 'extraordinary earths' that form the subject of the book. We deal first with potters' and brickmakers' clays, followed by pipe clay, sugar bakers' clay, ball clay, soap rock, china clay and china stone, as well as a manufactured material called Coade stone.

Chapter Three follows attempts to replicate Chinese porcelain across Europe, including those of the German Johann Böttger of Dresden. In Chapter Four, drawing upon recent researches, we describe the work of the Plymouth apothecary William Cookworthy and evaluate his pioneering contribution to the manufacture of English porcelain and the exploitation of West Devon and Cornish china clay and china stone.

The fifth and sixth chapters describe the discovery of important new uses for china clay in the production of cotton cloth and the making of paper. Other minor, less well-known applications discussed include cosmetics, wig powder, the manufacture of alum and ultramarine, the production of 'graphite lead' for pencils, the refining of sugar and the clandestine adulteration of flour for bread making.

Chapter Seven surveys the advances in land and sea transport which enabled the china-clay merchants to develop new markets, although ball clay exports were prohibited until the mid-1800s. Pit to port improvements included the establishment of turnpike trusts, the construction of canals and widening of rivers and the introduction of horse-drawn trams and steam-powered locomotives. We also discuss the building of new ports and the enhancement of existing harbours in Cornwall, Devon and Dorset and the hazards of coastal and ocean-going shipping.

In Chapters Eight and Nine we turn to changes in the ownership and control of the clay industry. The role, in North and South Devon and Dorset, of local families who made most of the running in ball clay production is here contrasted with the part played by outsiders in the early growth of the china clay and stone trade in Cornwall and West Devon. Within this context, we explore some hitherto neglected issues: why outsiders withdrew from china clay and stone production in Cornwall, but dominated granite and slate quarrying; why powerful Cornish metal mining adventurers, with strong interests in South Wales collieries, did not exploit their position to combine clay extraction and pottery manufacture; and why, for the best part of a century, no significant ceramic producers were sited in the South West.

Chapter Ten focuses on the wages and working conditions of male, female and child labourers in the clay industry, compared with workers in other relevant trades. It corrects the surprising exaggeration of the numbers involved in Cornish china clay production made by many previous writers on the subject. The chapter also surveys worker/employer relations in a turbulent period of rural history.

The second part of our work covers the period from the mid-1800s to the eve of the 1914 War — the golden age of the independent local clay merchants — and Chapter Eleven describes their origins and development. Chapter Twelve examines further industrial uses for clay: high class Parian Ware and statuary; Goss china seaside souvenirs and collectables; and a vast expansion in the output of sanitary ware, drainpipes, garden ornaments and terracotta decorations. The continued use of clay in cheap cotton cloth for the British Empire is mentioned, as well as a massive growth in the demand for high-class paper for illustrated books and magazines, and the introduction of wallpaper for middle-class homes. Further applications in alum, ultramarine and paint are discussed, as well as some unexplored links between linoleum manufacture, the clay industry and artists' colonies in the West Country.

In Chapter Thirteen we outline changes in the geographical pattern of clay sales that resulted from these new uses, including the lifting of the Port of Exeter interference with the export of ball clay and mounting competition from foreign clay producers. The continued advances in pit to port transport described include the construction of railways and pipelines and the enhancement of ports. Chapter Fourteen refutes the assertion by some previous writers that the Victorian clay merchants were technologically inert, and illustrates significant changes in the refining and drying of china clay.

Chapter Fifteen focuses on some of the adverse local impacts of the massive expansion of clay working: accident rates and health and safety hazards; river pollution, including the effects on fishing and shellfish breeding; and transport congestion, particularly in the St Austell area. The chapter ends with a discussion of the conflicting interpretations of early environmentalists and romanticisers of the clay landscape.

Chapters Sixteen and Seventeen return to the subject of industrial relations, and modify a conventional interpretation of the South West as a region of relative industrial harmony. The influence of successive waves of strike fever, spreading from the east, is described, including the 1875 and 1876 china clay disputes and the reasons why the first was a victory for the workers and the second a defeat.

Chapter Seventeen covers one of the most widely debated events in china clay history, the 1913 strike, and contrasts it with a parallel but little known strike of ball clay workers in South Devon. Explanations of why the strikers appeared to lose in Cornwall and won in Devon are offered, and opposing interpretations of the strike are evaluated – an unpremeditated action by local workers discontented with their lot, or part of an international conspiracy to overthrow the entire capitalist system?

Chapter Eighteen challenges the widely expressed view that clay producers did not engage in pottery manufacture because it was cheaper to send clay to places near the coalfields than to bring the coal to the clay-bearing regions. It illustrates the large variations in the amount of coal used to fire different kinds of earthenware, chinaware and porcelain. These variations, it suggests, help to explain why clay areas in North and South Devon and Dorset (and even a remote part of Ireland) witnessed an upsurge in ceramic production, while Cornwall did not.

In the final two chapters we concentrate upon the achievements of the clay merchants.

Chapter Nineteen describes the way in which, in an age of monopolies and mergers, they maintained their independence despite the increasing efforts of industrial magnates, both British and foreign, to take over part or all of their industry.

Chapter Twenty highlights the misleading and uncomplimentary opinions offered by well-known authors who lived in the clay district. It contrasts these with the accomplishments of the clay merchants, both as entrepreneurs providing employment and as contributors to local community life.

The following abbreviations are used in the end notes to each chapter.

ABK	An Baner Kernewek	MM	Mining Magazine
BBS	British Brick Society	MQE	Mine and Quarry Engineering
BCW	British Clay Worker	NE	Newquay Express
BHA	British History Association	NWLHS	North West Labour History Society
CCHSN	China Clay History Society Newsletter	PDNHAS	Proceedings of the Dorset Natural History and
CCTR	China Clay Trade Review		Archeological Society
CG	Cornish Guardian	PDRO	Plymouth and Devon Record Office
CRO	Cornwall Record Office	PWS	Proceedings of the Wedgwood Society
CT	Cornish Times	QJS	Quarterly Journal of Science
CTel	Cornish Telegraph	RCG	Royal Cornwall Gazette
DCNQ	Devon and Cornwall Notes and Queries	RCPS	Report of the Royal Cornwall Polytechnic
DH	Devon Historian		Society
DeRO	Devon Record Office	SAS	St Austell Star
EHR	Economic History Review	SDNQ	Somerset and Devon Notes and Queries
HS	History Studies	TDA	Transactions of the Devonshire Association
JBCS	Journal of the British Ceramics Society	TDS	Transactions of the Devon Society
JCALH	Journal of the Cornwall Association of	TECC	Transactions of the English Ceramic Circle
	Local Historians	TIMM	Transactions of the Institution of Mining and
JCH	Journal of Ceramic History		Metallurgy
JPMBT	Journal of the Paper Maker and British Trade	TPIBG	Transactions and Papers of the Institute of
JPMMC	Journal of the Plymouth Mineral and		British Geographers
	Mining Club	TRGSC	Transactions of the Royal Geological Society
JRCS	Journal of the Royal Ceramic Society		of Cornwall
JRIC	Journal of the Royal Institution of Cornwall	VCHD	Victoria County History of Dorset
JTS	Journal of the Trevithick Society	VCHL	Victoria County History of Lancashire
LHR	Labour History Review	WB	West Briton
MDA	Mid Devon Advertiser	WDM	Western Daily Mercury
MJ	Mining Journal	WMN	Western Morning News

Chapter One
Potters Clay, Ball Clay, Coade Stone and Soapy Rock

An early and greatly respected treatise on mining is *De Re Metallica*, written by Georgius Agricola from Saxony and published in 1556. Agricola spent much of his life as a doctor of medicine in Joachimsthal, then one of the most prolific metal mining regions of Central Europe and his book is outstanding because of its comprehensive coverage and wealth of woodcut illustrations of the mining methods and machinery used. In 1912 it was translated into English from the Germanic Latin used by Agricola by no less a person than Herbert Hoover, the mining engineer who later became President of the United States, and his wife Lou. The book's relevance to our own study stems from Agricola's observation that miners looking for metallic ores did not 'neglect the digging of extraordinary earths' wherever they were found.[1] He was referring to the working of non-metallic minerals such as clays and ochres, then regarded by other writers as of much less significance than metals such as gold, silver, iron, copper and tin.

Georgius Agricola, 1490–1555.

Over the years it was realised that whilst metallic minerals owed their value to the percentage of metal they contained, the worth of non-metallics depended on their properties such as whiteness, plasticity and fine particle size. Many kinds of men have joined in the search for metallic and non-metallic minerals: profit-motivated prospectors who tried to keep their discoveries a closely guarded secret and gentleman scholars who made public their findings in learned journals. Prominent among the latter were clergymen who had the education and resources to indulge in a fashionable hobby as amateur geologists and chemical analysts, such as the Cornish-born Revd William Gregor, who first identified the ore of the metal titanium in 1791 on the Lizard in Cornwall.[2]

Although originally known by a variety of names, the 'extraordinary earths' alluded to by Agricola include those now generally referred to in the UK as potters' and brickmakers' clays, ball clay, soapstone, china stone and china clay. In this chapter we concentrate upon three of these materials from the West Country which were largely developed before china clay and stone. We also deal with one of the earliest and most significant uses of ball clay in the manufacture of the ceramic product known as Coade stone.[3]

Potters' clays

The earliest form of pottery used local clay that varied in composition and colour, crudely fired to make what is called 'common pottery'. The later products from potteries can be classified depending upon the ingredients used and the different firing processes. Earthenware, fired at about 1,000°C, is granular in texture, opaque and porous. Somewhat similar is stoneware, which is fired to a higher temperature, resulting in a compact and generally non-porous finish. The most highly regarded product is porcelain, which is translucent, white and almost glass-like in texture. Three main kinds of porcellaneous ware can be recognised: soft paste, a rather brittle product made initially with china clay and ground glass and originally manufactured outside the United Kingdom; hard paste, made of china clay and china stone, requiring a higher firing temperature, first produced in China, followed by Europe and later by Cookworthy in England; and lastly bone china, discovered by Josiah Spode around 1800, using 50 per cent burnt and ground bone and 50 per cent china stone and china clay. Bone china is the only truly English ceramic invention.

Crudely decorated common pottery was made in Cornwall, Devon and Dorset before much mining took place, and relics of this early industry, chiefly sun dried, have been found, including funeral urns. Most English clays had to be cleaned before they could be used to make pottery, and all traces of coarse sand and pebbles were removed to leave only red or off-white clay. In earlier times the clay was softened by soaking in water, after which stones and gravel were picked out by hand. By the seventeenth century, small pits were dug in which heavier particles sank to the bottom of the water in the pit. The fluid mass was then poured through a sieve into a large, shallow area called a 'sun kiln' to a depth of three to four inches and dried in the sun. Further layers of slurry were added until the clay was 12–18 inches thick, after which it was cut out in blocks and stacked for use.[4] A similar method was later employed in the china-clay industry, which will be discussed in the next chapter.

West Country potteries, situated near beds of local clay, used timber from neighbouring woods for fuel. Barnstaple claimed an unbroken ceramic tradition from Roman times and was known for its great range of harvest jugs, while Bideford potters have been identified from the fourteenth century onwards, selling church tiles and clay ovens, as well as oil lamps and jars for pickling pilchards.[5] A Devonshire lease of 1762, for example, gives a 'licence to deal with earth, clay and sand' to make 'bricks, tiles and earthenware vessels'.[6] In surveys of seventeenth- and eighteenth-century Cornwall, however, Carew and the geologist and antiquarian Dr William Borlase made no mention of local potters, merely remarking that the area lacked good pottery clays.[7]

Some sedimentary clay for common pottery came from estuaries along the south coast of Cornwall, and it was here that nearly all the potters set up. According to Bernard Leach, at least 16 potteries clustered in small towns and villages from St Germans to Lostwithiel, Truro, Constantine, Mawgan in Meneage and Madron, near Penzance.[8] Their main products were basic earthenware kitchen utensils, flowerpots and ridge tiles for roofs but as their standards improved they found that the local Cornish clays were not good enough and they imported some ball clay, probably from Dorset, while a couple of potteries on the north coast near St Columb and at Boscastle bought in ball clays from North Devon.[9]

Brick making

Writing in 1758, Borlase commented that in Cornwall 'there are strata of clay for making bricks in so many places that there is hardly a parish, seldom a large tenement, without it'.[10] Here he was referring to bricks that, like local common pottery, required a mixture of clay and sand, although Borlase does not mention sand, which may have occurred with the clay when it was dug up, or else was readily available locally. Bricks came late to Cornwall because of its abundance of slate, granite and other building stone, and relatively few brick-built ancestral mansions or town houses of any distinction predate the later 1700s.

With the upsurge of metal mining in Cornwall and West Devon and an increased demand for better housing, bricks were needed in great numbers. Lighter than stone, they were used for the upper stages of mine chimneys, as well as for the window arches of engine houses, the floors of stables and firebricks in furnaces.[11] The first brickworks associated with Cornish china-clay operations did not appear until later in the nineteenth century and will be discussed in Part Two, but in the 1840s, at Lee Moor in Devon, the residue of mica and fine sand from china-clay works was used to make bricks. Firebricks were exported for use in metal smelters, as well as sold locally to builders in the rapidly expanding Plymouth area.

Competition from Staffordshire

After the Napoleonic Wars, coastal transport became much less hazardous and costly, so better and cheaper chinaware from potteries in Staffordshire and elsewhere began to permeate local markets. As we shall see in Chapter Two, this did not mean the end for local pottery production in Plymouth or North or South

Devon, but it seems to have reduced the number of Cornish potters. This is suggested by the growth in numbers of 'earthenware men' or 'hucksters', enumerated in Cornish and Plymouth trade directories, who dealt in imported china. In 1812 eight were listed in Plymouth and ten years later their numbers had grown to 13, together with Falmouth (six) and St Austell (three). By 1830 another six were operating at Redruth, four in Truro, two at Helston and one each at Penryn and Stratton.[12] Itinerant salesmen and women were also common through the nineteenth century. In one of Thomas Hardy's novels, set near Boscastle, a character buys 'a new basin and jug off a travelling crockery-woman who came to our door'.[13]

In the mid-eighteenth century a 'very fine clay' had been located by John Trehawke of Liskeard at a site in St Stephen-by-Saltash and also at St Neot, near Bodmin, which was sent to a pottery at Calstock.[14] This was probably the pottery seen by the Revd Doctor Pococke on a tour in 1750, which he described as making coarse earthenware but also as experimenting 'with things of a finer sort with a yellow clay from St Stephen near Saltash, from Hollowmore Bay, near St Germans and also from Kelly'.[15] Trehawke showed samples of these clays to Richard Horn, a Staffordshire potter, and to Borlase. Horn was rumoured to be setting up a pottery in Penryn around this time, but using 'soapy rock' (discussed later) from the Lizard. Neither he nor Borlase, who had identified deposits of white clay and stone in his *Natural History of Cornwall*, took any further action,[16] perhaps sharing a preference of that time for soapy rock as an ingredient in porcelain.

Tobacco Pipe Clay, Sugar Bakers' Clay and Ball Clay
Before these discoveries were made, however, a very fine-grained, highly plastic

sedimentary clay was being dug further east in the Petrockstowe Basin of North Devon, the Bovey Basin in South Devon and the Wareham Basin in Dorset.[17] The origin of these clays is not fully understood, but they appear to have resulted from the erosion, many millions of years ago, under tropical conditions, of earlier rocks some distance away. This was fol-

Sir Walter Raleigh, an early smoker, with a pipe made of fired ball clay.
(COURTESY THE BALL CLAY HERITAGE SOCIETY)

lowed by the transportation of fine and sandy weathered material which was deposited in lakes. Also found in these basins are layers of lignite, a low-grade brown coal derived from vegetation which flourished in tropical climates. Some of this lignite, as we shall see later, was tried as a fuel to fire potters' kilns. The extreme fineness of these clays makes them more plastic than china clay, but they are less white because they are usually tinted by decomposed vegetable matter in various shades – brown, blue, black and light grey – due to the presence of small amounts of iron and other oxides. This colouring disappears when the clay is fired to give a white or whitish shade

In Dorset, and probably in Devon as well, clay was utilised in Roman times to make common pottery, but it was the introduction of tobacco to England in the sixteenth century by Sir Walter Raleigh and others that led to the use of this material in making 'tobacco clay pipes', or simply 'clays', as they were called at the time. Later, new fashions brought fresh applications: 'sugar-bakers' clay' to refine muscovado or brown sugar into white sugar, fine chinaware from which to drink tea and powder for wigs. The way in which clay was used for these purposes will be discussed in Chapter Six.

In 1619 the first Charter of Tobacco Pipe Manufacturers was drawn up, one of the signatories being Swithin Bonham of Poole, who was granted a lease in that year by Sir Thomas Brudenwell to dig clay from his estate at Canford, near Poole. Another merchant, Thomas Brown, one of a family of tanners in Wareham, sold clay from land that his antecedent, also called Thomas Brown, had purchased at East Creech, near Wareham, in 1578.[18] Other merchants, including Thomas Cornwell of Poole, were sending clay to Plymouth, Lyme Regis and other South Coast ports, as well as to London and destinations on the East Coast, including Colchester, Yarmouth, Norwich, Boston, Hull and Newcastle.

From North Devon, the Bideford potters were using local clays to make an 'engobe' of equal parts of ball clay and china clay as a coating layer for jugs, bowls and other tableware. The Greening family was despatching clays from the Petrockstowe area in the 1650s. Recorded in 1691 were shipments of clay from Bideford to Padstow in Cornwall and to Bristol, Liverpool, Chester, Gloucester, Bridgewater and Swansea. In 1720 this trade amounted to 925 tons, and by the 1730s the annual figure was between 1,000 and 2,000 tons, but then declined, whereas shipments from South Devon increased rapidly.[19]

The first recorded shipment from South Devon dated from 1691, when small quantities of white-firing 'Chester clay', as it was called in Staffordshire, had been shipped to these potteries via Chester. In about 1700 John Osland sent '20 tons of Tobacco Pipe Clay' from Teignmouth to London. It was this substance, mixed with broken glass and sand from Alum Bay in the Isle of Wight, that was used, half a century later, by London potters in Bow, Limehouse and Chelsea to make their own version of soft-paste porcelain. However, the first successful clay producer in South Devon appears to have been a Dorset man, William Crawford of Poole, who leased land in the Bovey Basin and sent around 500 tons a year to London from the late 1720s. By the mid-1700s most of this clay was being applied as decoration to earthenware based on local discoloured clays. Clay exports through Exeter rose from only 400 tons in 1750 to nearly 2,500 by 1776.

Methods of working ball clay

Extraction processes varied over the years. Initially, outcrops of suitable clay were dug from shallow trenches or hillsides using various kinds of hand tool. An early description was given by Polwhele in 1793. These trenches grew into small open pits, often with neatly terraced, sloped edges, dug by men known as 'clay cutters'.[20] Hand-operated pumps came into use to remove water, and these were succeeded in larger pits by Cornish plunger pumps. Rain and ground water could cause the sides of large open pits to subside, and so in South Devon criss-cross timbering was used to support pits that were from 18 to 24 feet square. Alongside these pits an unusual type of crane called a 'crab crane' was erected to hoist clay out of the pit in a wooden bucket. The crab was a pivoted 'gallows' crane with two legs called tie backs, and was powered by hand winches or horses. These square pits produced only a few hundred tons of the most desirable potters' clay before the gear was moved to a new site, but the square pit system, as well as the large open pit system, continued in use until well into the twentieth century. However, in many locations the square pit system gradually developed into underground or 'shaft mining', described later in Chapter Eight.

White Ware and Queen's Ware

As public taste for 'white ware' grew, the potters demanded increasing quantities of white-firing ball clay, and the quantity shipped up the River Weaver near Liverpool to the Staffordshire Potteries rose from 200 tons in 1740 to 1,800 tons two decades later.[21] In the early 1760s Josiah Wedgwood produced his 'cream ware', a lead-glazed amalgam of ball clay and flint which was later marketed as 'Queen's ware' after Queen Charlotte was presented with a breakfast set in 1765. Wedgwood had already styled himself 'potter to the Queen' in 1763 through usage rather than any formal agreement, and no clear patent was ever awarded. This ware became immensely popular, and other potters in Staffordshire, Liverpool, Leeds, Castleford, Derby and elsewhere rushed to copy it.

Sets of Queen's ware made of Purbeck clay from the Calcraft Estate in Dorset are mentioned in letters of 1773 and 1775, and a map of 1772 of the Corfe Castle area shows, in a place called Threshers Heath, 'clay pitts... wherein is dug a very fine clay which makes what is named the Queen's Weare'.[22] The substitute for the more expensive porcelain for many middle-class families, Queen's Ware was not, however, as translucent as true porcelain, nor as white, and so Wedgwood and his rivals developed 'pearl ware', involving a ball clay or 'china glaze', which appeared whiter than cream ware.[23] This was achieved by incorporating a very small amount of cobalt blue pigment. Later on, china-clay producers were to 'whiten' china clay in a similar manner using various other 'blueing' agents.

Shipments of clay from Teignmouth rose from 3,600 to 15,000 tons a year between 1774 and the end of the century, and in 1801 the local traders petitioned for the status of an independent port, instead of being controlled by Exeter, because of the 'great increase in Coals and Culm brought back by the ships carrying clay to the several Ports of Liverpool, Newcastle, Sunderland, Wales etc'.[24] Shipments from Dorset totalled 10,000 tons in 1796,[25] and it was about this time that Josiah Wedgwood, sometimes with other Staffordshire potters, signed contracts with a number of Dorset producers for regular consignments of clay. They included Thomas Hyde of Poole, as well as the Brown family and the Pike brothers from Wareham.[26] By now many clay pits were being worked on what were called 'the Wastes of Purbeck', from the Arne Peninsula in the north east to Slepe, Stoborough and Green Island, and the 'celebrated Norden clay pits', as they were called at the time, near Corfe Castle.[27]

The naming of ball clay
In the next chapter the year 1783 is suggested as the date of the first written mention of the term 'china clay', but although what is now known as ball clay was in common use well over a century before that time, the leases and agreements located in Devon and Dorset do not contain that term. It is only by noting the district in which clay is found, and the uses to which it is put, that we can determine whether or not it is ball clay. For instance, a grant in 1836 of 'the liberty to dig for clay' near Polsue in Devon gives the right to erect chimneys and kilns for making bricks and tiles,[28] while an agreement of 1843 mentioning 'clay pits' near Fremington in North Devon can be assumed to relate to potters' or brick clay. The same could be said of sales of 'brick clay' pits in 1891 and 1894 near Pinhoe in Devon.[29]

Among the men who worked the clays, different grades were recognised colloquially by a variety of names, usually referring to the colour when dug or the sources – for example 'Purbeck Clay' – or the uses made of the clay. In 1907 experts analysing samples from the ball-clay areas of Bovey Tracey in South Devon and Torrington in North Devon identified Fire Clay, Pipe Clay, Paving Clay and Brick Clay. Whereas these experts employed the term 'china clay' for materials from Cornwall and West Devon, they never utilised the expression 'ball clay'.[30] Kelly's Directory of 1910 listed over 30 clay merchants in Cornwall and Dartmoor as 'china clay' merchants, but the five ball-clay producers of North and South

Devon simply described themselves as 'clay merchants'.

A common explanation of the word 'ball' is that it was cut out from the clay beds in cubes, which rounded into balls after being handled. As early as 1720, entries in the Port Books sometimes recorded shipments by the number of 'balls' as well as by the total weight, a ball averaging near to 40 pounds.[31] Another explanation is that a special spade, called a 'tubal' in Dorset, was used to dig out the clay and this led to the clay being called 'tubal clay', shortened to 'bal clay'. However, in Devon this implement was known as a 'tubil' or 'two bill', a farming tool, which makes this derivation doubtful.[32] On average, 70 clay balls weighed 22.5 cwt, which became the clay merchants' 'ton', from which piece rates for the workers and dues for the claylords were calculated.

Soapy rock: a forgotten Cornish mineral

The extraction of china clay and china stone in Cornwall has so overshadowed the story of the earlier exploitation of Cornish 'soapy rock' in the Lizard Peninsula that it has largely escaped the notice of today's local historians. Indeed, one of them went so far as to say that: 'No minerals of any commercial value have ever been found in the Lizard.'[33] Yet in the mid-1700s what was variously called 'soapy rock' or 'soapstone' or 'steatite' was seen for a time as an elusive stone used for centuries by the Chinese to make porcelain. Mineralogists and antiquarians flocked to the Lizard to inspect it, and potters from Liverpool, Bristol, Worcester, Staffordshire and London vied with one another to quarry it.[34]

The material is a magnesium silicate mineral formed as a result of the hydration of serpentine, found and worked on the Lizard Peninsula of Cornwall. It can be translucent or opaque and has a soft, waxy feel with a pearly lustre or a dull appearance. It is either milk white or veined and marbled with various tints of grey, yellow, green, blue, brown or red. Softer varieties may be called 'saponite', another term derived from a word for soap.

The discovery of soapy rock

An early reference to the material dates from 1667, when the celebrated poet, naturalist and Professor of Astronomy Dr Walter Pope showed fellow members of the Royal Society a specimen of 'the rock called the soapy rock in Cornwall'. By 1681 a 'Catalogue of Rarities' published by the Society included, under the heading of 'Stones Irregular', two items called 'Soap Stone' and 'Steatites'. These specimens were described as 'Part white, red, purple and green mixed together, as in Castile-Soap; and seeming, like hard Suet, greasie *(sic)* to the touch'. The term 'steatite' comes from the Greek for suet or tallow, and according to Borlase the Cornish word 'Soa' also meant suet or tallow.[35]

These samples had been sent to the Royal Society by the physician Dr Richard

Enlarged advertisement from the Sherbourne Mercury, *15 October 1770 for a soapstone lease at Gew Graze.*

Lower from St Tudy, near Bodmin, and later Cornish scholars who took an interest included the Revd William Robinson of Landewednack on the Lizard and Borlase. From 1735 onwards Borlase sent specimens of a large variety of Cornish rocks, or 'fossils' as they were called, to collectors, including members of the aristocracy and royalty. Most of the samples were of metals such as copper and mundic (iron sulphide).[36] The Catalogue of Rarities had earlier described a quality of soapy rock that was of key importance to potters: although it could be crushed by hand it became harder and whiter when fired. Yet for the best part of a century it remained of little interest until the race began in the mid-1700s to produce the first English porcelain.

The arrival of the potters
Just as we shall see with china stone and china clay, much of the early running in the extraction of soapstone was made by outsiders. The observations of a French missionary, Père D'Entrecolles, about the workings of Chinese potters had been translated into English, and ceramic producers were experimenting with different materials to make 'true' porcelain. Their quest soon led them to the Lizard Peninsula. First in the field in 1749, three years after William Cookworthy famously discovered china clay at Tregonning Hill, a few miles west of Helston, was Benjamin Lund from Bristol. Borlase seemed to have obtained his samples from

Kynance Cove, a mile or so north-west of Lizard Point, but Lund took out a lease on a recently discovered source at Gew Graze (today called Soapy Cove), a mile further north. A year later Dr Richard Pococke, later Bishop of Ossory and Meath in Ireland, was taken to the Lizard, where 'a rivlet *(sic)* runs over a vein of soapy rock into the sea... it is about four feet wide, most of it mixed with red'.[37]

In 1751 Nicholas Crisp of the Vauxhall Pottery in London took out two leases: one south of Kynance Cove from Lord Falmouth and one north of Gew Graze from the Hunt family of Lanhydrock. By the end of 1752, approaching 120 tons of soapstone had been quarried from various sites, and the record of a shipment of the material to the Port of Liverpool showed that its use had spread. In 1755 Richard Chaffers and his partner Phillip Christian of Liverpool entered into an agreement with Robert Podmore of Worcester, who passed on the secret of making porcelain from soapstone, and Chaffers leased two new sources south of Mullion from Sir Richard Vyvyan. Some five years later Chaffers won Wedgwood's admiration. 'Mr Chaffers,' Wedgwood asserted, 'beats us in all his colours and can produce for two guineas what I cannot for five.'

Problems in quarrying and firing

At this time the future of soapstone as a major ingredient of porcelain production seemed bright. New sources were being discovered and the material appeared to be abundant. A London mineralogist of Portuguese extraction, Emmanuel Mendes da Costa, in his *Natural History of Fossils* of 1757, asserted that 'no finer or fitter' material could be found to make 'elegant porcelain', while in the same year Borlase, in his *Natural History of Cornwall*, proclaimed that English potters would surpass European porcelain producers, who were not fortunate enough to possess such 'fine steatites'. However, despite these commendations from learned observers, all was far from plain sailing for the potters who quarried and used the material.

Lund had paid landlord's dues of 10s. (50p) a ton, but Pococke had seen men discarding the red patches so that the white rock could be sold at £5 a ton in Bristol; he was told that the men had been ordered not to take such care in removing all the red stains, since it made the material too expensive. Nevertheless, Lund appeared to have found the cost of hand-sorting and removing discoloured portions too great and sold his licence to Richard Holdship of Worcester. By the time Crisp had arrived the going rate for landlord's dues had more than doubled to one guinea (£1.05) a ton.

The usual method of extraction had been for half a dozen men to dig out the rock, which was sorted into different grades by small groups of women. To begin with, easily available rock from the surface or the cliff face was dug, but the location of deposits was becoming more scattered throughout the peninsula, while veins were often only one or two feet thick and varied greatly in their yield of saleable stone. Already by the 1760s adits or deep tunnels had to be dug and the existence of miners in the area suggested that they might have been employed, possibly at higher rates of pay. The cost of one of these underground workings was £7 per ton.

Apart from these extractive problems, the firing of porcelain was a hit-or-miss affair. In 1760 Richard Holdship of Worcester became bankrupt and in the same year Crisp of Vauxhall gave up his lease to two Staffordshire potters, John Baddeley and William Yates. However, they too were declared bankrupt in the following year and Crisp's business failed in 1763.

Despite these setbacks, two famous potters were attracted to the area: Thomas Turner of Caughley and Josiah Wedgwood. From 1775 onwards Turner attempted to acquire a lease at Gew Graze and also Wedgwood visited the Lizard in 1775, viewing workings at Gew Graze and another site on the other side of the peninsula at Gwavas Barton, run by Christian & Son, whose original partner Chaffers had died in 1765. Both Turner and Wedgwood, however, switched their attention to china clay and china stone. Turner tried to lease the Carloggas pit near St Austell that Cookworthy had discovered.

By now a few local men had shown an interest. In 1773 John Nancarrow, senior mine captain of the Godolphin group of mines nearby, licenced Boscawen land near Gew Graze, and four years later John Lean, a Redruth tinner, with William Thomas and William Trounson from Mullion, formed a group to extract soapy rock. In a letter of 1810 William Jenkin, steward of the estates of C.B. Agar of Lanhydrock, informed his master that John Williams of Scorrier House had applied for 'the grant of a spot... and one Wm Bennetts of Camborne hath since applied for another spot'.[38] Williams was a member of one of the richest mining and smelting families in Cornwall.

Although a few local adventurers showed an interest, most of the quarries that continued were in the hands of the Worcester potters. Thomas Flight of Hackney, the London agent of the Worcester factory, had acquired that firm in 1783 for his sons Joseph and John. Soapy rock continued to attract distinguished visitors to the Lizard. In 1792 James Boswell (without Dr Johnson) was taken to see a collection of minerals including 'the soapy rock'.[39] In the same year the Cornish geologist Majendie identified soapy rock extraction in a number of places on the Lizard. A decade later Britton and Bayley, in their tour of the 'Beauties of England and Wales' inspected a site of the 'celebrated soapy rock'.[40]

The decline of soapy rock
By the early 1800s, however, both the supply of, and the demand for, soapstone was on the wane. Production had fallen to under 100 tons a year,[41] compared with

over 1,000 for china clay and stone and 15,000 tons for ball clay. The original sources in Kynance Cove already appeared to be exhausted by the 1760s. A Swansea potter, Lewis Dillwyn, acquired a lease on Gew Glaze in 1814 at the high price of £75 per annum, plus £5 for every ton above 15 tons a year, and four years later the Revd G.C. Smith of Penzance, in a visit to the area, observed 'the soft clay dug out from the sides of this immense mountain to form our British China'.[42] However, these deposits seemed to have petered out before 1820.[43] Bone china, using bone together with china clay, china stone and a little ball clay, had become the standard item on English middle-class tables. It was less risky to fire, whereas potters had found that large items such as dinner plates, dishes and tureens made from soapstone tended to warp during firing.[44] Nevertheless, visitors were still taken to the quarries. In 1824 one of them recorded that three miles from Mullion was 'the celebrated Steatite or Soap Rocks, which to the geologist are highly interesting, and have been of great use to the china manufacturer'.[45]

Other uses were found for soapy rock, though in small quantities. De la Beche mentioned that 'two or three cargoes of it are shipped annually to Bristol for the purpose of furnishing magnesium for the carbonate of magnesia there manufactured'.[46] In 1853 Thomas Jackson of London licensed land from Lord Falmouth to extract soapstone for 21 years at £50 per annum for only 4s. (20p) per ton.[47] In 1864 a draft licence to search for soapstone in a different location at Veryan in Mid Cornwall was granted for an even lower price of 30p per ton,[48] but may not have been exercised. This seems to be the last recorded attempt at the commercial use of soapstone. By 1884 Brenton Symons, mining engineer and metallurgist, in his geological survey of Cornwall, only mentioned the material as 'formerly quarried for the manufacture of porcelain'.[49] However, mineral collections in Cornwall and elsewhere still contain samples of the stone, as well as souvenirs made from a material 'whose grace, beauty and charm', as Barry K. Hobbs has said, 'will forever be a delight to the eye and a balm to the soul'.[50]

Serpentine on the Lizard

It is perhaps worth mentioning the presence of steatite in another product of the Lizard. Ornaments and souvenirs of serpentine had been boosted by the patronage of Queen Victoria and displayed in the Great Exhibition of 1851 in London. By the early 1860s over 70 men were involved in extracting the stone and carving serpentine objects at Poltesco and other sites, even recruiting workers from the Blue John mines of Derbyshire.[51] Some of their products involved white steatite dotted with coloured fragments of serpentine, or serpentine spotted with pieces of steatite. Many examples of serpentine church furniture can be seen locally at Grade and Cadgwith, and there is a display of serpentine objects in an inn at Lizard village.

Mrs Coade's Stone

Ball clay was involved in the manufacture of Coade stone, a widely used material during the first half of the nineteenth century. Named after its manufacturer, Eleanor Coade, it is of particular interest because of its long and continuing record of stability and resistance to weather compared with most kinds of natural stone and with rival artificial stones of the time. Eleanor's father, of Cornish extraction, had set up in Exeter as a wool merchant but, after becoming bankrupt, moved to London, where he again became insolvent and died in 1769.

Coade stone lion on Westminster Bridge, close to the Lambeth factory.

Meanwhile, his daughter had established her own independent business as a linen draper and from 1769 set up a 'burnt artificial stone manufactory' called 'Coade's Lithodipyra' in Lambeth, the site of many potteries. The word *Lithodipyra* is derived from the Greek *litho* (stone), *di* (twice) *pyra* (fire), because fired ball clay was ground to make 'grog' and this was added to unfired ball clay, together with ground flint, sand and glass. The pre-shrunk grog reduced the shrinkage of the fired product and enabled large pieces to be made with ball clay accounting for about one-half of the total constituents. Because the mixture was not very plastic it was cast in plaster moulds and fired in cylindrical muffle kilns for four days at 1,100°C or more. For a long time it was thought to contained china clay, but tests in the 1980s showed that ball clay was the clay component utilised.[52]

Coade stone could hardly be distinguished from natural stone, and its creamy buff colour blended in perfectly with Georgian architecture. Through aggressive marketing, Eleanor Coade sold nearly 800 different patterns, ranging from ornamental statues and busts to huge memorial friezes, capitals and columns to eminent architects, including Robert Adam, John Nash and Sir John Soane. Coade stone is found in Buckingham Palace, St George's Chapel, Windsor, and in a ten-foot statue of Nelson in the Royal Naval College at Greenwich, as well as in North and South America, Poland and Russia. Dorset examples include decorations at Belmont, Eleanor Coade's home at Lyme Regis,[53] and a statue of George III at Weymouth.

In Devon we find numerous keystones at Southernhay East and West in Exeter, busts of Nelson, Raleigh and Wellington at Bicton Park and figures at the Egyptian Library in Devonport. Coade stonework can be seen in Cornwall at the Egyptian House at Penzance, in the Chapel parapet at St Michael's Mount, in vases and urns at Trelowarren, medallions and coats of arms on the Assembly Rooms at Truro, a statue at Caerhays and a monument 14 feet high at Mount Edgcumbe.[54] Eleanor Coade died in 1821 aged 89, but the firm continued for some years, run by William Croggan from Grampound, near Truro, a distant relative. However, the fashion for simple classical design during the Greek Revival period reduced the demand for Coade stone and the works closed down in the early 1840s.[55]

China clay overtakes soapstone and ball clay

At the beginning of the nineteenth century, while the ball-clay industry was well developed, the total output of Cornish china-clay pits, to be discussed in the next chapter, had yet to reach 2,000 tons a year. However, the demand for even whiter potting clays led to the growth of the china-clay industry and by the mid-1800s the output of china clay from Cornwall and West Devon was reaching the level of ball-clay production in Devon and Dorset. Teignmouth was then shipping out approaching 40,000 tons a year[56] and Poole perhaps as much as 60,000 tons, compared with Cornwall and West Devon's output of around 100,000 tons a year of china clay. While soapstone production petered out, ball clays remained of great importance to the potters because they were more plastic than china clay and therefore more easily worked by the potter. Eventually, the production of whiter china clay greatly outstripped that of ball clay when other significant uses, for which ball clay was not suitable, were found outside the ceramics industry.

References

[1] G. Agricola, *De Re Metallica*, translated edition, New York, 1950, p.115.
[2] R.J. Cleevely and C.M. Bristow, 'Revd William Gregor; The Contribution of a Cornish Cleric and Analytical Chemist', *JRIC*, 2003, pp.85–108.
[3] W.G. Maton, 'Observations on the Western Counties of England', in R. Pearce Chope, *Early Tours in Devon and Cornwall*, 1918, repr. New York, 1968, p.276; E. Morton Nance, 'Soap rock Licences', *TECC*, 1935, 3, pp.73–84; Joan Jones, *Minton*, Shrewsbury, 1993.
[4] Peter C.D. Brears, *The English Country Pottery*, Newton Abbot, 1971, pp.88–89.
[5] Helen E. Fitzrandolph and M. Doriel May, *The Rural Industries of England and Wales*, Oxford, 1927, pp.130–31.
[6] DeRO, 2974-A-1/P2LZ in Payhenbury Parish.
[7] H.L. Douch, 'Cornish Earthenware Potters', *JRIC*, VI, 1969, pp.33–64.
[8] Bernard Leach, *A Potter's Book*, London, 1940, p.113.
[9] Douch, 1969, pp.33–64.

[10] W. Borlase, *The Natural History of Cornwall*, Oxford, 1758.

[11] For a comprehensive treatment of this topic, see John Ferguson and Charles Thurlow, *Cornish Brick Making and Brick Buildings*, St Austell, 2005.

[12] *The Picture of Plymouth*, 1812; *The Plymouth Directory*, 1814, 1822; *Pigot & Co's London and Commercial Directory*, 1823, 1830.

[13] Thomas Hardy, *A Pair of Blue Eyes*, 1873, repr. London, 1952, p.255.

[14] John Penderill-Church, *The China Clay Industry*, n.d., and *Bodmin Moor*, 1981, Wheal Martyn Archives.

[15] Dr Richard Pococke, 'Travels through England', 1750 in Chope, 1968, p.205. There is a place called Kelly in West Devon, but the location that Pococke referred to is more likely to have been Kelly Bray, near Callington in Cornwall, since it was here that a fine altered killas, a white talc-like mineral, was found. It was used in the Second World War to seal a layer of bitumen in the construction of advanced aircraft landing strips. J.P.R. Polkinghorne, 'Slate Mining, Redmoor Mine, Kelly Bray, Callington Counthouse Section', *TRGSC*, XVII, pp.269–70.

[16] Penderill-Church, n.d.

[17] Apart from the references cited in the text, the information for the section on ball clay comes from The Ball Clay Heritage Society, *The Ball Clays of Devon and Dorset*, Newton Abbot, 2003; Robert Copeland, *A Short History of Pottery Raw Materials and the Cheddleton Flint Mill*, Cheddleton, 1972; M.J. Messenger, *North Devon Clay*, Truro, 1982; L.T.C. Rolt, *The Potters' Field: A History of the South Devon Ball Clay Industry*, Newton Abbot, 1974, pp.24–27, 39–40, 106; *WDM*, 12 January 1871, p.4; J.H. Collins, *The Hensbarrow Granite District*, Truro, 1878; J.A. Bulley, 'The Beginnings of the Devonshire Ball-Clay Trade', *TDA*, 1955, 87, pp.191–204; R.W. Kidner, *The Railways of Purbeck*, Oakwood, 2000.

[18] The land remained in the family's possession until 1874.

[19] Bulley, 1955, pp.194–95.

[20] For instance John Valance is recorded as a 'clay cutter' near Kingsteignton in 1815. DeRO, 3009A-99/PO16/820.

[21] Weaver Navigation Tonnage Books, quoted in Lorna Weatherill, 'Technical Change and Potters' Probate Inventories', *JCH*, 3, 1970, p.3.

[22] DRO, D/RWR/E20 and map E/16/5.

[23] Anthony Burton, *Wedgwood*, London, 1976, p.176.

[24] H.J. Trump, *West Country Harbours*, Teignmouth, 1976, p.32.

[25] *The Clay Mines of Dorset*, 1960, p.9.

[26] See for example the Draft Legal Case re Corfe Castle Clay Pits and Joseph Pike's Agreement with Wedgwood. DRO, RWR/L/8 0f 1790-1. Rodney Legg, *Purbeck's Heath*, Sherborne, 1987, p.30; Terence Davis, *Wareham: Gateway to Purbeck*, Wincanton, n.d.

[27] DRO, D/SEN/1/16, 1/3/1, 1/82. For details of leases see DRO, Family Papers of the Ryders of Rempstone; Pitt-Rivers Family Archive; Lester and Garland Family Archives; Encome Estate Archive; Papers of John Bishop and Thomas William.

[28] DeRO, 4671 F/L6-7.

[29] DeRO, 62/9/2/Box 7/5, 6.

[30] *GWR Company versus Carpalla United Clay Company*, London, 1907.

[31] Bulley, 1955, p.199.

[32] *The Clay Mines of Dorset*, London, 1960, p.7; L.T.C. Rolt, *The Potters' Field*, Newton Abbot, 1974, p.73.

[33] Jill Newton, *Bygone Helston and the Lizard*, Chichester, 1987.

[34] Apart from the sources cited in the text, the information upon which this article is based comes from Joseph Mayer, *On the Art of Pottery*, Liverpool, 1871, p.41; E.G. Harvey, *Mullyon: Its History, Scenery and Antiquities*, Truro, 1875, pp.40–41; A. Majendie, 'A Sketch of the Geology of the Lizard', *Transactions of the Royal Geological Society of Cornwall*, 1818, p.32; R. Pearce Chope, *Early Tours in Devon and Cornwall*, 1918 repr. New York, 1968, pp.197, 276; E. Morton Nance, 'Soap rock Licences', *Transactions of the English Ceramic Circle*, 1935, 3, pp.73–84; E. Morton Nance, *The Potteries and Porcelain of Swansea*, London, 1943, pp.94, 474; Barry K. Hobbs, 'New Perspectives in Soapstone', *Transactions of the English Ceramic Circle*, 15, 3, 1995, pp.368–92. We are indebted to Colin Hanley for his valuable assistance.

[35] William Borlase, *Antiquities, Historical and Monumental, of the County of Cornwall*, 1754, repr. 1973, p.455.

[36] P.A.S. Pool, *William Borlase*, Truro, 1986.

[37] Pococke in Chope, 1968, p.197.

[38] Royal Institution of Cornwall, HJ/J/10. The first 'spot' referred to disused workings at Predannack, south of Mullion, the second to a vein discovered in an old adit near Kynance Cove. We are indebted to Jim Lewis for these references.

[39] Edwin Jaggard, 'James Boswell's Journey through Cornwall', *Journal of the Royal Institution of Cornwall*, 2004, p.13.

[40] John Britton and Edward Bayley, *The Beauties of England and Wales*, London, 1801, II, p.331.

[41] Revd Richard Warner, *A Tour through Cornwall*, Bath, 1809, p.224.

[42] G.C. Smith, *Wreckers: or a Tour of Benevolence from St Michael's Mount to the Lizard Point*, 1819.

[43] Henry de la Beche, *The Geology of Cornwall, Devon and West Somerset*, 1839, p.511.

[44] Pat Halfpenny, *Penny Plain, Tuppence Coloured*, Stoke-on-Trent, 1994, p.22.

[45] F.W.L. Stockdale, *Excursions Through Cornwall*, London, 1824, repr. Truro, 1972, p.70.

[46] De la Beche, 1839, p.499.

[47] Nance, 1935, p.81.

[48] CRO/DD/CF 3880, draft licence drawn up by Shilson Coode, St Austell, from Thomas Blamey to W.L.S. Clark.

[49] Brenton Symons, *A Sketch of the Geology of Cornwall*, London, 1884, p.26.

[50] Hobbs, 1995, p.391.

[51] Michael Sagar-Fenton with Stuart B Smith, *Serpentine*, Truro, 2005, pp.34–35.

[52] Traces of titanium oxide were found in it, which are present in ball clay but not in West Country china clay.

[53] Belmont has recently been acquired by the Landmark Trust.

[54] A massive 'statue of Neptune and seahorse' was sold for £44 to James Goldsworthy, proprietor of the Exeter Waterworks, in 1821 to crown the entrance of Southernhay Baths, later destroyed. C.G. Scott, 'A Coadestone Statue in Exeter', *DH*, 70, 2005, pp.22–23. For illustrations of Coade Stone in Truro buildings, see Ferguson and Thurlow, 2005.

[55] Alison Kelly, *Mrs Coade's Stone*, London, 1990; George Savage, *Pottery through the Ages*, Penguin, 1959, p.188; Hans Van Lemmen, *Coade Stone*, Princes Risborough, 2006.

[56] Trump, 1976, p.106.

Chapter Two
West Country China Stone and China Clay

T he search for whiter forms of ball clay and soapy rock and the use of artificial 'whiteners', described in the previous chapter, was part of a general social and economic trend in eighteenth-century Europe: the commercialisation of luxury for the bourgeoisie.[1] Just as medieval alchemists sought ways of transmuting base metals into gold, so apothecaries, potters and aristocratic *dilettanti* – arcanists, as they were called – experimented with different substances in a variety of locations to try to reproduce the purity, the translucence and the elegance of Chinese porcelain. At the beginning of the eighteenth century, porcelain imported from China was known in Britain as 'Chinaware', later simplified to 'china'.

By the early decades of the nineteenth century the china made in Britain was mainly 'bone china', incorporating a substantial proportion of ox bone as an additional flux and whitener, which was not a component

Probus Church tower, built in 1522.

of the genuine Chinese hard-paste porcelain.[2] It was in the course of this transition in usage from 'Chinaware' to 'china' that the white clay and stone of Cornwall came to be called china clay and china stone, and the first bone china was known as 'English Cornish china'. As Colin Bristow has remarked, 'The massive occurrences of china clay in Cornwall and Devon are unique, in terms of their combination of quantity and quality, in the world'.[3] We begin, however, by discussing china stone because, although china clay later became much more important, it was initially china stone that was more widely used.

St Stephen's Granite and Petunse

Stone known at various times by these names is found largely in the western part of the St Austell china clay area, but in much smaller quantities than china clay. It was sometimes referred to as St Stephen's granite because the main quarries were located in that parish. A largely white granite, it was used as a building material for hundreds of years, for instance in Probus Church tower, built in 1522 and admired by Dr Pococke in his 1750 tour of Cornwall. This structure is now partly covered with lichens and moss but would originally have had a white appearane. The interior of Colan Church, near Newquay, with pillars, capitals and arches of stone from St Stephen, shows an original, unweathered effect. In practice, St Stephen's granite can be divided into two types, 'china stone' and 'shell stone'. The first type is a 'topaz granite' which contains some fluorine minerals and a silvery mica. It is slightly whiter and has a lower iron content than the second type, known locally as 'shell stone', a lithionite granite that contains bronze-coloured lithium-bearing micas and occurs in patches alongside topaz granite. Both types can be used as building stone, since they are relatively easy to cut. Lumps of all types of granite found on the surface in Devon and Cornwall, the result of weathering or glaciation, were referred to as moorstone.

Much of the granite in the St Austell area is wholly or partly decomposed, including St Stephen's granite. Topaz granite contains some undecomposed feldspar, which acts as a flux in a ceramic body when ground to the same fineness as china clay, with which it is mixed and fired. It fuses the more refractory (heat resisting) china-clay skeleton into a homogeneous ceramic article without the discoloration or specking that occurs if shell stone is used. It is this ability to fuse that led the Chinese to refer to petunse (china stone) as the 'flesh' of porcelain, whereas kaolin (china clay) was the 'bones', expressions which at first seem to be the opposite of what one might expect, confusing some commentators over the years.

The quarrying of china stone was aided by the fact that it was generally located in areas of ground with pronounced vertical and horizontal joints known to the quarrymen as 'heads' and 'beds'. These assisted quarrying but had to be followed by the careful selection of stone suitable for ceramics, the topaz granite. The use of hammers, wedges and heavy bars allowed much stone to be broken without explosives, but some hand drilling and blasting was noted by Dr William Fitton of Northampton, writing in 1813 and describing a trip to Cornwall in 1807.[4] In the early years of the industry the stone was ground by potters close to, or in part of, their works and it was some time before grinding mills were established in Cornwall.

Early prospectors found it difficult to distinguish between different types of granite. Wedgwood, for example, noted a heap of stone during a journey through Cornwall in 1775, which was being used for road-mending because it contained too much iron for pottery manufacture. However, he observed that a stranger would not notice these impurities 'without being minutely and particularly shown'. This stone was in St Stephen's parish, but on Tregonning Hill, near Helston, he took samples which he judged to be good but which on further analysis on his return home he also found to contain too much iron.[5] It was topaz granite that was used by the Plymouth apothecary Cookworthy for his porcelain experiments, and it has been widely utilised up to recent times in a variety of ceramic bodies. Different grades of stone were quarried according to the degree of decomposition of the feldspar, and softer grades were favoured in the early years of the industry. Demand for topaz granite led to the enforced extraction of some lithionite granite, unsuitable for ceramics, which was used in a number of nineteenth-century buildings in Cornwall, an important example of which was, as ashlar, for the internal walls of Truro Cathedral in the 1880s. The buff colour blends well with the Bath stone used for window tracery and roof vaulting.

Cornish crucibles

An early reference to topaz granite as 'china stone' appears in Pococke's account of his 1750 tour: 'East of the Fal above Grampound,' he wrote, 'the Cornish china-stone (as it is called) is principally procured. At Truro this substance has been manufactured into retorts and corcibles *(sic)* of so excellent a quality as to stand the fire with uncommon success, and it contains so small a quantity of iron, that the porcelain made from it in Worcester and Staffordshire is very

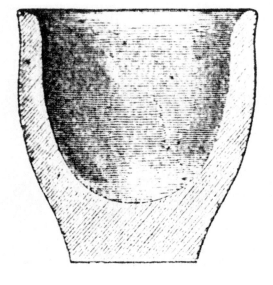

Section of a Cornish crucible, made in various sizes by Juleff of Redruth.
(Source: *Metallurgy* by Percy, 1875)

little discoloured.' [6] It is likely that a local clay, as well as stone, was also used as an admixture for these articles. The making of crucibles for assaying metals had developed earlier in Germany, but in 1766 the organisation which is now the Royal Society of Arts paid a premium to Jacob Lieberich to open a factory at Calenick, near Truro, on the recommendation of Henry Rosewarne, Vice Warden of the Stanneries, Recorder and MP for Truro. It was to this kiln that Cookworthy sent some cups to be fired in 1768.[7]

The Calenick works were possibly the factory visited by Pococke in 1750, since in 1778 William Pryce mentioned Lieberich's crucible factory as 'set up within these few years'. In the same year Pryce remarked that the ability of the crucibles to 'resist the most intense fire' proved the 'strength of our clays, when mixed with granite'.[8] In 1839 Sir Henry de la Beche, the first Director of what was called the Ordnance Geological Survey, remarked that 'the elvan on the west of Killiganoon is so decomposed it has been worked as crucible clay'.[19] Another firm that carried on the tradition of Cornish crucible manufacture was the Pednandrea works at Redruth, established about 1760 by John Juleff and continued by his son and grandsons. Both Michell of Calenick and the Juleffs of Redruth exhibited 'Cornish crucibles' at the Great Exhibition in London in 1851.[10] Cornish producers had built up such a fine reputation for reliability that their products were exported world-wide for many years.

Since Cornwall was a major world centre of metalliferous mining at the time, it made economic sense to produce crucibles locally. However, most china stone was exported to potteries elsewhere, and as early as 1799 a trade directory for St Austle (sic) reported that 'there are several quarries in the neighbourhood, which produce what is commonly called China Stone. Sometimes not less than one thousand tons per year are shipped at Polmear (later called Charlestown) and conveyed to Bristol, Liverpool or Wales, and from those places to Staffordshire, where it is manufactured into porcelain'.[11] Exports of china stone rivalled those of china clay in importance for several decades until, as we shall see, important additional uses for clay were developed but few for stone.

Kaolin, growan, 'porcelain earth', 'china earth' and china clay

Various names have been given to a fine white powder, formed by the decomposition of feldspar in granite which, under a powerful microscope, is seen to consist of hexagonal platelets of an aluminium silicate mineral called kaolinite. It is found in association with granite in many parts of the world. In Cornwall and Devon the exact nature of the decomposition process of feldspar was debated for many years, but it is now thought that a complex process began with hydrothermal action in the granite caused by ascending hot gases and fluids. This was followed by weathering when surface water soaked through cracks in the rocks and was warmed at depth by the unusually high content of radioactive elements in the granite. The slow convective circulation of water, over hundreds of millions of years, altered feldspar to china clay. The presence of radioactive elements was reflected in local mining activities. A mine at South Terras near St Stephen, on the edge of the granite area, produced ores from which salts of uranium and radium were prepared in the late-nineteenth century until 1928.[12]

Sixteenth-century 'claye pittes'

A very early reference of 1548 records the boundaries of tin workings, where a 'claye pitte' marks the northern side of the tin works of Stenackgwyn, near the present-day village of Foxhole.[13] It is unlikely, though not impossible, that this was worked for a crude form of china clay, but more probably it was a source of building material. Richard Carew's *Survey of Cornwall*, published in 1602,[14] refers to 'the poor cottager' who 'contenteth himself with cob for his walls and thatch for his covering'. A mixture of clay and sand can be used to make cob, and clay might also have been used as a mortar in stone buildings.[15]

In 1705 a sale of items from a ship coming from the Far East listed 'china earth',[16] and Père d'Entrecolles, a Jesuit missionary, in his letters describing the manufacture of porcelain in China from 1712 to 1722, referred to china clay as kaolin and china stone as petunse.[17] In a memorandum of around 1765, William Cookworthy, the Plymouth-based apothecary, mentioned his discovery of two materials he called 'Caulin' and 'Petunse' but in his patent of 1768 he used the terms 'Moorstone' or 'Growan stone' and 'Growan Clay', and in correspondence of this period with Thomas Pitt he also alluded to 'china ware clay'.[18] 'Hard growan' was the expression used by Cornish miners at the time for granite, while 'soft growan' was their term for decomposed granite.[19] When the Staffordshire manufacturers Wedgwood and Turner visited Cornwall in 1775 in search of suitable materials to make porcelain, they enquired where 'Growan clay and stone' could be found.[20]

In 1783 the term 'china clay' appeared in a letter from Rudolph Erich Raspe, a renegade German mineralogist, addressed to Phillip Rashleigh of Menabilly, an avid mineral collector. Professor of Antiquity and Curator of the mineral collection of the Landgrave or Prince of Hesse, he was discovered selling items from the collection to pay off his debts. Raspe fled to Holland and then to England, where he was employed from 1782 onwards in Redruth by Matthew Boulton, of the famous Midlands engineering partnership of Boulton & Watt, to inform them of affairs in Cornwall. He suggested to Boulton that the strength and casting properties of iron might be improved by the addition of tungsten, a metal that appeared to have been produced, perhaps by accident, in blowing houses or reverberatory smelters such as those at St Austell, making them possibly the world's first producers of tungsten.[21] Although primarily concerned with metallic minerals, Raspe took an interest in china clay. While in Redruth he also found time to write *The Surprising Adventures of Baron Munchausen*, a collection of tall stories, published anonymously, which was translated into many languages.[22]

His letter to Rashleigh was accompanied by samples of Cornish copper ore together with some of what he called 'china clay', suggesting that the term was already in use by that time. However, Raspe referred to 'soft steatite or china clay', and the term 'steatite' (a variety of soapstone) implies that he did not properly understand the fundamental mineralogical difference between china clay and soapstone.[23] Potters were using the terms 'porcelain earth' or 'porcelain clay', and Dr Fitton, in 1807, came across 'porcelain earth or china clay, as it is denominated by the workmen'.[24] Dr Fitton, talking of stone, referred to 'granite, the feldspar of which is in a disintegrated state'. This description would fit the 'soft white' or 'hard white' grades of china stone used by the potters at that time. It seems likely that at some point the workmen had named the materials 'china stone' and 'china clay' because their early application was for pottery. However the term 'kaolin', a Chinese word meaning 'high ridge', is in general use in America, France, Germany and other countries.[25]

The preparation of china clay

As will be discussed later, William Cookworthy, a Plymouth apothecary, found decomposed granite containing china clay in Cornwall and in a memorandum written before 1768 he made the preparation of china clay seem quite straightforward. 'Nothing more is required but pouring a large quantity of water onto it so that it may not, when dissolved [suspended] be of so thick a consistency as to suspend the mica. Let it settle about ten minutes and pour off the dissolved clay into another vessel. Let it settle, pour off the water and dry it'.[26] In a simplistic fashion, Cookworthy had identified the three main stages for china clay production: breaking up clay bearing ground with water (a method once used by the ancient Romans for working gold in Spain); a refining stage where unwanted sands are settled out to leave china clay in suspension with water; allowing the clay to settle and a drying process.

An early account of china-clay working

Dr Fitton, in his visit of 1807, describes a more detailed process. First, topsoil and stained clay ground, called 'overburden', was moved to one side of the area to be worked. The clay-bearing ground was then dug out and transferred to a cutting known as a 'strake', where water was allowed to run over it and the break-down was assisted by men with one-sided picks called 'dubbers' and shovels. The resultant suspension of clay, water and sand was run into a series of pits where coarse and fine sands settled out and china clay finally flowed into a pond. It appears that as pits developed the strake became part of the pit face rather than a fixed site.

According to Fitton, the settling of clay in ponds was sometimes assisted by the use of alum as a flocculating agent, and the thickened clay was run or pumped to 'sun pans' using simple lift pumps. These pans were shallow tanks where clay stood for months before being cut out as nine-inch cubes. These were set out on the ground, probably on short wooden boards, to harden in the sun and wind before either being transferred to an open-sided store, referred to as an 'air dry' or being stacked under 'reeders', simple panels of thatched rushes, to give some protection from the weather. These operations were probably based on the 'hacks' used by brickmakers. When dry, the cubes were scraped clean of any sand and lichen by

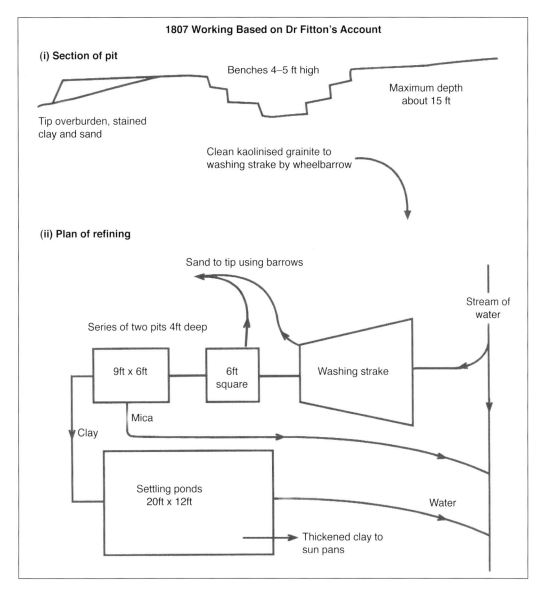

1807 Working Based on Dr Fitton's Account

(i) Section of pit

Benches 4–5 ft high

Maximum depth about 15 ft

Tip overburden, stained clay and sand

Clean kaolinised grainite to washing strake by wheelbarrow

(ii) Plan of refining

Sand to tip using barrows

Stream of water

Series of two pits 4ft deep

9ft x 6ft

6ft square

Washing strake

Mica

Clay

Settling ponds 20ft x 12ft

Water

Thickened clay to sun pans

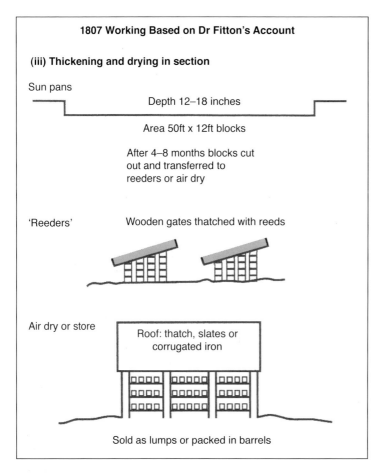

1807 Working Based on Dr Fitton's Account

(iii) Thickening and drying in section

Sun pans

Depth 12–18 inches

Area 50ft x 12ft blocks

After 4–8 months blocks cut
out and transferred to
reeders or air dry

'Reeders'   Wooden gates thatched with reeds

Air dry or store

Roof: thatch, slates or
corrugated iron

Sold as lumps or packed in barrels

women using triangular-headed tools like oversized paint scrapers. A visitor to what he called the 'dreary district north of St Austell' in the 1850s gave an account of these methods, describing the scrapers as 'instruments resembling small Dutch hoes'.[27] Later, in the mid-nineteenth century, these lengthy drying operations were speeded up by the use of coal-fired drying kilns, to be described later.

The St Austell china-clay district

The earliest china clay pits in the Mid Cornwall area were operated solely for the production of materials for ceramics and were located in the western side of the area. This reflected geological differences within the local granite. In simple terms the decomposed granite in the eastern half contained more iron minerals than in the western half. Fortuitously, early experiments with clay and stone from Carloggas, near Nanpean, and its immediate region used materials that were the most suitable for ceramics, and this area still produces most of the china clay sent to the potteries. The same ground

Lift pumps operated by women and children to transfer thickened clay from settling ponds to sun pans. Sketch by J.M. Coon, 1927, based on his recollections.

also contained topaz granite or china stone, and in many cases the clay and stone suitable for potting were found adjacent to each other in a single pit.

Areas in the eastern side, where china clay contained iron impurities that made it unsuitable for ceramics, also contained other metalliferous minerals alongside china clay. At some old tin-streaming sites such as Carclaze the china clay was disregarded and allowed to drain into local rivers. Underground mines such as Rocks and Beam hoisted tin ores from areas that were not worked as clay pits until the latter half of the nineteenth century. By this time, china clay producers were finding other uses for which clay from the eastern side were suitable. This gradually led to a concentration of china clay working around the St Austell area and the development of an infrastructure of engineering, transport and administrative services.

West Country china clay outside the St Austell district

Although Cornwall has always produced at least 80 per cent of china clay output from the South West, many places on Dartmoor in south-west Devon, such as 'Whitehill', derive their name from the presence of the material. In 1755 Cookworthy is said to have come across this source in his travels, but rejected it as inferior to the deposits he had observed in Cornwall.[28] It was not exploited until John Dickins, a Plymouth china merchant, began to explore Dartmoor in the 1830s, and by 1839 De la Beche observed clay works 'in the south part of Dartmoor', as well as in Cornwall.[29] De la Beche regarded the clay he saw as 'artificial china clay', whereas he called ball-clay 'natural china-clay of inferior quality'. To him ball clay was 'natural' because it was dug out as a 'pure' clay without sandy contaminants from a stratum of ground,[30] whereas china clay was 'artificial' because, as described earlier, it needed washing, the removal of much sand, settling and drying before it could be used. Even so, he considered ball clay to be 'inferior' because it was not as white.

After the closure of the Calstock pottery mentioned in Chapter One, its stocks of clay were transferred to Bovey Tracey, where Staffordshire potters had constructed kilns using lignite, or 'brown coal', found nearby, as a fuel. Throughout the areas worked for ball clay, small amounts of lignite were encountered, but in the Bovey Basin some stratified beds up to eight feet thick or even more were located, together with ball clay and sand, and the lignite was often referred to locally as 'Bovey coal'. However, it was inferior to coal as a fuel, having a lower carbon content, intermediate between peat and coal, and was first used locally in lime-burning kilns in the seventeenth century. Geologists studied the area, including the lignite, in the eighteenth century and at this time the largest lignitic excavation was known as the 'Coal Pit'.[31]

This made industrial-scale pottery production seem possible, and Cornish potters John Edyvean and Edmund Carthew were active in the area. Unfortunately the lignite proved unsatisfactory because its sulphurous fumes discoloured the chinaware and gave out only a fifth of the heat of bituminous coal, making it difficult to reach the high firing temperatures of around 1,450°C needed for true porcelain.[32] However, an account of 1820 tells that 'Bovey coal' was used for the first, or biscuit, firing and 'North Country' coal for the second firing after glazing. The offensive, sulphurous fumes limited general domestic use to poorer cottages in the district. Cookworthy's unsuccessful links with Nicholas Crisp at Bovey Tracey will be discussed in the next chapter.

Eventually, Dartmoor became an important source of china clay, as did Bodmin Moor, and these developments will be considered in Part Two. In addition to these sites, clayworks opened in various parts of Cornwall, but most of them never yielded large quantities and were only worked for a short period. In West Cornwall the most celebrated site was Tregonning Hill, identified by Cookworthy in the mid-1700s, rejected as too iron-stained a quarter of a century later but then reworked briefly around 1830 and again in 1850.

Further to the west, pits were dug at Bedlam Green, near Georgia in Towednack Parish, St Ives, Sancreed near Penzance and Leswidden near St Just. The pit at Bedlam was possibly worked from the 1820s and it may have yielded around 5,000 tons in total,[33] while Sancreed output was probably smaller[34] and Leswidden continued to produce until the 1920s. Several small pits turned to brickmaking after their clay was found to be unsuitable for use by potters. An example was Wheal Amelia at St Day, near Redruth, which started life as a china-clay pit before concentrating upon brick making.

Is china clay a mineral?

As a postscript to this discussion, it is curious to note that, a century and a half after its discovery in Cornwall, the legal status of china clay as a mineral had not been established. This was more than a matter of academic interest, but had important financial consequences if clay-bearing ground adjoined a railway line. According to the Railway Lands Clauses Act of 1845, if a mineral owner was prevented from working land because of a railway development then the railway had to pay compensation. If, on the other hand, china clay was not a mineral but merely a compound or mixture, then no compensation was involved. In 1885 this Act was tested when the Great Western Railway served a writ on Candy & Co. to stop it extending a pit producing potters' clay and brick clay towards the Mortonhampstead branch line near Newton Abbot, alleging that it threatened the stability of the railway track, and the clay company complied.[35] Perhaps they did not have sufficient resources to fight a case against such a powerful opponent, or possibly they felt their clay, used mainly for brickmaking, did not justify a special dispensation from the GWR. In 1908, however, in what was described as 'the most important china clay case ever fought', the Great Western Railway Co. brought an action against the Carpalla United China Clay Co. clayworks at Foxhole, near St Stephen's Beacon, and Lord Clifden, the landowner.[36] The clay company was extending its pit towards the railway and applied to continue digging under it. After a hearing lasting nine days, in which many expert witnesses on both sides were called, the judge held that china clay was a mineral and dismissed the GWR's action. The railway company then appealed, and once more geologists, chemists and other experts gave their evidence. The GWR's case was that china clay was not a mineral but rather a manufactured substance that could be made from common clays that were found in many places.

'Men Working in a China Clay Pit', c.1912. This watercolour painting is the earliest of two by Dame Laura Knight. The pit is a small working, possibly at Leswidden, near St Just. China clay is being broken in the stream seen clearly on the left-hand side. In the foreground are two sand pits, that on the left being filled with clay and sand while that on the right is being emptied.

However, the geologist Professor Gregory contended that china clay was a mineral that was worked in much the same way as other well-known minerals, such as alluvial gold and tin, and Dr Marshall, a chemist, argued that common clays such as brick clay were 'of a different clay altogether from china clay' and that 'no man in his senses could mistake the one for the other'. J.M. Coon, with 30 years' experience as a Cornish china-clay merchant and clay works manager, asserted that china clay had always been known as a mineral in the St Austell district. Samples were examined and analysed of 'common' and 'best' potters' clays and 'bleaching' clays, as well as of mica and sand from the Carpalla works, and compared with potters' clays from Newton Abbot, Bovey Tracey and Birmingham. The GWR again lost the case. China clay really was an 'extraordinary earth'.

References

[1] Maxine Berg, 'From imitation to invention', *EHR*, 2002, pp.1–30.

[2] George Savage and Harold Newman, *An Illustrated Dictionary of Ceramics*, London, 1974, repr. 2000.

[3] Colin Bristow, *China clay, a Geologist's View*, St Austell, 2006, p.1.

[4] William Fitton, 'On the Porcelain Earth of Cornwall', *Annals of Philosophy*, 2 November 1813, pp.348–51.

[5] Geoffrey Wills (ed.), 'Josiah Wedgwood's Journey into Cornwall', *PWS*, 1986, pp.81–82, 86, 90.

[6] Dr Richard Pococke, 'Travels through England', 1750, in R. Pearce Chope, *Early Tours in Devon and Cornwall*, 1918, repr. New York, 1968, p.252.

[7] Cyril Staal, 'Calenick Crucibles', *RCPS*, 1957, pp.44–54. The Calenick works was later acquired by the James family and closed after the death of John James in the early 1900s. Its stock in trade was acquired by the last survivor of the early Cornish potters, Lake's of Truro.

[8] William Pryce, *Mineralogia Cornubiensis*, 1778, repr. Truro, 1972, p.32.

[9] Sir Henry de la Beche, *Report on the Geology of Cornwall, Devon and West Somerset*, 1839, p.177. Elvan is a fine-grained granite which, when decomposed, becomes a mixture including sand, and Killiganoon is located near Truro.

[10] *Official Descriptive and Illustrated Catalogue of the Great Exhibition*, 1851, II, p.725.

[11] *Universal British Directory of Trades, Commerce and Manufactures*, London, 1799, p.363.

[12] Courtenay V. Smale, 'South Terras: Cornwall's Premier Uranium and Radium Mine', *JRIC*, 1993, pp.304–21.

[13] We are indebted to Allen Buckley for this information.

[14] Reprinted by Tor Mark Press in 2002.

[15] John McCann, *Clay and Cob Buildings*, Shire Publications, 2004. In recent years there has been a revival of interest in 'earth building', including a bus shelter of novel design at the Eden Project, near St Austell, using a blend of Cornish china clay and a red Devon clay.

[16] Geoffrey A. Godden, *New Hall Porcelain*, Woodbridge, 2004, p.69.

[17] William Burton, *Porcelain*, London, 1906.

[18] Albert Douglas Selleck, *Cookworthy 1705–80 and his Circle*, Plymouth, 1978, pp.243, 248.

[19] Pryce, 1972, p.322.

[20] Lynn Miller, 'Wedgwood in Cornwall', *JTS*, 15, 1988, p.32.

[21] Colin Bristow and Bryan Earl, 'Did the earliest production of tungsten take place in Cornwall?' *Materials World*, February 2004, pp.12–13.

[22] John Carswell, *The Prospector*, London, 1950.

[23] Colin Bristow, 'China Clay — the origin of the term', *CCHSN*, 2007, 15, p.8.

[24] Fitton wrote two articles on the subject: 'On the Porcelain Earth of Cornwall', *Annals of Philosophy*, 2 November 1813, pp.348–51 and March 1814, pp.180–84. In the first he did not perceive the difference between china clay and stone. He talked of a 'curious rock' that was 'sent to Staffordshire in two states', either in lumps or washed out and then dried. In the second article he referred to quarries that were 'annexed to every porcelain earth works' from which material was 'extracted in a solid form'.

[25] The American town of Sandersville in Washington County, Georgia, calls itself 'The Kaolin Capital of the World'.

[26] Douglas Selleck, *Cookworthy: 'a man of no common clay'*, Plymouth, 1978, p.243.

[27] Walter White, *A Londoner's Walk to the Land's End*, London, 1879, pp.126–30. His visit took place around 1854.

[28] John Penderill-Church, *William Cookworthy*, Truro, 1972, p.46.

[29] De la Beche, 1839, p.257.

[30] De la Beche, 1839, p.511.

[31] W. Pengelly, *The Lignites of Bovey Tracey*, London, 1962, p.19.

[32] Brian Adams and Anthony Thomas, *A Potworks in Devonshire*, Bovey Tracey, 1996, pp.10–14.

[33] John Tonkin, 'Bedlam Green', *CCHSN*, 6, August 2004, pp.2–3.

[34] William Jory Henwood, 'On the Preparation of China Clay', *JRIC*, 1839, p.56.

[35] Ian Turner, *Candy Art Pottery*, Melbourne, Derbyshire, 2000, pp.11–12.

[36] The House of Lords, *On Appeal from His Majesty's Court of Appeal between the Great Western Railway Company and the Carpalla United China Clay Company*, 1909, I, Ch. 218; SAS 17 July, 26 November 1908.

Chapter Three
'The True Art, Mystery and Secret of Porcelain'

So far we have focused upon the discovery in the West Country of various kinds of 'extraordinary earths'. However, in order to put these developments in context, and before examining the role of William Cookworthy, we need to widen the discussion and consider the spread of porcelain production in Europe. Centuries earlier the Japanese and the Koreans had challenged the Chinese monopoly in the manufacture of porcelain, and experimentation spread slowly northwards and westwards through Europe from Italy where, around 1470, the Venetians, long familiar with the product, attempted to reproduce it.[1]

Johann Böttger, the alchemist turned potter.

A century later the Grand Duke of Florence also tried his hand.[2] Because of the translucence of porcelain, he assumed that glass might be an ingredient, along with sand and powdered rock, but could only produce a greyish finish. A hundred years after this, factories were established at Rouen and Paris producing what is known as 'soft paste' porcelain, by mixing together a variety of materials including sea salt, soda, plaster-of-Paris, alum, saltpetre, white sand, glass, chalk, lime and china clay. These ingredients were fired at around 1,150°C, whereas 'hard paste' porcelain, to be discussed later, required a temperature of around 1,450°C.

Such combinations were also sometimes known as 'frit porcelains', because the materials were 'fritted' (melted) and then mixed with clay or plaster before firing. A leading English innovator in this field was John Dwight, who applied for patents in 1671 and again in 1684 for his discovery of 'the Mistery of Transparent Porcelanne... or China or Persian Wares'. He set up a factory in Fulham to produce it and between 1693 and 1696 took legal action for infringements of his patents against a number of other London potters, including John Chandler, who had worked for him and presumably stolen his secrets. Although these products sometimes achieved whiteness and supposed translucence, they were extremely costly to produce and much more fragile, requiring several firings, each potentially disastrous. As a result, they gradually went out of favour when more easily manufactured recipes were developed.[3]

'God our Creator has turned a gold-maker into a potter'

German princes also competed in the race to produce a true porcelain, and went to enormous lengths to devise and preserve their formulae. Augustus the Strong, King of Poland and Elector of Saxony, was so eager to expand his porcelain collection that he exchanged 600 of his cavalrymen with King Frederick William I of Prussia for the latter's collection of Chinese porcelain. He had apprehended and imprisoned an alchemist, Johann Böttger (1682–1719), in Dresden, but he became so enraged by Böttger's costly failures to produce gold that he transferred him first to the king's old fortress of Albrechtsburg, near Meissen, and later back to Dresden, warning him that he would not be released until he achieved success in making porcelain, an equally desirable commodity.

Böttger experimented for several years and was inspired by a nobleman, Ehrenfried Walter von Tschirnhaus, a mathematician and scientist, until his seath in 1708. He progressed to using mixtures of local clays from Colditz and Aue, together with small proportions of chalk or alabaster. It is possible that the Aue clay contained some unaltered felspathic material to provide a fluxing action. In 1710 a factory was opened at Dresden producing hard-paste porcelain, and shortly after the death of Böttger it began to use feldspar as a flux, with local clays. 'God our Creator has turned a gold-maker into a potter', was the inscription in German emblazoned above Böttger's threshold.[4]

Despite the most rigorous efforts to conceal the formula (the clay was delivered by armed guard), news of it spread, possibly through skilled workers who were enticed away from Meissen by other nobles, although it might well have been developed independently by Greiner in Thuringia or in Vinograduff in Russia. Then, in 1736, a book by J.B. du Halde was published in Paris, based upon the observations of the French missionary Père d'Entrecolles, mentioned in Chapter Two, that contained an account of the workings of Chinese potters east of Shanghai, where 18,000 were said to be employed using assembly line methods. The primary ingredients of porcelain were identified as what we now call china clay and china stone, although 'soapstone' was also mentioned erroneously.[5]

Tea drinking and the need for hard paste porcelain

By now the possession of porcelain was such an essential symbol of royal, princely and aristocratic status that rulers vied with one another to set up porcelain producing enterprises. Most of the German states had one. The Russian Imperial

China stone working in China

These sketches were drawn in the nineteenth century on the basis of letters from D'Entrecolles about Chinese potters in the eighteenth century. z

Showing the working of china stone. Stone is being broken by hammers in a largely underground working.

The broken stone is wet ground by tilt hammers (stamps) and is made into briquettes in a covered building to the rear. Here the ground stone is covered with cloth and stacked so that some water is squeezed out. Sketches by the same author of china-clay working are more conjectural.

Source: *J. Allen Howe Handbook to the Collection of china clay and china stone*, 1914 (p.87).

Porcelain Factory, founded in 1744 by the Empress Elizabeth at St Petersburg, achieved importance under Catherine the Great, whilst among several factories in Moscow was one run by an Englishman, Francis Gardner. Potteries opened in Belgium, Spain and Sweden and the Royal Danish Porcelain Factory was formed at Copenhagen. Then, during the Seven Years' War from 1756 to 1763, the Meissen works were severely damaged and European leadership in porcelain production passed to the Sèvres factory in France, which had come under the patronage of Louis XV in 1753. Sèvres had reached the height of fashion with its range of highly decorative and gilded wares, initially made of soft-paste porcelain using ground glass.

It was, however, much less efficient at withstanding hot liquids, and this was a serious drawback, for porcelain had to be practical as well as beautiful. Tea drinking had become increasingly fashionable, an important social occasion for the display of wealth and refinement among the English bourgeoisie, and porcelain had to take boiling water in order to brew tea. For this purpose hard-paste porcelain, fired at a higher temperature than soft paste, had been brought to England from China in the holds of cargo ships by the Honourable East India Co. It also made a useful ballast that could be placed towards the bottom of the hold, since it would not be spoiled by seawater, unlike tea, silk and spices.

By this time du Halde's book on Chinese methods had been translated into English, so that in the mid-1700s a whole host of potters claimed to have discovered how to make 'true' porcelain. Workshops were established in London at Bow, Limehouse, Isleworth, Vauxhall and Chelsea (the last named moved to Derby in 1770) and Fulham (from where one of the potters moved to Lowestoft), as well as in Liverpool, Bristol, Swansea, Worcester and Bovey Tracey in Devon. Fast growing centres of ceramic production were also located around Burslem in North Staffordshire. In contrast to continental practices these factories were not established by royal or aristocratic patrons but by merchants, apothecaries and artisans. Skilfully exploiting the snobbery of the rising bourgeoisie, they first presented their finest work to the aristocracy and nobility, such as Catherine the Great of Russia, and then, having secured royal approval, sold copies to the middle classes.[6] Wedgwood received the approval of Queen Charlotte for his 'Queensware', and after Worcester potters were visited by King George III and Queen Charlotte in 1788 they marketed their products as 'Royal Worcester Porcelain'. While British producers never received the high levels of subsidy enjoyed by continental potters, they were protected from foreign competition by tariffs on imports of over 100 per cent during the Napoleonic Wars. The French blockade and the actions of privateers added to the hazards of importing rival products. British potters were also offered premiums to create employment for local artisans by copying a wide range of foreign articles.

Porcelain producers in England

In England, as elsewhere in Europe, industrial espionage was rife, and rival potters stopped at nothing to steal each others' secrets. John Astbury, a Burslem potter, was said to have posed as a half-witted labourer and worked in a competitor's firm to spy on their processes.[7] Fear of the undercover methods of porcelain manufacturers is illustrated by the example of a doctor of medicine, John Wall, and an apothecary, William Davis, of Worcester, who set up a business in 1751 to develop 'the real, true and full art, mystery and secret' of making porcelain. They had 13 other partners, including Edward Cave, a London printer and publisher of the *Gentlemen's Magazine*. Wall and Davis promised that they would not divulge their secret to any other person, on penalty of paying the other partners the sum of £4,000.[8] The Worcester potters soon opened their first showroom in London and later successfully developed a product using Cornish soapstone. The following year another partner, the Quaker Richard Holdship, negotiated the purchase of the Bristol manufactory of Benjamin Lund, which was moved lock, stock and barrel to Worcester.[9]

The technique of transfer printing, 'a particularly English form of ceramic decoration',[10] was developed at this time by a number of manufacturers. John Sadler of Liverpool unsuccessfully applied for a patent for printing on earthenware tiles. His story was that he had discovered it by accident when his children, playing with some old engraving plates for printing on paper, applied them to some broken china.[11] A delightful tale, but his application came five years after the Worcester experiments in printing on porcelain and another application for a patent to print

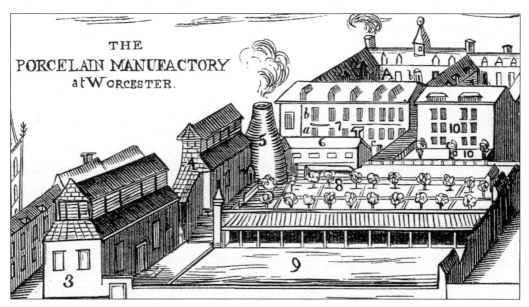

The Worcester porcelain factory in 1752, from Ceramic Art in Great Britain.

on enamel by John Brooks of Birmingham, who later worked at York House, Battersea, an enamel works in London.[12] Whatever the origin of his innovation, Sadler was certainly a skilled pottery printer, for in 1761 he received the supreme accolade when Josiah Wedgwood sent his creamware by pack horse to the firm of Sadler & Green to be printed. Liverpool was then a thriving pottery centre, with over 400 potters recorded in one district alone, of whom 117 were wealthy enough to qualify for a vote in the Parliamentary elections.[13]

Other successful Liverpool potters of the period included Seth Pennington, whose Oriental decorations were of such a high quality that a Staffordshire potter offered him 1,000 guineas for the secret recipe for a certain blue colouring, an offer which he refused.[14] A further claimant, or plagiarist, was a Staffordshire potter, William Jenkinson, who in 1751 advertised his discovery of the 'art, secret and mystery of making a certain porcelain ware in imitation of china'. His product, however, seemed remarkably similar to that of yet another potter who had set up a works in Chelsea in 1745.[15]

Any attempt to produce porcelain at that time was a very hit and miss affair. Clay only fused under intense heat, but if the temperature rose too high in any part of the kiln the result was a shapeless heap on the floor, known as a 'waster'. The early kilns were liable to this kind of problem, and the proportion of 'wasters' could sometimes be as high as, or even higher than, 90 per cent.[16] Perseverance and patience were needed, as well as a scientific approach, exemplified in the experiments not only of Josiah Wedgwood but also of a wholesale chemist of Plymouth, William Cookworthy, who had become interested in the manufacture of porcelain.

References

[1] The name comes from the Italian porcelino, or 'little pig', after the translucent, pearly little-pig-shaped cowrie shells once used as currency.

[2] Apart from the sources cited in the text, information for this section comes from the porcelain collections at the Bowes Museum, Barnard Castle, Birmingham Art Gallery and Museum, the Royal Worcester Porcelain factory, Pottery Museums in Stoke on Trent, Preston Art Gallery, Holborne Museum of Art at Bath, Plymouth Museum and the RIC Museum at Truro.

[3] A.R. Mountford and F. Celoria, 'Some examples of sources in the history of 17th century ceramics', *JCH*, I, 1968, pp.12, 17; Paul Rado, *An Introduction to the Technology of Pottery*, Oxford, 1969, pp.170–71; 'The Evolution of Porcelain', *JBCS*, 1963, I, 3, pp.419–23; George Savage, *Pottery Through the Ages*, Penguin, 1959, pp.178–79; William Burton, *Porcelain*, London, 1906. Chapter IV has letters of du Halde in full. Dwight's recipes, as given in his notebooks and found by Lady Charlotte Schreiber, exist as copies in the British Museum.

[4] Janet Gleeson, *The Arcanum*, London, 1998, pp.45–56.

[5] The mineral that they called 'hua-stih' was not soapstone at all but an impure kaolin containing a large proportion of white mica.

[6] Chris Freeman and Luc Soete, *The Economics of Industrial Innovation*, London, 1997, pp.43–48.

[7] Willaim J. Claxton, *In the Potteries*, London, 1913, p.71. This tale appears in other histories of the potteries.

[8] Henry Sandon, The *Illustrated Guide to Worcester Porcelain*, London, 1980, pp.3–5; E. Morton Nance, *The Pottery and Porcelain of Swansea*, London, 1943, pp.473–74.

[9] Ray Jones, *Porcelain in Worcester*, Worcester, 1993.

[10] Pat Halfpenny, *Penny Plain, Twopence Coloured*, Stoke-on-Trent, 1974, p.7.

[11] Sebastion Cradelwith, 'When Liverpool was the Potteries', *Liverpool and Merseyside Illustrated*, March 1961, pp.22–23.

[12] Savage,1959, p.30; Halfpenny, 1994, pp.11–12..

[13] *VCHL*, II, 1908, repr. London, 1966, p.406.

[14] Joseph Mayer, *On the Art of Pottery*, Liverpool, 1871, p.41.

[15] *Plymouth Porcelain*, Plymouth Department of Lifelong Learning, 2002.

[16] Savage, 1959, p.23.

Chapter Four
What did William Cookworthy Achieve?

A few miles west of Helston, on Tregonning Hill, a stone commemorates the 300th anniversary of the birth of 'William Cookworthy, a Plymouth Quaker chemist, who discovered China Clay' on that spot around 1746. Since he has acquired an iconic status in the china clay story as the virtual founder of the Cornish industry, his career is worth examining in some detail. It vividly illustrates the triumphs and disasters that were common among would-be porcelain producers of the time, reflects the secrecy in which they operated and, in his case, the value of the Quaker connection among merchants and industrialists.

Cookworthy (1705–80) was born near Kingsbridge in Devon, son of a prosperous weaver, but his father's death in 1718 ruined the family finances and he was apprenticed to Quaker pharmacists in London, who eventually set him up in business in Plymouth.

A man of wide scientific interests, he developed an overriding curiosity about the manufacture of porcelain at a time when sources of what later became known as china clay were being identified in various parts of Europe and also in North America. Darnet, a French surgeon, had come across it near Limoges, where a porcelain factory later opened in 1761.[1] Earlier, the proprietors of a factory in Bow had taken out a patent to make porcelain using a clay imported from the Cherokee territory of what is now North Carolina, and clay from this source had also been used by Wedgwood and shown to Cookworthy in 1745.

Experiments in porcelain manufacture had been proceeding in America at least since 1738,[2] and the delivered cost for clay was £13 a ton. Even at this price, Cookworthy believed, it could be used to make 'china as cheap as common stone ware'.[3] However, during trading visits in Devon Cookworthy had come across deposits of china clay on southern Dartmoor, but rejected them because they did not fire as white as those he had been shown by a Quaker mine captain at Tregonning Hill.[4] This clay, found between Helston and Penzance, was being used by a mine captain and fellow Quaker, John Nancarrow, to mend his furnaces.[5]

Cookworthy was not without knowledge of the subject. He had read *De Re Metallica*, the pioneering treatise on mining by Agricola mentioned in Chapter One, and had studied the techniques developed by potters at Bow and Worcester, both run by fellow Quakers. He had also experimented with a mixture of the china clay and stone he had spotted at Tregonning Hill, and he carried out a short-lived attempt, with others, to produce hard-paste porcelain in 1765 at a factory at Bristol, a city that had long been a thriving pottery centre and where fuel was more readily available. However, he ceased operations after only a few weeks because he was unable to fire the chinaware without smoke-staining. This venture was financed partly by his brother-in-law, Thomas Were, another Quaker, who was a Somerset woollen merchant. Between them, the partners managed to lose £600, but, like some earlier German and French potters, Cookworthy eventually found support from a rich and aristocratic collector of porcelain.

William Cookworthy and Thomas Pitt

This was Thomas Pitt (later Lord Camelford), of Bocconoc in East Cornwall, nephew of Prime Minister William Pitt the Elder. Cookworthy's correspondence with Pitt, identified in 1980 by Colin Edwards of the Cornwall Record Office, contains small samples of the porcelain that Cookworthy was making, and also offers valuable insights into both Cookworthy's and Pitt's knowledge of porcelain production. In this one-way correspondence, from Cookworthy to Pitt only, we can discover details of his preparation of clay and stone, and his development of firing methods, which were not available to earlier clay historians and which are worth outlining here.[6]

Cookworthy had also discovered stone on Pitt's lands at Tregargus and clay and stone at Carloggas, both near St Stephen-in-Brannel. He believed the Carloggas stone to be 'immense in quantity' and superior in quality to the Tregonning Hill deposits. Apprehensive that Pitt might ignore a proposal from an unknown chemist as 'mere fancy and chimera', he asked Lord Edgcumbe, a member of one of the greatest Cornish families, with whom he had some acquaintance, to act as a go-between. Emphasising that he had found the 'genuine materials used by the Chinese' on Pitt's land, Cookworthy mentioned his intention to build a small pottery in Plymouth to experiment with them. Since, he claimed, other potters sometimes had to fire their materials three times, while he often only needed one firing, he would be able to 'undersell them greatly'. Cookworthy was comparing a single high-temperature firing to produce glazed ware in hard-paste porcelain with soft paste requiring separate biscuit and glazing firings

At this stage Cookworthy did not ask for financial assistance but simply for an assurance that he could use as much of Pitt's materials as he required. When Pitt approved of this, Cookworthy proposed a licence lasting 'three lives', a common form of leasing land and buildings, offering £0.50 for every ton of washed clay produced and £0.125 per ton of stone. China clay and china stone were used in roughly equal proportions in hard-paste porcelain. Should his experiments fail, Pitt would receive half the profits that could have been made on sales of clay and stone to other potters. However, he was soon negotiating the raising of finance, asking for one-tenth of any profits made by the pottery to cover his own start-up costs of about £80. He also asked for ten of the 20 shares in the business for himself and his friends. A sum of money of the order of £400 was raised from Pitt and other subscribers, again including Thomas Were. One of the contributors was Richard Champion, a young Bristol merchant from a Quaker family, who had taken an interest in Cookworthy's Bristol venture and who had sent samples of china clay, acquired from his brother-in-law in America, to a Worcester pottery to be fired, but with indifferent results.

Problems with china stone grinding and the making of saggars

In his pioneering work, Cookworthy met problems with the grinding of china stone. A finished sample of glazed porcelain had been contaminated by 'a Mill of the common foul moorstone used to reduce the [china] stone which composes the glaze to a fine powder, whereas the mill ought to have been made of the same stone'. The contamination probably referred to brown specks on the sample caused by iron impurities. The mill may have been a traditional bread flour mill with two flat, circular grinding stones, one fixed and one rotating. These were quite often cut from local granite moorstone, which contained dark, iron-bearing micas. It was soon recognised, however, that when these mill stones were used to grind china stone for use in chinaware, they would wear, contributing irony particles to the final product with deleterious results.

In February 1767 Cookworthy was planning clay and stone production at Carloggas and gave instructions for 'forming proper Conveniences for washing the clay', ordering a stone mason at St Austell to make a 'Mill for Grinding the Stone'. The Potteries had experienced health and contamination problems when dry grinding calcined flint in iron mortars and stamp mills, and as early as 1726 Benson had patented a wet grinding mill to avoid this hazard. In essence, the lower millstone was replaced by a shallow watertight pan, paved with non-contaminating stones called 'pavers', while the rotating stone was replaced by rough cubes ('runners') of similar stone. These were pushed around the pan by wooden arms on cast-iron brackets driven by gearing from a water-wheel. It was probably this type of mill that Cookworthy had built using china stone runners and pavers, with a water-wheel to drive the sweep arms. There are no records of Cookworthy's mill, but it may have been on the site of a group of mills built at Tregargus.

Another of his difficulties was in making 'saggars', (a term derived from 'safe-guards'), the ceramic containers in which pottery was fired in a kiln to prevent it from being affected by smoke and gases from the kiln fires. In 1767 Cookworthy ordered some 'pipe clay' (ball clay) from Teignmouth for use in saggars, and in 1768 he experimented to find the 'best composition' for saggars. He later obtained Stourbridge clay and reported that he could make satisfactory saggars, although we do not know the final recipe he developed.

Experiments at Bovey Tracey and Truro

Correspondence with Pitt also revealed more about his contacts with a pottery at Bovey Tracey. The presence of deposits of lignite close to beds of ball clay had encouraged potters from Staffordshire to move there but unfortunately the lignite proved unsatisfactory as a fuel. Originally Cookworthy had intended to use 'sea coal' from Newcastle, 'vastly cheaper than wood' to fire his kiln, but he later decided that 'coal will not do for us', perhaps because of the fumes. Reconsidering wood, he thought that Lostwithiel (close to Bocconoc) was 'without doubt the proper place for carrying out our Manufacture'.

However, wood also proved unsatisfactory, so he next experimented with a mixture of coke and wood from various sources, including Bovey Tracey, Plymouth Dockyards and Truro. Hoping that an experienced potter might produce better results than his own men, he sent his clay and stone mix to Nicholas Crisp of Bovey Tracey. Crisp, formerly a partner of the Vauxhall pottery in London, seemed a useful associate for experimental work on porcelain. Of scientific bent, a founder member of the Society of Arts, he was also a member of the Haberdashers' Company and, until 1751, a jeweller. He received a Society of Arts premium of £50 for work in connection with cobalt in 1764. Yet, like Cookworthy, while always imagining he was on the brink of great success, he never achieved it. He had been declared bankrupt in London in 1763 before moving to Bovey Tracey, where he died in poverty in 1774. In the 1990s excavations took place at the main Bovey pottery sites. The quality and range of pottery found was a surprise, and has led to a reappraisal of the significance of this area.[7]

They only worked together for a year or two, because they had different views

Pottery shards from the Indeo site at Bovey Tracey, unearthed in 1992. These revealed a wide range of colours and types previously unknown in this area.
(ILLUSTRATION COURTESY OF BRIAN ADAMS)

on firing techniques. Relations between the two men also deteriorated because Cookworthy appeared to poach some of Crisp's skilled workmen, including a decorator, a modeller and a 'burner', or expert in firing techniques. Cookworthy, on the other hand, suspected Crisp of passing on his secrets to John Bolton, a London enameller, who had also previously been employed at the Vauxhall pottery, leaving there in 1755 to set up a factory in Kentish Town in London.

Cookworthy also used Jacob Lieberich, the crucible manufacturer at Calenick, near Truro, mentioned in Chapter Two, to try out his mixture of clay and stone, hoping that he would achieve better results. Yet still success eluded Cookworthy, and he was plagued by worries that someone else would produce porcelain using the same ingredients before he did. One such rival was a Frenchman, the Duc de Brancas, Comte de Brancas-Lauraguais, who, Cookworthy learnt, 'was pushing very hard to get a patent'. Brancas was said to have discovered kaolin at Alençon in Normandy, to have worked at Sèvres and visited England and Scotland in the 1760s looking for raw materials for hard-paste porcelain.

He took out a provisional patent in 1766, two years ahead of Cookworthy, for a 'new method of making porcelain ware in all its different branches'. However, he gave no technical details or recipes for making porcelain and so the patent did not take effect.[8] One historian of porcelain, however, has doubted whether Brancas ever developed a viable process,[9] and in any case it was Cookworthy who succeeded in taking out Letters Patent which gave him the monopoly throughout England and Wales. His formula, he claimed, had 'all the Characteristics of the True Porcelain... in a degree equal to the best Chinese or Dresden Ware'.[10]

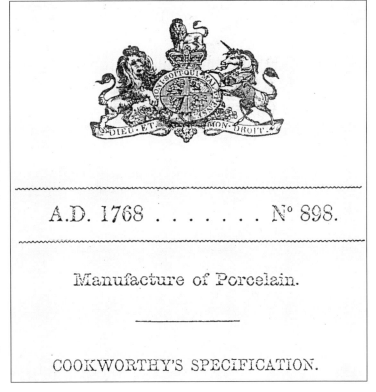

A.D. 1768 . . . . N° 898.

Manufacture of Porcelain.

COOKWORTHY'S SPECIFICATION.

Title heading for Cookworthy's patent of 1768. The full patent has three pages.

Pitt's involvement in pottery

Thomas Pitt was now taking an increasing interest in Cookworthy's experiments and had visited potteries in Worcester and Musselburgh in Scotland and passed on what he felt were helpful comments on their kiln construction and firing methods, as well as sending a quotation from du Halde's work. Cookworthy, however, was politely dismissive of these comments, convinced that his own kiln design and porcelain formula were markedly superior, and condemned du Halde's methods as 'an unintelligible piece of nonsense'. To Cookworthy, as he made clear in his patent application of 1768, all other so-called English porcelain was a poor imitation of his own true product. Some of his output, though, continued to be flawed, with blurred glaze, smoke stains, specking and fire cracks, due to irregularities in the materials, the firing process and the flat topped kiln design.[11]

By this time Pitt seemed to have been carrying out his own experiments into a cobalt-based blue pigment for painting on chinaware, and Cookworthy was advising him how to construct and line a kiln. Perhaps Pitt was becoming disillusioned by Cookworthy's slow progress, but Cookworthy now took a decisive step, opening a factory, possibly at Coxside, near the Barbican at Plymouth, and engaging experienced designers, decorators and mould makers from Crisp at Bovey Tracey, as well as a French painter from Sèvres. The factory employed between 30 and 60 workers and at first made relatively simple domestic ware. These used an under-glazed decoration which fired to an inky blue-black or pale grey-blue, the designs being inspired by Chinese porcelain. Soon, though, the factory was making more decorative articles with over-glaze coloured enamels, many applied with over-glaze polychrome enamels.

Champion takes over

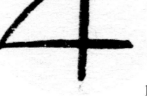

Cookworthy's alchemical pottery trademark.

In 1770, however, Cookworthy closed his Plymouth manufactory and production was transferred to Bristol. According to one version of events, it was sold to a consortium of Bristol merchants in which Richard Champion, his earlier partner, was involved.[12] In an alternative version, Cookworthy retained a controlling interest but Champion ran it.[13] The moulds, designs and the Plymouth 'alchemical' trademark, together with some workers, went to Bristol.[14] In 1774 Cookworthy, perhaps because of his advancing years, assigned his Letters Patent to Champion, who claimed to have 'been at a very considerable Expense and at great Pains and Labour... by reason of the great Difficulty attending a Manufacture upon a new Principle'. In order to 'receive an adequate Compensa-tion', an extension of the monopoly for a further 14 years was requested, and granted.[15]

Champion's porcelain was of high quality, with more ornate designs and lavish use of gilt, yet he still failed to make ends meet, partly perhaps because of the legal costs incurred in fighting a challenge by Wedgwood to break his monopoly. He may also have found it difficult to sell to potential royal or aristocratic patrons because of his robust support of the opposing Whig party, detested by George III. Champion had applied to renew Cookworthy's 99-year lease on Pitt's land, but this had been resisted by Wedgwood. The latter did not want china clay to mix with china stone to make porcelain in the way that Cookworthy did, but rather to brighten the appearance of his 'cream ware', made from ball clay. For although Wedgwood's 'Queen's ware' achieved popular acclaim, he realised that he would be able to charge higher prices for a whiter product.[16]

Wedgwood, with the aid of other potters, succeeded in limiting the licence so that they could utilise china clay and china stone if they did not use the Cookworthy formula, which Champion, running short of funds, eventually sold with his moulds to a syndicate of ten Staffordshire potters, later reduced to four. He took up a post in London as Assistant Deputy Paymaster General for a few months before emigrating to America in 1784. The syndicate, which became the New Hall Co., continued to make porcelain, using a much modified Cookworthy-Champion formula incorporating animal bone, which is known today as 'hybrid porcelain'.[17] There is no evidence that they sold compounds suitable for making a porcelain body before 1796, when the patent expired, or at any time afterwards. However, they may well have done, since this was a logical step, as 'body' and 'glaze' must be carefully matched for successful porcelain production.

Hybrid porcelain is twice fired in its undecorated state and requires relatively high firing temperatures, with a greater risk of failure than products made from later bone recipes. Although the hybrid formula was used by Chamberlain of Worcester and also at Caughley and at Coalport, it was effectively eclipsed by a new and specifically English product. Around 1800, Josiah Spode II introduced china made of 50 per cent bone ash, 19 per cent china stone, 19 per cent china clay and small amounts of ball clay and flint. This recipe had been developed by his late father before his death in 1797, and the new product, originally called 'Stoke China', later became known as Bone China. An advantage was that large plates could be fired without warping at lower firing temperatures than the Cookworthy formula. This made it suitable for firing in the 'bottle' ovens of Staffordshire and it soon became the standard form of English china. Its manufacture was almost exclusively limited to England, with just one other firm in Sweden. The New Hall

Co. stopped making porcelain using their modified Cookworthy-Champion formula in about 1810, and continued with bone china manufacture until 1835.

Cookworthy: 'a man of no common clay'

To sum up, how important was the contribution of Cookworthy? As one authority on his work has remarked, 'many fictional embellishments' have been added to his story, and contemporary opinions about him varied.[18] William Pryce, Redruth surveyor and author, predicted that Cookworthy's Bristol factory would produce porcelain 'not less elegant than the best Asiatic China'. Jabez Fischer, an American visitor who met Cookworthy, asserted that he was the 'most sensible, learned, kind man I ever knew, the history of every nation is familiar to him, he has explored every country, understands many languages...'[19]. Yet Cookworthy's obituary in *The Gentleman* confined itself to praising his good works as a Quaker,[20] whereas Champion, who had done less to merit recognition, was referred to in his obituary in the same magazine as the proprietor of a 'China Manufactory.[21] Not until a century later did the Revd C.M. Edward Collins of Blisland, an advocate of potteries in Cornwall, as we shall see in Chapter Seventeen, praise Cookworthy for his 'rare but still beautiful Plymouth Ware which, for grain and texture, has seldom been excelled'.[22]

William Cookworthy, 1705–80, from Wheal Martyn Museum Guide, 1983.

Cookworthy was not the first to use china clay, and the problem of emulating the ancient Chinese makers of porcelain was solved in Saxony around 1708 and then in various parts of the Continent. Over half a century elapsed before Cookworthy was technically successful in producing hard-paste porcelain and then only for a dozen years, during half of which he had retired. Moreover, during their lifetimes, both Cookworthy and his successor, Richard Champion, were dismissed as commercial failures. Although connected by marriage to the highly successful Fox family of Falmouth, they still managed to lose considerable sums by investing at the wrong time in the Fox's normally profitable Polgooth tin mine near St Austell.[23]

However, Cookworthy and Champion were far from alone in failing to make money out of porcelain production, and the history of the industry is littered with firms going into liquidation. Did Cookworthy have a negative influence, though, in delaying progress in porcelain development by holding on to his monopoly of china clay in Cornwall? James Watt, the great pioneer of steam power, has been accused of 'clogging engineering enterprise for more than a generation' by extending his licence in steam technology by 25 years.[24] However, the same could not be said of Cookworthy, because Wedgwood and others were not interested in using Cookworthy's formula and experimented with china clay from Cornwall, as well as from other sources, even when Cookworthy held his licence.

Cookworthy's true claim to fame is not that he was the first to discover china

Cookworthy's grave in the Quaker cemetery at Efford, following reinterment after the postwar development of Plymouth.

clay in England, nor that he founded the Cornish china clay and stone industry, which would have prospered without him. His achievement was in pioneering the relatively complex process of refining china clay for use in firing at high temperatures at a time when most potters simply dug up clay and used it more or less in its raw state, like ball clay. The making of porcelain was not just a matter of digging up china clay and stone and converting them into chinaware. As we saw in Chapter Three, as late as 1839 De la Beche was remarking how china clay was being 'artificially prepared' by washing, screening and drying, suggesting that these practices were still a novelty in England, and praising Cookworthy, 'a person of much talent', for his trail-blazing preparation of these 'artificial clays'.[25]

In doing so, Cookworthy raised the whole standard of British ceramics by exhibiting the whiteness of Cornish china clay and the fusibility of Cornish china stone, which were used with great success in other chinaware recipes. Cookworthy had the honour of being the first potter to replicate in Great Britain, albeit with indifferent commercial success, 'the true art, mystery and secret of porcelain' invented by the Chinese.

References
[1] Paul Rado, 'The Evolution of Porcelain', *JBCS*, 1963, pp.419–23.
[2] Alice Cooney Frellinghusen, *American Porcelain*, New York, 1989.
[3] F. Severne Mackenna, 'William Cookworthy and the Plymouth factory', *TECC*, 1982, II, 2, pp.84–98.
[4] E.A. Wade, *The Redlake Tramway and China Clay Works*, Truro, 1982.
[5] A.D. Selleck, *Cookworthy, a Man of No Common Clay*, Plymouth, 1978, p.56.
[6] For Cookworthy's letters to Pitt, see Cornwall Record Office, DDF (4), 80/1-30; for a commentary on these letters see Cyril Staal, 'William Cookworthy and his Plymouth Porcelain Factory', *JRIC*, 1981, VIII, 4, pp.267–74; Geoffrey Wills, 'The Plymouth Porcelain Factory', *Apollo*, 1980, CXII, 226, pp.377–85 and 1981, CXIII, 227, pp.29–37. The letters were reprinted in Geoffrey Wills, *The Plymouth Porcelain Factory Letters to Thomas Pitt*, Canterbury Ceramic Circle, 1998.
[7] Brian Adams and Anthony Thomas, *A Potwork in Devonshire*, Devon, 1996.
[8] A.H. Church, *English Porcelain*, London, 1885, 1898, revised HMSO, 1911; Mackenna, 1982, pp.90–91.
[9] George Savage, *Porcelain*, Penguin, 1959, p.12.
[10] CRO, DDF(4) 86/28.
[11] Some examples of Cookworthy's work, showing these faults, are displayed in Plymouth Museum.
[12] Lady Radford, 'Plymouth China', *The Devon Year Book, 1920*, pp.30–41.
[13] *City of Plymouth Museums and Art Gallery Catalogue*, 2005.
[14] The Plymouth pottery works may have continued, though, since a pottery is listed at Coxside in 1812 and 1814, operated by Fillis & Co. and in 1822 by William Alsop. *The Picture of Plymouth*, Plymouth, 1812; *Plymouth Directory*, Plymouth, 1814, 1822.
[15] An Act for Enlarging the Term of Letters Patent granted... to William Cookworthy, CRO, DDF (4) 80/28.
[16] Rado, 1963, p.443; J. Allen Howe, *Handbook for the Collection of China Clay and China Stone*, HMSO, 1914, pp.7–8.
[17] Geoffrey A. Godden, *New Hall Porcelain*, Woodbridge, 2004, pp.82–86; David Holgate, *New Hall and Its Imitators*, London, 1971, p.100.
[18] Dr F. Severne Mackenna, 'William Cookworthy and the Plymouth Factory: An Updating', *TECC*, 1982, II, 2, pp.84–98.
[19] William Pryce, *Mineralogia Cornubiensis*, 1778; Jabez Fischer, 1775, quoted in Godden, 2004, p.72.
[20] Wills, 1980, p.377; Selleck, 1978.
[21] Godden, 2004, p.86.
[22] C.M. Edward Collins, *RRCPS*, 1868, p.33.
[23] R.L. Brett, *Barclay Fox's Journal*, 1865, pub. London, 1979.
[24] T.S. Ashton, *An Economic History of England*, London, 1955, p.102; Peter Mathias, *The First Industrial Nation*, London, 1983, p.35.
[25] Sir Henry De la Beche, *The Geology of Cornwall, Devon and West Somerset*, London, 1839, pp.257, 509, 510, 512, 513.

Chapter Five

Sizing, Bleaching and Filling Cotton Cloth and Paper

Until the 1830s, sales of china clay and china stone kept roughly in line, with china stone making up about 40 per cent of the combined total, but from then on clay sales accelerated much more rapidly, as the graphs illustrate.[1] The ceramics industry was the only significant customer for china stone, so the divergence in production suggests that demand from potters for china clay and china stone continued to grow at a slow rate, whereas demand by non-ceramic users for china clay alone was rising much more swiftly. Cotton textiles and paper-making were the two main outlets where clay's smoothness and fineness made it useful for sizing, whitening and filling, whereas china stone, even when ground into powder, was too abrasive and not as white.

The development of cotton cloth manufacture and paper-making have several features in common, apart from their use of china clay as a filler and whitener. Both utilised forms of cellulose as their main raw material and both benefited from the discovery of chemical bleaching. Cotton fibres are the hairy covering of cot-

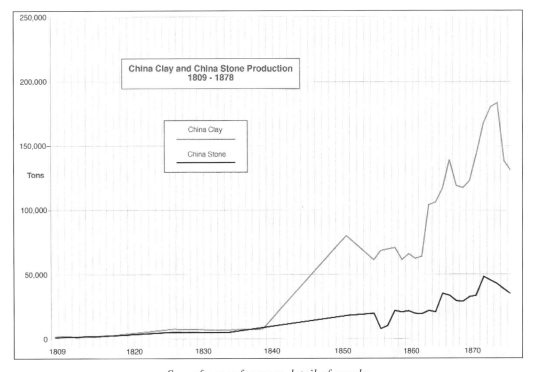

See reference for more detail of graphs.

ton seeds, and after cotton cloth was downgraded into rags, these became the main constituent of paper. When demand for paper overtook the supply of rags, other forms of cellulose were substituted, including straw, esparto grass and, most importantly, wood fibres. However the timing of the growth of the two industries differed. While the boom in cotton cloth manufacture occurred from the later 1700s, the massive increase in paper production took place in the later 1800s, when it became the main outlet for china clay.

When we try to estimate the relative amounts of clay used by cotton cloth or paper makers we run into two sets of difficulties. In the first place, just as the potters kept some of their operations a closely guarded secret, so some cotton textile and paper manufacturers, for reasons that will be discussed later, were unwilling to disclose that they used materials like china clay, and because of this the clay merchants were similarly reluctant to admit that they were producing clay for these markets.

This may account for the fact that, although production of cotton cloth expanded massively before 1830, recorded sales of china clay to cotton manufacturers do not appear to be reflected in the production graph shown earlier in this chapter. While china clay producers referred to the material they sent to textile and paper makers as 'bleaching clays', they were not used for 'bleaching' in the usually accepted meaning of the term, but for other processes such as the 'sizing' of cotton threads and the finishing of cotton cloth using calendering rolls and as a constituent in the mixing 'vat' used to feed paper-making machines.

'Sizing', 'bleaching' and 'filling' in cotton textiles

Traditionally, cotton textile manufacture has been regarded by historians as the major propellant force of the early Industrial Revolution: between 1760 and 1800 it exploded from a relatively small and localised activity into the most dynamic sector of the British economy. Despite the vast distances over which the raw material had to be transported compared with wool or linen, advances in technology and engineering reduced the cost of a piece of cloth by 1830 to one-seventh of its 1780 price, leading to a vast increase in demand.[2] This rapid growth was largely due to the fact that cotton fibre was more versatile and easier to handle mechanically than wool or flax, while cotton fabric was more readily printed with patterns. The share of cotton goods in the total value of British merchandise exports escalated from only 1 per cent in 1750 to 46 per cent by 1801, whereas that of woollen goods fell from 47 per cent to 14 per cent.[3]

However, the vast increase in the scale and speed of production created problems of its own, and it was to help solve some of these that china clay and similar substances were needed. The manufacture of cotton goods involved a range of processes, beginning with spinning to convert cotton fibres into thread, which was then woven on a loom to make cloth. In the 1760s a major advance in spinning cotton mechanically was made by James Hargreaves, a Lancashire hand weaver, with his 'spinning jenny'. Until then the yarn had to be spun by twisting it by hand, but the multiple rotating spindles of Hargreaves allowed the operator to spin a number of threads at the same time.

In 1769 Richard Arkwright patented an improved spinning machine and set up a mill at Cromford on the River Derwent that was driven by water power, and for this reason the name 'water frame' was given to his innovation. Ten years later Samuel Crompton invented a machine combining features of the spinning jenny and the water frame which, being a hybrid, was named a 'mule'. In weaving, a series of longitudinal threads called warps are criss-crossed by other threads called wefts. The warp threads are subject to flexing and abrasion as cloth is woven, and to minimise damage to them, and a consequent loss of production time, a 'size', or coating of a flour and water paste, was applied by hand in the early days of the industry.

In 1803 machines were introduced to speed up this process, but they also increased the risk of damage, and the sizing of warps became a critical operation. Other components were added to the size to increase its efficiency such as oils, Epsom Salts (magnesium sulphate) and china clay. One advantage of clay was that it was three times as heavy as flour, and the value of sizing mixtures was often judged by weavers on their weight, apparently on the doubtful principle that 'heavier is better'. At this time heavy drapes and materials were the fashion.[4] For fine

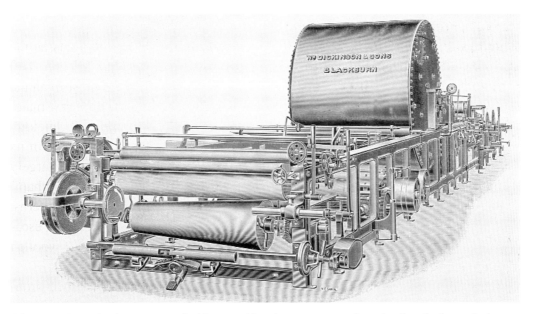

The tape sizer, also known as a slashing machine, incorporates a 'sow box' at the far end where a heated mixture of size is put on the yarn. A heated cylinder dries the yarn, which is finally wound onto a beam for use in the weaving process,

materials, hand-spun and woven Italian muslins had always been preferred, but mechanically produced British cloth was now just as delicate. This was made possible by a wide range of special machinery for coating and embossing cloth.

As we have seen, china clay producers tended not to differentiate between use for sizing or other purposes, simply referring to sales of 'bleaching clays'.[6] Thus, in the severe trade recession of the 1830s, a number of leading clay producers, including John Lovering, asked for a rebate on port charges for what they called 'bleaching clays' which sold for £0.50 less than best-quality potting clays but nevertheless represented a useful source of revenue.[7] In the 1850s, a clayworks was recommended for sale on the grounds that it was eminently suitable for 'bleaching and calico, for which china clay is extensively used',[8] and a well-known guidebook of the time recommended tourists to visit a clay pit and see how a material was extracted 'for bleaching paper and calico to give them weight and body'.[9]

That clay was widely used for non-ceramic purposes is evident from the *Catalogue* of the Great Exhibition of 1851 at the Crystal Palace in London. Although clay producers displayed samples of china clay and china stone as well as chinaware made from these materials, they mostly, in their written entries to the *Catalogue*, referred to other uses. The West of England China Stone & Clay Co. alluded to clay 'being sent to the potteries and bleaching manufactories in various parts of the old and new world'. Charles Truscott and Elias Martyn mentioned 'bleaching and paper-making', Sarah Michell wrote of 'bleaching paper and calico', William Phillips of Lee Moor referred to 'calico dressing and paper-making' and Philip Wheeler to 'bleaching clay used in the cotton and paper manufactories'.[10] On the other hand, William Browne and Thomas Thriscutt made no mention of uses other than in ceramics.

The *Catalogue* entry by Jenkins & Courtney, who worked the Bodilla china stone quarries between Treviscoe and St Dennis, and who only supplied potters, included a derogatory memorandum about other uses. The author, 'R.H.', who might have been the authoritative Robert Hunt, Government Keeper of Mining Records, claimed that 'a large quantity of china clay of an inferior quality is used by the paper makers and calico dressers for the purpose of giving weight and body to their fabrics'.[11]

Chemical bleaching and 'bleaching clay'
This use of the term 'bleaching' referred to the ability of a material such as china clay, alum or plaster of Paris to make cotton cloth or paper look whiter and smoother, not to the use of a chemical agent to take the colour out of it. Until the later 1700s, cotton cloth was bleached by 'whitsters', who washed it and spread it out to dry on grassy fields to fade in the sun. This slow process, taking as long as

six to eight months, was speeded up by soaking the cloth in 'lye', a mixture of plant ash (potash) and water before exposing it to dry. Sometimes sour milk, a weak acid, was also used, or even sulphuric acid. Like many important discoveries of the time, chemical bleaching was invented in France but applied in Britain. In 1784 the Frenchman Berthollet discovered the bleaching properties of chlorine, which greatly reduced the working capital needed to finance the 'lead time' during the bleaching process, which dropped to just a few days. It also eliminated the cost of leasing land on which to spread the cloth.[12] To put the impact of this innovation on the price of cotton into perspective, bleaching costs made up about 4 per cent of the total sales price of cotton cloth.[13]

By 1830 the principal chemical bleaching centres of Britain were Bolton, Blackburn, Manchester and Glasgow.[14] However, the predominance of the Lancashire cotton trade has overshadowed the contribution of West Country producers. As early as 1695 the spinning and weaving of cotton was advocated as a means of providing employment in workhouses for the poor at Bristol and, a century later, when Champion was operating his porcelain factory, at least four cotton mills were working in the neighbourhood. In the first half of the nineteenth century there seemed no obvious reason why the Bristol area should not become as significant a cotton manufacturing centre as Lancashire. After all, high-quality coal from South Wales was cheaper than in Lancashire and the port was just as accessible for American raw cotton imports and cotton cloth exports.

In the year 1836 alone, prospectuses appeared for the Bristol Cotton Co. with a proposed capital of £2,500,000, for the Bristol & Bitton Cotton Twist Manufactory with £300,000 capital and for Bristol Cotton Twist & Power Loom Cloth Co. for £200,000. In 1840 one firm, the Great Western Cotton Works, employed over 900 hands, rising to nearly 2,000 by 1899. Yet Liverpool cotton merchants proved more dynamic in forging links with New York, so that cotton imports and exports tended to go between Bristol and Liverpool by coaster, and then by ocean-going vessels across the Atlantic.[15]

One problem with chemical bleaching, though, was that it left cotton cloth with a rough and crumpled appearance, and something was needed to make it look whiter and also fill the interstices of the material, giving it a smoother and richer appearance. Historians of the textile industry have underestimated the importance of this process for, as early as 1788, the filling process required one-sixth of the total capital invested in cotton manufacture, some £300,000,[16] and fillers eventually made up to 70 per cent by weight of the cheaper cotton cloth. The first stage of finishing consisted of mechanical treatments to produce various surfaces. This was followed by 'calendering' or mangling processes involving various combinations of rollers which picked up a layer of the clay mixture from troughs below the rollers and coated the cloth with it.

China clay and paper-making

From the first century AD, the Chinese pulped fibres such as the bark of trees, hemp and rags for paper-making.[17] A frame with a mesh called a mould was then dipped into a vat of macerated fibres and water until a thin, uniform layer of pulp formed on the mould. This layer was then taken off the frame and placed between layers of felt cloth, where the water was squeezed out. The sheets of paper formed were then dried in lofts equipped with canvas or hessian cloths to support them. A final

Paper mill, c.1662. In this drawing various processes have been compressed into one room. The water wheel (A) powers stamps B, C, D, E. A vat (G) holds pulp for the hand frames. The vat man formed a sheet of paper which was passed to the coucher, who pressed water out with a screw press (F).

stage for some paper was to coat or 'size' sheets to minimise smudging when printing ink was used. Gelatine derived from animal carcasses was often employed as the coating, but starch or gypsum was also used by the Chinese.[18]

From the traditional home of the craft, knowledge of paper-making slowly followed much of the same route as that of porcelain, through the Middle East and the Eastern Mediterranean to Italy, France, Germany and eventually into England, arriving there at the end of the sixteenth century. Most of the output at that time was wrapping paper, or pasteboard made by pasting layers of paper together for book covers or playing cards. The invention of printing then provided the stimulus for whiter grades, which replaced the more expensive parchment made out of animal skins. During the seventeenth century most of the growing British demand for paper was met by imports, but by the 1720s British mills were making about two-thirds of the total required.

At the end of the eighteenth century these simple methods were still in use in thousands of small artisanal, water-powered mills across Europe.[19] Almost all the paper mills in the West Country at that time were on sites converted from corn or woollen mills, and although some produced writing and printing paper, most made wrapping paper or pasteboard for local use. While the woollen industry was still important in Devon, some mills manufactured glazed paper for the hot-pressing of cloth. Of the 52 mills identified in Devon, all but two at Barnstaple were situated in the south of the county, concentrated along the coastal areas from Plymouth to the lower Exe valley and its tributaries. Twelve paper mills operated at various times in Cornwall, again spread along the South Coast from Penzance to the Tamar valley. In Dorset only four mills have been identified, two near Wareham and two near Wimborne.[20]

Their location was largely determined by two factors. A fast-running clear stream was essential to power the stamping mills used to pulp fibres and also provide pure water to clean the raw materials. A local supply of these materials was another important advantage. Cotton and linen cloth was needed to produce white paper, but woollen rags, old rope and canvas sails were adequate for making the cruder sorts of brown wrapping paper. A sizeable local town, especially a port, would both provide these commodities and a public that demanded paper, which helps to explain the concentration of mills around Exeter in the South West. This city also had the advantage of cheap labour compared with London and the expanding towns of the North, where other industries competed for workers. Paper mills in the West Country employed a larger proportion of young girls than, for example, Lancashire, where higher-paid work in cotton mills was available. To put these figures in perspective, however, there were over 4,000 paper mills in England at the end of the eighteenth century, which meant that the South West only made up 4 per cent or less of the total.

The mechanisation of paper-making

In the 1790s a Frenchman, Nicholas-Louis Robert, invented a process to produce paper in a continuous roll instead of in separate sheets. This was developed during the Napoleonic Wars by an English mechanic, Bryan Donkin, and evolved into the Fourdrinier machine, named after two London brothers with paper-making interests who helped to finance its development. A new process for sizing was devised from 1807, when Illig from Switzerland proposed the use of rosin and alum instead of gelatine. Rosin was made by distilling resin from trees. Because of this, the year 1807 is sometimes suggested as the year in which china clay was also first applied in paper-making.[21] However, Illig's process was not widely used until 1830, when drying cylinders had been added to continuous paper-making machines, improvements linked to the names of a London stationer, John (later Sir John) Dickinson, and T.B. Crompton. Both became eminent paper-makers, Dickinson in the South of England and Crompton in Manchester.[22] It is perhaps at this point that china clay began to be used, which helps to account for part of the upsurge in clay sales shown earlier. Rosin and alum made a good 'size', which worked well in paper filled with china clay. Because the alum was slightly acidic, an inert filler such as china clay was essential.

Paper-makers were impressed by the potential of china clay as a filler for white paper of relatively fine size which gave opacity to paper by filling the voids between fibres. In addition it was heavier than rags or, later, wood pulp and also cheaper, characteristics that mirrored the advantages found when china clay had been introduced into the cotton industry for sizing and filling. Some evidence suggesting its use in Cornwall dates from the 1830s, when the paper mill at

A Fourdrinier paper-making machine, c.1835. The wooden vessel onthe right holds pulp. On the left are three steam-heated drying cylinders.

Coosebean, west of Truro, was sold in 1839, after the death of the owner, with various items, including a quantity of china clay, alum, rosin and two early Fourdrinier paper-making machines.[23] Richard Pearse has also suggested that in the mid-1800s 'someone, somewhere, reputedly in Holland, had discovered that the cost of producing paper could be appreciably reduced by incorporating china clay' but, as we have seen, this application was known earlier.[24]

In the earlier decades of the nineteenth century, the number of Fourdrinier machines in operation increased relatively slowly. By the time his patent ran out in 1822, 42 of them had been sold to a widely scattered group of paper-makers from Aberdeen to Cork, from Norwich and Dover to Coosebean in Cornwall. Mechanisation did not become widespread until the 1830s, but by 1860 hand-made paper only accounted for 4 per cent of total UK output, which had escalated from 14,000 tons at the end of the Napoleonic Wars to nearly 100,000 tons a year. From being a net importer, Britain was said to be the world's biggest exporter of paper, although this amounted to less than 10 per cent of total British production.[25]

The main stimulus for this growth came from internal demand. The British population nearly doubled between 1800 and 1860, and Britain was the world's greatest exporting nation for goods of all kinds, increasing demand for wrapping paper. Demand was further stimulated by a fall of 60 per cent in the average price of paper between the Napoleonic Wars and 1860, partly due to a fall in costs through mechanisation, partly because the government reduced and finally abolished taxes on newsprint and paper. By the end of the 1850s the average price of books had halved. However, while Britain may have been the workshop of the world, the literacy level of its workers was low compared with some other western nations, mass-circulation newspapers did not exist,[26] and so the main thrust for growth was in brown, not white paper.

By the mid-1800s, the geographical contraction of the paper-making industry had set in. The spread of railways benefited mills close to the main lines and led to closures in more remote locations, while the introduction of steam-powered machinery meant that mills with easier access to coal were more competitive. Of the four Dorset mills identified earlier, three had disappeared. In North Devon, only one survived at Barnstaple. In South Devon the number fell from a peak of 39 in 1820 to 24 by 1860. Those that remained tended to specialise, either by providing coarser wrapping paper mainly for local use, or by concentrating upon the production of high-quality writing and printing paper, where hand-crafted products always enjoyed a certain cachet.

A shortage of raw materials

A shortage of clean cotton rags, still the basic material for paper-making, was eased when the method of whitening rags by chemical bleaching, mentioned earlier, became available. This involved the use of bleaching powder, calcium oxy-chloride, which was added to rags and water in stirred tanks in a batch process. The use of bleach not only whitened stained rags but it also whitened some materials that had already been dyed, and made quantities of coloured rags suitable for paper-making.

Chemical bleaches were widely used but not always carefully applied. If excess bleach was used and some of it remained, paper could disintegrate over a period of time. For instance, in 1816 the British & Foreign Bible Society found a quantity of bibles 'crumbling to dust' after two years. Bibles were sometimes printed on potters' tissue, very thin paper made at mills including Hanley, Cheddleton and Newcastle under Lyme and used for transferring decoration onto pottery[27]. In another case, in 1834, some 30,000 copies of a book printed in 1818 fell to pieces through 'excessive bleaching with chlorine'. This kind of problem might have discouraged potential users of china clay in paper if they could not differentiate between the result of over-bleaching by chemicals and the effect of the china clay filler.[28]

By the mid-1800s there were signs that the rapid growth in British paper production would be halted because of an increasing shortage of the main raw material, rags. Domestic demand was outstripping supply and imports were becoming hard to find. France, Germany and Belgium were expanding their own production of paper and, in true protectionist fashion, imposing duties on, or even prohibiting, their exports of rags. The USA was also becoming a substantial importer of European rags, and between 1845 and 1859 its imports, mainly from Italy, rose three and a half times. The prices of various grades of rags rose by 20 to 30 per cent in the 1850s and, since rags accounted for half of the cost of paper, this was reflected in its price.[29]

Paper makers were looking for alternative materials, and over 100 patents were taken out for the use of different substances, with suggestions ranging from thistles and nettles to rhubarb. The two most practical substitutes were straw and the sweepings of textile mills. Some Lancashire paper manufacturers made extensive use of cotton waste from neighbouring factories, while Irish and Lancashire firms utilised waste from flax spinners. However, their demand drove up the price of these materials to make them just as costly as rags, and the same happened to straw. Although widely available, three tons of straw were needed to make one ton of paper, and larger quantities of chemicals and coal had to be used. Specialist equipment was also required to process it, and the end product was of poor colour and quality.[30]

Another substitute attracting the attention of paper-makers was esparto grass, which grew wild in the western Mediterranean. Patents were taken out on methods of processing it during the 1850s, notably by Thomas Routledge, an

Oxfordshire mill owner, who was joined by John Dickinson, a leading paper-maker, in setting up a works near Sunderland to prepare it. Later, esparto grass was to become an important source of material, but around 1860 it was not in widespread use. Another alternative was wood pulp. In Germany, F.G. Keller had patented a method of grinding timber logs to make pulp, and in the 1850s a number of Continental paper-makers were employing this process. Chemical processes to break down cellulose fibres in wood were also developed in the second half of the nineteenth century as an alternative to mechanical treatment. In 1860, however, wood pulp was little regarded by the majority of British paper-makers, and one of the most prominent of them, Sir John Evans, regretted that, 'there is no probability of any substitute for rags being discovered'.[31] In fact, the British paper industry had reached a turning point and, as will be discussed in Chapter Twelve, it was about to embark upon a period of massive growth through the use of wood pulp.

Comparative demand from pottery, cotton cloth and paper makers

How did the growing demand for clay for cotton textiles and paper compare with the traditional use in ceramics? Hunt's official statistics, given for the first time in 1858, suggest that over half of the pits produced exclusively for pottery, about a quarter solely for either cotton textiles or paper-making, and the rest for more than one of these uses.

POTTERY, TEXTILE AND PAPER USES OF CHINA CLAY IN 1858

Specialised production			Mixed production		
	% pits	% tonnage		% pits	% tonnage
Pottery only	56	56	Pottery, text, paper	2	3
Textiles only	16	14	Pottery, textiles	8	11
Paper only	10	10	Pottery, paper	5	3
			Textiles, paper	3	3
Total	82	80		18	20

Calculated from Robert Hunt's Mineral Statistics, 1858.

Popular suspicions about the use of china clay

The figures in this table, however, depend upon the accuracy of the information given by the clay producers, and there were commercial reasons why they might understate sales for non-ceramic uses. In addition, there was some public suspicion about its use in other materials, such as cotton textiles and paper. Apart from this, chemical knowledge was rudimentary and some textile manufacturers kept their preparations involving china clay a close secret, because they believed them to be superior.[33] It should also be remembered that china clay was often used in small amounts when first used in paper-making

According to Richard Pearse, paper mills sometimes obtained supplies of china clay via third parties so as to prevent rival paper manufacturers knowing that they were using it. Some also decided not to take out patents on processes involving the application of clay to keep their use of it a secret. Such clandestine practices explain why clay producers themselves were not always aware of the uses to which their clay was being put.[34]

In keeping quiet about sales to paper manufacturers, the clay merchants may have been influenced by government policy. For many years Her Majesty's Stationery Office continued to restrict the 'mineral' content of their paper to a level 'not exceeding 16 per cent' or even to make it 'free from added mineral matter'. Indeed a technical paper written as late as 1936 commented that: 'Even today there are paper-makers who would assume an attitude of injured innocence if it was sug-gested that they put china clay into their manufactured product; of course they do, but they do not want their customers to imagine that they would do such a thing.'[35] While HMSO specifications were not obligatory for other users, they probably influenced perceptions of china clay as some kind of adulterant.

Public apprehension about the use of china clay, however, was not confined to its application to paper. Its use outside the pottery industry continued to be attacked as unnatural and even immoral, as we shall see in the next chapter. Yet this seemed to have little or no effect upon demand, and sales of china clay con-tinued to increase, as illustrated in the Production Graph.

Adit working for china clay

As clay sales grew, small-scale labour-intensive clay workings needed to expand their output, and some pits used mining expertise to assist them. The technique of

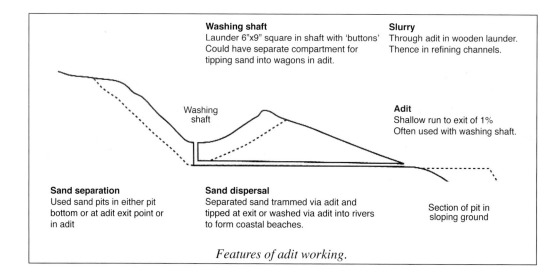

Washing shaft
Launder 6"x9" square in shaft with 'buttons'
Could have separate compartment for
tipping sand into wagons in adit.

Slurry
Through adit in wooden launder.
Thence in refining channels.

Washing
shaft

Adit
Shallow run to exit of 1%
Often used with washing shaft.

Sand separation
Used sand pits in either pit
bottom or at adit exit point or
in adit

Sand dispersal
Separated sand trammed via adit and
tipped at exit or washed via adit into rivers
to form coastal beaches.

Section of pit in
sloping ground

Features of adit working.

'adit' working used an almost horizontal tunnel cut from ground to one side of the pit at a lower level close to a site where clay refining could be located, to the bottom of a working pit. The adit would slope slightly upwards to the pit to allow a steady flow from the pit. In most cases a short shaft was 'raised' to intercept a low point in the pit down which washed clay and sometimes sand could pass in a stream regulated by a vertical wooden launder (a square pipe) with holes to allow discharge through it.

Arrangements varied, and coarser sands could be settled in sand pits within the pit area or leave the pit with the clay slurry and be separated at the outlet of the adit.

Pits using adits for handling washed clay included Lee Moor and Carclaze, while Littlejohns and Dubbers shared an adit which took their clays to Nanpean for refining. Many other pits used adits, sometimes referred to as 'levels', usually on a smaller scale, to collect water for use in clay washing and stored in pools adjacent to the works.

Adit working used techniques from the mining industry, and these served as a basis for the increased tonnages required by customers in the second half of the nineteenth century.

Detail from a china stone quarry, photographed c.1880. In the upper half two men using hammers and a boy holding a drill steel are drilling holes for blasting. The jointing of stone in the quarry with horizontal 'beds' and vertical 'heads' can be seen. In the foreground sorted stone has been piled to await removal to the surface by an inclined railway.

References

[1] The graphs are based on figures collected by J.H. Collins from various sources in *The Hensbarrow Granite District, Truro*, 1878. Intermediate totals for dates not covered by Collins are estimated by the authors on the basis of trends.

[2] Deirdre McCloskey, '1780–1860', in Roderick Floud and Deirdre McCloskey, *The Economic History of Britain Since 1700*, Cambridge, 2000, I, p.250; Janvier Cuenca Esteban, 'Factory costs, market prices and Indian calicos: cotton textile prices revisited 1779-1831', *EHR*, 1999, p.755.

[3] N.F. Crafts, *British Economic Growth during the Industrial Revolution*, Oxford, 1985, p.143.

[4] J.T. Marsh, *An Introduction to Textile Finishing*, London, 1957.

[5] C. Knick Harvey, 'Cotton textile prices and the industrial revolution', *EHR*, 1998, p.67.

[6] H.M. Stocker, 'An Essay on the China-Stone and China-Clay of Cornwall', *RRCPS*, 1852, p.88.

[7] *WB* 17 June 1831; CRO, Johnstone Records.

[8] *WB*, 19 March 1858.

[9] *Murray's Handbook for Travellers in Devon and Cornwall*, London, 1859, p.244.

[10] The word 'calico', derived from Calcutta cotton, was often used to describe cotton goods from any part of the world.

[11] *Catalogue of the Great Exhibition*, London, 1851.

[12] *VCHL*, 1908, pp.398–99.

[13] Knick Harvey, 1998, p.71.

[14] Edwin Baines, *History of the Cotton Manufactures of Great Britain*, London, 1835, repr. 1966, p.420.

[15] S.J. Jones, 'The Cotton Industry of Bristol', *TPIBG*, 1947–48, pp.61–79.

[16] Maxine Berg, 'Factories, workshops and industrial organisation', in Floud and McCluskey, Cambridge, 2000, I, p.133.

[17] The information in this section comes from Edwin Sutermeister, *Chemistry of Pulp and Paper Making*, 1929, p.352; R.H. Clapperton, *Modern Paper Making*, 1952, p.185; F.H. Norris, *Paper and Paper Making*, 1952, p.95; J.P. Casey, *Pulp and Paper*, 1960, II, p.985.

[18] Dard Hunter, *Papermaking*, New York, 1947, repr. 1978, pp.62, 194, 468.

[19] Michael Havinden, in Roger Kain and William Ravenhall (eds), *Historical Atlas of the South West*, Exeter, 1999, pp.336–38; Alfred H.Shorter, *Papermaking in the British Isles*, Newton Abbott, 1971.

[20] Alfred H Shorter, 'Paper Mills in Devon and Cornwall', *DCNQ*, 1947, pp.97–103; 'The Paper Making Industry in Cornwall', *RRCPS*, 1948, pp.30–41; 'Paper Mills in Dorset', *SDNQ*, 1948, pp.144–48.

[21] For instance, by Shorter, 1971.

[22] Later the Crompton family supplied paper to nearly all the Northern and London newspapers, producing an annual output in the later 1800s of 1,400,000 tons. *VCHL*, 1908, p.407.

[23] In a mill owned by Antony Plummer, see *JTS*, 2, 1974, pp.91–94. In the catalogue of items for sale, rosin was referred to as 'oil', the term used by paper-makers at that time.

[24] Richard Pearse, *The Land Beside the Celtic Sea*, Redruth, 1983, p.64.

[25] D.C. Coleman, *The British Paper Industry 1495–1860*, Oxford, 1958, pp.201–20.

[26] Michael Sanderson, *Education, Economic Change and Society in England, 1780–1870*, Cambridge 1995, p.18.

[27] Shorter, 1971.

[28] Richard I. Hills, *Papermaking in Britain 1488–1988*, London, 1988; Harry Carter, *Wolvercote Mill*, Oxford, 1957, p.43.

[29] Coleman, 1958, pp.327–38.

[30] Coleman, 1958, pp.338–42.

[31] Coleman, 1958, p.344.

[32] Hunt's statistics of clay output first appeared in 1858 but not again until 1868, and annually thereafter.

[33] Percy Bean, *The Chemistry and Practice of Sizing*, Manchester, 1910, pp.3–4.

[34] Pearse, 1983, pp 64–65.

[35] J.E. Aitken, 'The Mineral Constituents of Paper', *The Paper-maker and British Paper Trade Journal*, November 1936, pp.167–71.

Chapter Six
Some Reputable, and not so Reputable, Uses

We have already seen how some china clay customers kept their usage a secret, in case the public thought they were hoodwinking them by employing a cheap and somehow 'unnatural' filler. Applications in pottery would seem to be quite acceptable, indeed sanctified in the Bible, but even this usage sometimes met with disapproval. For instance, when, in the later eighteenth century, the Revd George Thomson, vicar of Jacobstowe, was dining with friends at Camelford and was served with food on an earthenware plate, he politely asked for a wooden platter. 'The poor wood turners are being starved by these new plates,' he explained.[1] The reaction of Cornish tin miners as china replaced pewter was more violent. In 1776 100 of them descended on Redruth market and smashed all the chinaware on display because it was displacing pewter ware, made of four parts of tin to one part of lead. Moving to Falmouth, they were about to break all the pottery there, but an astute pair of notables bought quantities of pewter ware and promised to discourage the sale of chinaware.[2]

Tobacco pipe clay

As mentioned in the Introduction, some early applications of clay were associated with changes in style and fashion, often originating from the middle and upper classes. Examples are the use of ball clay for the manufacture of tobacco pipes and the production of elegant chinaware for drinking tea sweetened with white sugar, which might have been refined using clay. As discussed in Chapter One, the craft of making clay pipes for smoking tobacco began in England in the 1580s after the introduction of tobacco from America, and sources of 'tobacco pipe clay', as it was called, can be identified in Dorset from the early 1600s. Other small local deposits of white firing clay, such as those found at Broseley in Shropshire, were also used. Although the basic shape of the clay pipes, probably adapted from that of American Indians, remained virtually unaltered for centuries, changes in fashion led to variations in size. Longer stems were favoured by the gentry in the mid-1700s and 100 years later stems a yard long enjoyed some popularity.[3] Already by 1650 at least 100 pipemakers operated in London, with many others in towns such as Bristol and Chester.

Bristol was the chief source of trade in pipes to Virginia, where originally the main customers were the native inhabitants. Some idea of the enormous number of pipes produced comes from the records of one Bristol firm, Israel Carey, which shipped over a million pipes a year to Virginia in the period before the American War of Independence in 1776. Another Bristol pipe maker, the Ring family, exported an average of a million pipes per year to Philadelphia, New York, Baltimore, Boston and New Orleans, for 77 years.[4] The eastern ports of England had a good coastal trade with the Low Countries and Germany, although the French preferred snuff-taking until the later 1700s. Some pipemakers added to their income by making other small fired clay objects such as hair curlers.

Except for the earliest pipes, which were moulded entirely by hand, the method of manufacture remained much the same after the introduction of two-piece moulds shortly before 1600. Initially the lumps of clay were broken down, washed to remove dirt and stones and partly dried. After this the clay was pulverised with a heavy iron bar, a process which took several hours, to expel air. The clay was then kneaded until smooth and plastic enough to be pressed into moulds. Next it was rolled by hand into rough pipe shapes and these were placed in groups of 12 on 'dozening boards' and passed to the moulder. A 'piercing rod' of steel or brass was inserted into the stem and the pipe was placed in a mould made in two halves. These were pressed together by a 'gin-press', which also formed the bowl of the pipe. The pipes were then trimmed by hand if necessary and placed in 'saggars' (as used in other pottery kilns) in a small

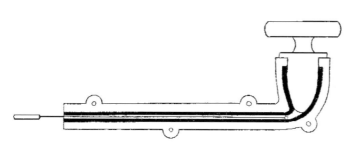

Half mould for tobacco pipes, showing the 'stopper' at the bowl end and the steel wire used to form the hollow part of the stem.

updraught kiln and fired at a temperature rising to about 950°C. The firing and cooling process took about three days.

Sugar bakers' clay

A perfectly legal application of clay in food production was referred to in 1768 in the will of Giles Brown, owner of a ball-clay pit at Furzebrook, south of Wareham, in Dorset. The will mentioned pits of 'tobacco pipe, pottery and sugar bakers' clay'.[5] These 'sugar bakers' were engaged in the refinement of white sugar from sugar cane. Over many centuries the cultivation of this plant had spread from its origins in the South Pacific through China, Japan, South-East Asia and the Middle East to the Mediterranean, and in the 1500s European settlers took it to the New World via Madeira and the Canary Isles. In 1600 Brazil was the major sugar producer but in the 1840s it was introduced to the Caribbean Islands, where production expanded massively. What had been a luxury item became a common commodity.[6]

The normal method of production, after harvesting the cane, was to crush it with rolling mills, yielding a juice which was heated in copper cauldrons, clarified and then strained. At this point the liquor remained a mixture of dark molasses and brown sugar called muscovado, and the refining process involved transferring this liquid into open-topped cone-shaped vessels about two feet tall which had a half-inch hole at the bottom. This was plugged during filling, and when the plug was removed the molasses ran off, leaving a granulated brown sugar loaf. One method of producing white sugar was to seal the drip pots with a layer of very moist white clay several inches thick, and as the clayey water gradually percolated down it washed off the remaining molasses. This process could be repeated until the required degree of whiteness was achieved.[7]

The technique of 'claying' sugar was said to have been brought to the Americas from Spain and Portugal and at the time that Brown's will referred to 'sugar bakers' clay', nearly all Brazil's output was 'clayed', as was the production from the French West Indies ('*sucre terré*'). Several accounts offer a curious explanation for the discovery of the claying process. Sugar bakers, presumably located in an area of white clayey soils, noticed a trail of white footprints left by chickens that had walked over brown muscovado sugar, and when the clay was washed off it took the remaining coating of molasses with it. Although one authority, N. Deerr, dismissed this story as 'silly', later research may have added some credence to it. Today, in Peru and Brazil, parrots and macaws lick white clay to purge themselves of toxins derived from juniper berries.[8]

The sugar from the British West Indies, however, marginally less than that from the French colonies (80,000 tons a year), was nearly all 'unclayed' brown muscovado, the last English record of imported 'clayed' sugar dating to 1721. The reason for this was quite simple: the 'sugar bakers' of Britain had lobbied the government to impose high duties on imported white sugar to protect their own claying businesses. In the 1720s over 500 British sugar importers existed and, for another century, cane sugar was the most valuable single British import. By 1800, however, the number of sugar merchants had declined to 85 as the trade concentrated in fewer and larger concerns. The main centres were London and Liverpool, but Bristol Quakers played a significant role.

It is highly unlikely that the clay referred to in Giles Brown's will was exported from England, since Brazil and the French colonies had local supplies of white clay (as did Barbados, used for the production of white sugar for local consumption). If the French colonies imported any clay it would probably have come from France, where Rouen and Saumur were cited as sources of fine white clay.[9] For half a century or so, sales of 'claying sugar' may have been a useful, if subsidiary, outlet for West-Country clay, but in 1812/13 an Englishman, Edward Howard, patented a technique for whitening muscovado by washing it with a sugar solution (Patent No. 3607) and refiners developed more sophisticated methods. These involved steam processes, centrifugal separators and charcoal cisterns to decolour sugar.[10] The claying process was still mentioned in a work of 1830,[11] but the claying of sugar fell out of use. Thus ended a West-Country connection in which clay as a whitening agent had served the needs of a middle-class public, just as it whitened the chinaware teacups in which they put their sugar.

Wigs, cosmetics and buttons

While Böttger, mentioned in Chapter Three, was trying to make porcelain for Augustus the Strong in the early 1700s, he is said to have come across china-clay deposits at Aue, 50 miles south-west of Dresden, discovered by Schnorr and sold as 'Schnorr's famous white earth' for powdering wigs. In the eighteenth and early-nineteenth centuries these were worn by many men and women, and British officers were required to wear them while other ranks had to grow a 'queue', or plait of hair, which also had to be powdered. Officers' wigs were usually dressed with a starch or flour mixture, whereas other ranks used ball clay, or pipe clay, as it was called at the time.[12] The fashion for periwigs further strengthened demand for ball clay. It was also easily moulded and heat resistant, which made it ideal for the manufacture of curlers used in making and maintaining wigs.[13]

Wig-making was a thriving business and no fewer than ten 'peruke-makers' were listed at Falmouth, at that time a naval base, in 1799, and others at St Austell, St Columb Major, Fowey and Penzance.[14] When flour was short during the Napoleonic Wars other substances apart from clay were introduced to powder military wigs, including rice, plaster of Paris and slaked lime. When the military queue was abolished in 1808 the fashion for wigs began to die out, but china clay was used later by hairdressers as an inert carrier in hair bleaches.[15]

An early application of china clay is said to be by a Cornish apothecary, William Stephens of St Austell, who marketed a 'Cornish Veterinary Powder', consisting of semi-refined clay from the Trethowel–Carthew area, where it had been exposed by tin-streamers. A sample of this clay may have reached a Staffordshire potter, Tom Astbury, and may have been used to whiten earthenware, but the secret of the source and the treatment of the clay died with Stephens.[16] China clay, probably from local sources, was later used as a face powder by the ladies of Georgia in the USA and in Australia, and as a complete body covering by certain African tribes.[17]

In the 1840s the *Royal Cornwall Gazette* reported a 'novel application of china clay', the manufacture of various shapes and sizes of buttons made from 'pressed china clay powder'. These were described as clear in appearance, and sold at only a third of the price of mother-of-pearl.[18] In fact buttons had been made in large quantities by potters since the 1770s. Buttons were fashionable items of gentlemen's clothing, and those who could afford it had them made of diamonds. For the less well-off Wedgwood produced brilliant Jasperware buttons. The chief producer of hard-paste porcelain buttons at that time was the Royal Porcelain Factory at Copenhagen, while soft-paste porcelain buttons were manufactured in small quantities in France at Sèvres and in Belgium at Tournai and Etterbeck. By the 1840s an important manufacturer of porcelain buttons was Prosser of Birmingham, in collaboration with the Minton pottery in Staffordshire.[19]

The Miners' Safety Fuse

In the eighteenth and early-nineteenth centuries, Cornwall's dominant metal mining industry used gunpowder as an explosive to break ground, a practice developed by German miners a century earlier. The gunpowder was commonly ignited by lighting a trail of goose quills fitted together, containing a thin core of powder. This crude method was far from reliable and the toll of deaths and serious accidents was high. In underground working, dirt sometimes infiltrated the frail tubes, or damp caused the gunpowder to smoulder, resulting in long delays. The miners would return, thinking there had been a misfire, only to risk being blasted by rocks in a late explosion.[20]

One humane soul, leather merchant William Bickford from Tuckingmill, near Camborne, was distressed by the sight of victims of blasting accidents, and set out

Bickford, Smith advertisement from the 1880s.

to find a solution to the problem. After several experiments he succeeded in 1831 in patenting what he called 'The Miners' Safety Fuse', whereby a small trickle of gunpowder was poured into the centre of a thin rope as it was being spun. This proved very efficient and its use spread rapidly to mining districts around the world, and was also adapted to military purposes.

As the process was refined, the safety fuse was coated with a waterproofing compound containing tar, gutta percha or rubber to enable it to be used in damp conditions. China clay was then used as a whitening agent in the final varnishing of the fuse, and more generally as a dusting to avoid stickiness on the hundreds of thousands of miles of fuse produced at the Tuckingmill works until 1970.[21]

The manufacture of alum

Alum was a substance with a symbiotic relationship with china clay. It was often applied in the settling of clay, while clay could be used in the production of alum. The first alums were based upon the mineral alunite, a potassium aluminium sulphate, which was refined in one of the oldest chemical industries, going back to ancient Egypt. In medieval Italy the main source of alum was alunite, but as demand grew other methods of preparation evolved, using silicates such as pyritic potassium shales. These were roasted and leached with water to give a crude alum which was then purified. This process was described by Agricola in 1556, and six years later William Kendall from Launceston was granted exclusive rights for alum making in the South-Western counties and the Isle of Wight.[22]

The raw material used in the process was essentially a ball clay containing some pyrite, later known as 'pinney' ball clay, but after some years Kendall lost interest and others tried, with limited success, to make alum in the Bournemouth and Poole area. Place names such as Alum Bay on the Isle of Wight and Alum Chine near Bournemouth owe their origin to attempts at alum manufacture. From the beginning of the seventeenth century, shales on the north-east coast of Yorkshire were found to be easier to treat and this area took over as the premier source of alum in England.[23] Alum has been used mainly as a mordant to fix dyes, as a size for sealing the surface of paper or walls and as a flocculant for settling fine solids in liquid suspensions, including beer and wine making.

In the year of the French Revolution, 1789, a distinguished French chemist, Jean Chaptal, gave an account of his process to produce aluminium sulphate as a substitute for alum, which was cheaper and served most uses. His method involved the roasting of a 'white' china clay and reacting it with sulphuric acid.[24] It was used in France until the mid-nineteenth century but was never patented. However, in 1839 a German chemist from Duisburg obtained a patent for converting potters' clay to aluminium sulphate, which also involved roasting and reacting china clay. This process also appears to have been carried out at Lee Moor for a time.[25] One application of alum, known since 1828, was for the settlement of china clay in thickening tanks. It was the patenting of a process for manufacturing an alum substitute by H.D. Pochin, which used china clay and sulphuric acid, that later led to his involvement in the china-clay industry, in which he and his family played an important part, as will be described in Chapter Twelve.

Ultramarine blue

From the 1830s, china clay was used as a component in substitutes for the blue pigment ultramarine. A pigment derived from the attractive and relatively rare blue mineral lapis lazuli (or bluestone), ultramarine was found in such far-away places as Afghanistan, Siberia and Brazil, hence its name. The mineral was usually found mixed with white marble, so laborious hand sorting was needed to pick out the blue lapis lazuli for use by artists as a colour and for jewellery. By the early-nineteenth century the chemical composition of the blue fraction was partly understood and attempts were made to produce an artificial ultramarine blue. This led to a competition sponsored by the *Société d'Encouragement pour l'Industrie Nationale* of France for a cheaper, artificial ultramarine. An eminent French scientist, J.B. Guimet, from Toulouse, succeeded and was awarded a prize of FF6,000 in 1828. His artificial ultramarine cost only a tiny fraction of the original blue mineral,[26] which was transformed from being the most expensive to the cheapest artists' colour.

Guimet used a mixture including china clay (30–50 per cent) sodium carbonate, sulphur and pitch or rosin. These were ground to a paste and

Victorian advertising for 'dolly' blue.

then roasted in a furnace at 800°C. Slow cooling took place, allowing the product to oxidise, before it was ground into powder, washed and dried. German manufacturers developed the production of artificial ultramarine on an industrial scale, followed by the French and British. The purest blue-coloured product was used to make artists' colour. The French Impressionists were among the earliest to employ the finest grades of the pigment to impart a new brilliance to their paintings.[27] Lowgrade ultramarine was sold as 'dolly blue' for household laundering and some was used in the china-clay industry as a tint for china clay, added discreetly to the clay product from the mica drags to make it look whiter.[28] In the 1860s blue aniline dyes were also used for this purpose, and clay companies had to be careful to add appropriate 'blueing' agents to avoid unwanted interactions in their customers' processes.

Graphite and pencils

Pencils as we know them today were made, from the beginning of the sixteenth century, using graphite, a form of carbon, from mines in Cumberland. This was cut into lengths of rectangular cross section and encased in wood. By the early-nineteenth century, however, these mines were exhausted and the pencil makers looked for substitutes. Although overseas sources of graphite were found, they were not suitable for direct use as pencil leads, and from 1832 graphite was ground to a fine powder and mixed with a plastic ball clay. By using different proportions of graphite and clay, circular section pencil leads of varying degrees of hardness are made after drying, baking and encasement in cedar or similar wood. For example, the most commonly used grade of pencil is 'HB', of medium hardness, which contains about 10 per cent clay, baked at about 850°C.[29] Coloured pencils were later made by using china clay and pigments in a similar process.

Adulterating flour with china clay for bread making

While the public accepted the use of china clay to make pottery as eminently respectable, some dealers kept quiet about non-pottery uses such as cloth or paper because they were regarded as unethical or even physically harmful. At about this time there was public disquiet about the development of new industries and suspicions that products were sometimes adulterated to increase profits. Some of the most notorious cases involved the mixing of china clay with flour. The first recorded malpractice of this kind appeared during the Napoleonic Wars. As already discussed, flour was widely used during the later eighteenth century in the form of starch for non-food processes such as sizing and filling in cotton cloth and paper manufacture, as well as for bookbinding, linen printing, trunk-making, paper hanging and the powdering of wigs. However, when maritime blockades during the Napoleonic Wars cut off grain supplies and the price of flour and bread more than doubled,[30] food riots broke out and the government was forced to intervene.

The urban workforce spent a large proportion of their wages on basic food, so flour millers and bakers were easily identifiable scapegoats for high prices, whether they were guilty of profiteering or not. In rural corn-growing areas they were accused of diverting corn and flour to the towns, where they could make bigger profits, and violent demonstrations occurred when famine threatened. In West Cornwall grain was so scarce that, from time to time, flour almost disappeared from local markets, and the militia had to be summoned to protect corn merchants from angry crowds of would-be purchasers.[31]

During the food shortages of 1795 and 1800 manufacturers were prohibited from using wheat and barley in the production of starch and powder for hair. The British output of starch, which had risen from around two million pounds in weight in 1750 to over eight million by 1794, fell sharply to under two million in 1795. It then recovered, after prohibition was lifted, to over seven million by 1798, only to drop back to below three million when prohibition was reintroduced.[32] Faced with these disruptions in supply, users of flour for non-food applications turned to other sources, such as chalk, plaster of paris or china clay.

However, rumours that food merchants had been mixing other materials with flour naturally caused a great uproar. In 1814 William Jenkin, steward to the Agar-Robartes of Lanhydrock, reported that: 'This neighbourhood is now greatly agitated by... a nefarious act supposed to be committed by some persons in Truro and Redruth.' They had been selling flour at much below the market price which contained: '... an alarming quantity of clay... exactly similar to what a gentleman by the name of Wedgwood showed me at Lanhydrock three or four years back'. According to Jenkin it was: 'perfectly white, and so very fine, that in a dried state it feels like balls of very fine flour.' Following complaints from Truro to the Mayor of Plymouth, a vessel had been seized there with 60 tons of adulterated flour and clay on board.[33]

The British Government, in its anxiety to keep down the price of bread and stop the poor from rioting, may even have inadvertently encouraged adulteration. Until 1797 they allowed local magistrates to fix the price of bread according to the quality of flour that individual bakers used. But in that year they froze the price of bread, and also allowed bakers the same remuneration per sack of flour turned into bread, whatever the quality. Since flour was so expensive, bakers who used pure flour found it difficult to make a living, but in collusion with the miller, they could keep bread prices below the fixed limit and make bigger profits by adulterating the flour.[34]

Other adulteration scandals
Mixing flour with some other substance such as alum, plaster of Paris or china clay

was part of large-scale fraud involving the supply of many materials by army contractors during the Peninsular War. In 1813 a baker of Greenwich had been fined
£10 for adulterating bread with a substance identified as 'calcified stone finely pulverised', and shortly afterwards a biscuit maker of Deptford was prosecuted for
mixing flour with a similar material in the proportion of four parts of flour to one
part of the other ingredient. This mixture was also detected in army stores in Spain
and in a camp on Dartmoor. It did not take long to trace the illicit trade to St
Stephen-in-Brannel, where a Redruth man called Osborne was bringing cart-loads
of clay to Truro, supposedly for a potter on the Scilly Isles, a most improbable
journey for a highly unlikely destination.[35]

In fact in 1814 clay was taken to flour mills west of Truro, the property of
William Trahar, who had also sent adulterated flour to Plymouth, which had been
seized. Local feelings ran high, especially when Trahar got off lightly while his
employees were punished, and £50 was raised by mining adventurers to take him
to Bodmin Assizes, but again he got off scot-free after reducing the main prosecu-
tion witness to a state of drunkenness. These, though, were not the only shady
dealings of which Trahar was accused. In 1813 he had also been involved in forging tokens that the mining magnate Williams coined to pay his workers, for use in
his company shops.[36]

Nor was Trahar alone in mixing china clay with flour. Indeed, a Plymouth magistrate found it necessary to issue public notices putting all millers and bakers, as
well as their customers, on their guard against: 'Frauds of this Description, practised for a long Time past, and to very considerable Extent.' According to these
notices, John Rowe and Henry Rundle, owners of two flour mills in the Truro area,
had been convicted of 'adulterating Meal and Flour' and fined £10 each, and a
Plymouth miller, William Smith, had made several sacks of adulterated flour into
bread. Another Plymouth trader, John Bartlett, possessed 40 sacks consisting
entirely of 'pulverised China Clay, resembling Flour of the best quality', sent by a
Mr Ostler of Truro.[37]

The importance of this trade is impossible to estimate with any certainty because
of its clandestine nature. Newspapers claimed that thousands of tons of clay were
involved, but recorded china-clay output was under 2,000 tons a year, and most of
this went to the potteries, so if newspaper reports were correct, the illegal trade in
china clay for use in flour would have exceeded that for legitimate uses, which
seems highly unlikely. After the wartime blockade was lifted, the unlawful trade
diminished, but it probably did not die away completely, for flour with china clay
or chalk or plaster-of paris in it actually looked much more palatable than 'pure'
flour, which had a dingy grey appearance. In 1850 an 'Analytical and Sanitary
Committee' discovered that 49 loaves of bread from various London shops contained alum.[38]

Not until much later, with the introduction of roller-milled flour, which separated the wheat germ, did flour become whiter (but less nutritious), and so bakers were tempted to continue using china clay surreptitiously to enhance its appearance.[39] Of course the effect upon the digestive system of adding china clay to flour was more problematic. Some customers complained of internal pains and perhaps the clay felt heavy in the stomach, because its specific gravity was three times that of flour. Even so, china clay was later used extensively in medicines and toothpaste, as we shall see in Part Two.

Summing up this survey of lesser-known uses of clay, while some applications continue to this day, others, such as wig-powdering, button-making and sugar-clay-ing, are long forgotten. However, the scandal of the adulteration of flour with clay cast a long shadow on the industry and the general public also mistrusted its use in other materials, such as cotton textiles and paper. They suspected it was an attempt by manufacturers to hoodwink

TO THE PUBLIC.

WHEREAS, some Persons carrying on the FLOUR TRADE, near TRURO, have been convicted of having mixed Clay with the Flour which they have sold to the public, to the great injury of the health of those who have used the same —And whereas REPORTS have been circulated tending to injure the character and trade of other persons concerned in the manufacture and sale of Flour.

We, whose names are hereunto subscribed, feel that we owe it to our character to declare, that we are at all times ready to produce our present servants, as well as those who, at any former time, may have been employed by us, to prove on oath, that they have never known us to mix any substance whatever, with the Flour which we have manufactured or sold. And we are ourselves ready to make oath that we have never done so, or procured any person to do so for us.

And we do hereby offer a REWARD of TEN POUNDS to any person who shall give such information to any of us as may secure the conviction of any person or persons propagating reports injurious to the character of any of us, relative to the adulteration of Flour.

And further, we do hereby offer a REWARD of TWENTY POUNDS to any person or persons, who shall give information to any of the Magistrates in this neighbourhood, so as to secure the conviction of any person or persons guilty of adulterating Flour with Clay.

Given under our hands this 11th day of May, 1814.

WILLIAM MOORE, Ladock.
RICHARD CROSSMAN, Truro.
WILLIAM HAMM, St. Erme.
P. A. GREEVE, Probus.
RICH. WHITFORD, Tresillian-Bridge.
WILLIAM GLASSON, New-Mills.
SIMON HUGO, Probus.
NICHOLAS COURTENAY, Cornelly.
FRANCIS JOHNS, St. Clements.
EDWARD CROSSMAN, Truro.
JAMES CHARLES, Probus.
THOS. HEAN, Goodern-Mill, Kenwyn.

May 1814 notice in the West Briton.

the public into buying products that contained a high proportion of a cheap, use-less and possibly harmful substitute. Fortunately for the clay producers, however, this seemed to have little discernible effect upon the demand for china clay, the sales of which continued to increase by leaps and bounds.

References
[1] Mark Guy Pearse, *The Ship where Christ was Captain*, London, 1926, p.184.
[2] Peter C.D. Brears, *The English Country Pottery*, Newton Abbot, 1971, p.55.
[3] Eric G. Ayto, *Clay Tobacco Pipes*, Princes Risborough, 1994.
[4] Julian Lea Jones, *Bristol Curiosities*, Edinburgh, 2007, pp.20–22.
[5] *Blue Pool*, Wareham, c.1998, pp.15–16.
[6] Apart from the references cited in the text, the information for this section comes from N. Deerr, *The History of Sugar*, Vols I and II, London, 1944, 1950; R.S. Dunn, *Sugar and Slavery*, London, 1973; J.H. Parry, Philip Sherlock, Anthony Maingot, *A Short History of the West Indies*, London, 1987, pp.109–10; Alison Grant, *North Devon Pottery*, Exeter, 1983; J.H. Galloway, *The Sugar Cane Industries*, Cambridge, 1989; the staff of the Barbados Museum in Bridgetown, the Barbados Sugar Museum and the Holetown Library, Barbados.
[7] In some regions of Asia, wet vegetable matter was placed on top of the sugar instead of clay.
[8] Macaw and Clay Licks, http://www.inkanatura.com/macawclaylicks.asp
[9] Henri Duhamel de Monceau, *L'Art de Raffiner le Sucre*, 1764.
[10] Bristol refiners were unable to compete and the last closed in 1912. Donald Jones, *Bristol's Sugar Trade and Refining Industry*, BHA, n. d.
[11] G.R. Porter, *The Nature and Properties of the Sugar Cane*, London, 1830, p.9.
[12] During the siege of Gibraltar, 1779–83, officers were forbidden to powder their wigs with flour. *The Military Heritage Centre*, Gibraltar.
[13] *Blue Pool*, c.1998, p.13.
[14] *Universal British Directory of Trade, Commerce & Manufacture*, London, 1799.
[15] J.S. Cox, *An Illustrated Dictionary of Hairdressing and Wigmaking*, London, 1984; Paul Rado, 'The Evolution of Porcelain', *JBCS*, 1963, I, 3, p.427.
[16] John Penderill-Church, *The China Clay Industry*, Wheal Martyn Archives, n.d., p.1.
[17] J. Allen Howe, *A Handbook to the Collection of Kaolin, China-Clay and China-Stone*, HMSO, 1914, p.44.
[18] *RCG* 19 January 1844.
[19] Victor Houart, *A Collectors' Guide to Buttons*, London, 1977.
[20] Bryan Earl, *Cornish Explosives*, Penzance, 1978, p.67.
[21] Bickford Smith & Co. Ltd, *The Miners' Safety Fuse*, c.1931, repr. Trevithick Society, 2006.
[22] R.B. Turton, *The Alum Farm*, Whitehaven, 1938, p.37.
[23] I. Miller, (ed) *Steeped in History*, North York Moors National Park Authority, 2002, p.9.
[24] Charles Singer, *The Earliest Chemical Industry*, London, 1948, p.135.
[25] *Ures Dictionary of the Sciences*, 7th Edition, 1875, p.120.
[26] *Ures Dictionary*, 1875.
[27] D. Bornford, J. Kirby, J. Leighton and A. Roy, *Art in the Making, Impressionists*, National Gallery, 1990, p.57; G.H. Hurst and N. Heaton, *Painters' Colours, Oils and Varnishes*, London, c.1900, pp.228–35.
[28] David Cock, *A Treatise on China Clay*, London, 1880, p.93.
[29] Alfred Searle, *Clay and What We Get From It*, London, 1925, p.131.
[30] B.R. Mitchell and P. Deane, *Abstract of British Economic Statistics*, Cambridge, 1962, pp.97–99.
[31] A.K. Hamilton Jenkin, *Cornwall and Its People*, Newton Abbott, 1970, pp.574–75.
[32] Excise Revenue Accounts, quoted in T.S. Ashton, *An Economic History of England*, London, 1955, pp.55 and 244.
[33] Letter of 19 May 1814 to Pasco Grenfell, a Cornish merchant in London. We are indebted to Jim Lewis for this reference.
[34] Ashton, 1955, p.57.
[35] CRO AR/T/29.
[36] J.A. Mayne and J.A. Williams, *Coins and Tokens of Cornwall*, 1985, pp.49–54. We are indebted to Jim Lewis for this reference.
[37] PDRO, 1/684/10.
[38] Clifford Lines, *Companion to the Industrial Revolution*, London, 1990, p.5.
[39] Mary Mackinnon, 'Living Standards 1870–1914', in Roderick Floud and Deirdre McCloskey, *The Economic History of Britain Since 1700*, London, 2000, II, p.278.

Chapter Seven
Transport and Marketing

Since a defining characteristic of clay and stone is the low value for their bulk compared with metals, and since most customers are located far away from the West Country, a significant part of the final cost to them consists of handling, storage and transport charges. Improvements in materials handling can thus play a large part in reducing the final price of clay. We begin this chapter by discussing advances in pit to port transport, examining the part played by road improvements and Turnpike Trusts, canals, rivers, horse-drawn tramways and steam locomotives. We then describe the construction of new ports in the St Austell clay district and also the role played by other harbours in Cornwall, Devon and Dorset. Finally we consider improvements in coastal and ocean-going shipping. Since the production of clay and stone involves the import of coal, timber and other materials, inward transportation will also be touched upon.

Pit to port transport

In the earliest days of clay extraction there were no canals or railways in the region and the roads were generally in an appalling state. In wet weather, commodities had to be carried by mule or packhorse because loaded wagons would sink axle-deep in mud. The sea was the only long-distance highway, and since the average load of clay that animals could carry was just over 200 pounds, and the distance they would cover in a day was 18–20 miles, it might take up to a dozen animals to carry one ton of clay from a pit to a local port and bring back necessary supplies, such as coal.[1] Turnpike Trusts, which were monopolies granted by Parliament to build private highways and charge a toll to travellers, used improved methods of road making devised by Macadam and Telford. The period 1750–70 saw the greatest expansion of these roads, and, as a result, large four-wheeled wagons pulled by three or four horses could now carry 3–3.5 tons about 20 miles a day, the equivalent of 30 or more packhorses.[2] The toll roads only covered the main routes between towns, however, and country lanes remained in a poor state. Their construction, moreover, sometimes aroused opposition in the West Country and elsewhere. During periods of food shortage, workers feared that, by facilitating outflows of food, they might raise local prices, while other producers saw that they might intensify competition from outsiders.[3]

In many cases clay was only one, and not the most important, of the many commodities carried; mules and horses could crowd into a port on busy days. Transport by pack animals, moreover, was not merely cumbersome but also costly, because the price of animal feedstuffs rocketed during periods of food shortage, such as the Napoleonic Wars or the potato famine of the 1840s. In North Devon a Turnpike Trust built a road between Bideford and Barnstaple in 1763, but packhorses were still needed for inland journeys. According to a traveller in 1796, the only other means of carrying bulky loads was what were known as 'truckamucks', sledges made of two or more trees bound together and fastened to a horse. The high cost of packhorse transport from the ball-clay pits at Peters Marland to the port of Bideford a dozen miles north, compared with the shorter hauls in South Devon and Dorset, seems to have caused the clay works to close down for a time, until the construction of Lord Rolle's canal, discussed later.[4]

In the Bovey Basin, Kingsteignton and Newton Abbot were only a few miles apart, and the Totnes Turnpike Trust built a road connecting them in 1836. In Dorset, Turnpike Trusts built roads south of Wareham near to ball-clay land at Creech, Furzebrook and Corfe Castle in 1766.[5] The clay was stored in 'clay cellars' at Corfe Castle, Stoborough and Wareham Quay.[6] During the nineteenth century, however, the two leading clay producers, Joseph Fayle and the Pike Brothers preferred to construct their own tramways over shorter distances to Poole Harbour.

The construction of canals in Devon

Canals played a large part in the development of some English regions, such as the Staffordshire Potteries, and the most ambitious canal scheme involving the clay trade in the West Country was the proposed Dorset and Somerset Canal of 1793. The promoters aimed to avoid the difficult and dangerous sea voyage around Lands End and up towards Bristol or Liverpool by constructing a canal from Poole to Bristol via Bath, with a branch to the Somerset collieries. Clay was to be taken by barge and then transferred to sea-going vessels, the barges returning with a cargo of coal. The promoters hoped to raise a capital of £200,000, but although a start was made from the northern end, only £58,000 was actually paid up and the venture was abandoned.[7]

In the Bovey Basin, much land was of little agricultural value, the roads were poor and it was the ball-clay trade that stimulated improvements such as canals. The industry was entirely dependent in its earlier days on access to coastal trade along the tidal estuaries from Teignmouth as far inland as Newton Abbot, and to improve access to the clay pits, two short stretches of canal were built in the 1790s. The first, the Stover Canal, went from the tidal reaches of the River Teign as far as Bovey Tracey, upstream from Newton Abbot, through the estate of James Templer II, whose father had made a fortune in India.

The project was financed by Templer but only just under two miles of it were completed in 1792, finishing at Ventiford, near Teigngrace. It had no towpaths for horses and barges, which were either sailed or bow-hauled by men. Each carried 30–50 tons of clay to Teignmouth, where it was transferred to sea-going vessels. The number of barge loads increased from about 400 a year in 1790 to 1,000 by 1854, and clay shipments rose from around 12,000 to 30,000 tons per annum. Coal was sometimes transported on the return journey and to handle these cargoes, as well as Haytor granite, quarry owner George Templer built the New Quay at Teignmouth in 1827. Shortly afterwards he sold the quay to the Duke of Somerset, along with his canal. The rising

Top: *Clay barges under sail on the Stover Canal.*
Below: *Barges at Teignmouth.*

importance of the clay trade also led another landowner, Lord Clifford, to build a second canal, the short Hackney Canal, in 1843 from the head of the River Teign estuary near Hackney Wharf to the Newton Abbot–Kingsteignton road.[8]

From North Devon, shipments of ball clay were recorded in the 1650s by the Greening family of Bideford, who worked the Peters Marland clay pits, mentioned earlier. The port was sheltered so that barges could load clay and unload coal and limestone, but the problem was carriage along the hilly road to Bideford, and the cost of transporting clay made it uncompetitive, as previously discussed. However,

Aqueduct on the Rolle Canal over the River Torridge. Opened in 1827, closed in 1871.

this was eased in 1826 when Lord Rolle, the local landlord, at his own expense built a canal from Annery, near Bideford, to Great Torrington at a cost of £40,000–£50,000.[9] His aim was to carry coal from South Wales, as well as culm and limestone to Great Torrington, where he owned a foundry, a large mill and lime kilns, and special boats were built at Bideford for their transport.[10] Finally, although at least ten canals were projected in Cornwall, few were built and only one was used by the clay trade, at Carclaze.[11]

Horse-drawn tramways in Dorset
Another solution was to construct railways along which animals could pull trucks with much heavier loads than were possible on the rough roads of the times. The first of their kind in Dorset was built in 1806 by Benjamin Fayle, a London merchant who had bought existing clay pits at Norden, near Corfe Castle. He had learnt of this new form of transport which was used in the coalfields, and constructed a tramway just over 3 miles long following the line of an old cart track across the heath to a quay he built on an inlet of Poole Harbour called Middlebere. Three horses drew small carriages, each holding two tons of clay, making the return journey three times a day. Each rail trip carried the equivalent of 100 donkey loads, and the clay could be tipped directly into the trucks at the pit and emptied directly onto the barges at the quay.

The scheme cost £7,500, a large amount compared with the small expense of starting up a clayworks, but it soon paid back the investment. The saving on manpower enabled him to increase production from 14,500 to 22,000 tons a year and cut his labour force by a half to about 100 men. Fayle had first introduced himself to potters in 1804, when he appointed John Scarlet of Newcastle, Staffordshire, as an agent for blue and brown clays from the estates of William Morton Pitt and John Calcraft, and in 1806 he offered clay at 18s. (90p) per ton on board ship.[12] In 1813 he invited Wedgwood to visit his pits and by this time he was meeting strong competition from South Devon and also from Dorset, where the Pike family opened up new pits. Another firm, Watts, Hatherley & Co., also developed clay-works at Furzebrook and constructed their own narrow-gauge, horse-drawn railway to the estuary at Ridge Wharf, a mile east of Wareham, in 1838–40. This was acquired by the Pike family around 1850.[13]

The beginnings of rail haulage
The steam-powered haulage of clay took a long time to develop because the cost of constructing railways was almost always higher than the original estimate, whereas the revenue often proved lower than expected. The most enterprising early railway scheme concerned with clay extraction was the revival of a plan mentioned earlier to connect the ball-clay area of Dorset with the Somerset collieries and the Bristol Channel. In 1825 a meeting was held in Wareham, chaired by John Calcraft, the local MP and lord of the manor, to build a railway line. The aim was the same as that of the earlier proposal for a canal, to take ball clay to the channel and bring back coal, but once again the promoters could not raise the capital required, an estimated £300,000.[14] In the Bovey Basin the Mortonhampstead branch railway from Newton Abbot was opened in 1867 enabling clay to be loaded at Teignbridge sidings adjacent to the Stover Canal. In North Devon attempts were made in 1831 and in the 1840s to provide a railway, but it was not until 1855 that Bideford and Barnstaple were connected to Exeter at a cost of £44,000.[15]

In the atmosphere of cut-throat competition that prevailed, especially in Cornwall, spoiling tactics were common. As soon as one group of entrepreneurs got wind of another's plans to construct a road or a railway, they put up an alternative scheme to outdo it or to block it. In Mid Cornwall the main adversaries were local clay merchants and landowners on the one hand and the great Mid-Cornwall adventurer Joseph Austen Treffry on the other. In 1833 Austen obstructed a scheme to construct a railroad westwards from the clay country by proposing a railway from St Austell to Par. Two years later he undermined a project, backed by the clay landlords Sir Charles Graves Sawle and Henry Lambe and clay merchants Elias Martyn and John Martin, to build a turnpike road north from St Austell, by announcing plans for a railway serving the same area. A decade later he hindered another railway scheme of Elias Martyn, John Martin, Yelland, Truscott and others, which would have linked the clay country with Pentewan, by proposing his own railway to serve the same territory.[16]

In view of all these impediments to progress, it may seem surprising that vast improvements in the transport network were made, mainly with other products in mind, such as copper, tin, lead and iron ores, granite, limestone and sea sand. As a result, during the first half of the nineteenth century the average cost per ton of moving clay from pit to port in Cornwall fell from £1.25 around the year 1800 to only £0.25 by the 1850s.[17]

The Lee Moor tramway in Devon

In Devon, as in Cornwall, transport developments that were vital to clay producers often came about through the growth of other industries. Dartmoor was the largest granite district in the South West, the first significant quarries opening in 1820 at Haytor. A tramway was constructed to the Stover Canal, mentioned earlier, where barges took the granite to Teignmouth for shipment by sea. A much bigger complex of quarries was developed north of Plymouth on Dartmoor to supply the needs of its dockyards. The granite was transported by the Plymouth & Dartmoor Railway, completed in 1826 and promoted by a prominent local landowner, Sir Thomas Tyrwhitt, who also intended to open the moor up to Princetown for agricultural production.[18]

A single-track, horse-drawn line of 4feet 6 inches gauge, it passed through land owned by Lord Morley. Instead of paying him for the use of his land, the tramway company built him a branch line to his slate quarry at Cann Wood, about halfway between Plymouth and Lee Moor, where he owned china clay-bearing land. Unfortunately, when clay extraction began in 1834, progress in extending the line to Lee Moor was blocked by the refusal of George Strode, who owned some intervening land, to let the railway pass over it.[19]

In 1850 another effort was made to reach Lee Moor by rail, only to be abandoned when the South Devon & Tavistock Railway Co. wished to cross Morley's land and agreed in return to build a branch line for him to Lee Moor. Sadly, the cuttings, embankments, viaducts and bridges that they constructed were poorly made and, on the insistence of William Phillips, who had leased Lee Moor pit, had to be rebuilt. This delayed completion for two years which, according to Phillips, cost him £22,000, since the price of clay had dropped during the period by £0.10 a ton.[20] The rail track, moreover, stopped half a mile below the place where Phillips had constructed his workers' cottages, and to gain easy access to them a steep counter-balanced inclined railway was necessary, which continued on to another clayworks at Cholwichtown, leased by the Martins of St Austell. In 1858 the Earl and Countess of Morley attended the ceremonial opening of this incline, where they were impressed to see wagons travelling up and down in just two minutes and 30 seconds.

New ports in St Austell Bay

Until the 1790s, the nearest harbours to the St Austell clay pits were Mevagissey and Fowey, but access to them from the clay district involved laborious journeys. Fowey offered the best all-tidal harbour, but did not come into its own as a clay port until a branch railway arrived around 1870.[21] However, the landing point closest to the main clay and stone area was nothing more than a beach at West Polmear in St Austell Bay, or Tywardreath Bay, as it was called then. From time to time small vessels were beached there to unload coal and limestone and ship out copper ore, granite, china clay and stone and other materials. In rough weather this was a hazardous operation and vessels were sometimes wrecked on the shore.

Early photograph of Charlestown Harbour, where coal was unloaded and copper ores and china clay were shipped out.

Adventurers saw the need for a proper port, and over the next few decades they endowed the clay district with no fewer than three within a few miles of each other: Charlestown, Pentewan and Par. Again, as with toll roads or railway improvements, they did not owe their provenance entirely, or even mainly, to the china-clay trade. Apart from the export of pilchards and granite, large copper deposits had been discovered along the eastern part of the bay, and in the 1790s Charles Rashleigh built the small harbour of Charlestown, named after him. He owned the land at West Polmear, and augmented the pilchard fishing fleet, erected fish cellars, set up rope makers, lime-burners and brick makers, as well as a hotel, and designated open storage areas for china stone and copper ores. Since maritime locations at the time were in danger of attack from French privateers and potential invasion by the French navy, he installed a four-gun battery of 18-pounder guns on the cliffs to guard the harbour. If he had not been systematically swindled by his agents he would have had enough capital to expand still further, but he died, bankrupt, in 1823.[22]

By this time the volume of china clay and stone trade from Charlestown had greatly expanded. In 1810 some 1,900 tons of clay and 1,600 tons of stone were shipped out, and by 1826 the traffic had more than trebled, to 7,000 tons of clay and 5,000 tons of stone. In 1846 the directors of the Charlestown Estate, anticipating the construction of a railway from Plymouth to Truro, by the Cornish Railway Co., entered into an agreement with the company to build a branch line to the port. Realising that the gradient leading to the docks was probably too steep for conventional steam locomotives, the directors considered the 'atmospheric' system, a fashionable idea at the time patented by Brunel and tried out at Dawlish in South Devon in that year.[23]

Another option was a counter-balancing system, using a stationary steam engine to wind a cable that pulled wagons up and lowered them down an inclined railway from the top of Charlestown Road to the docks. A system of this kind had been installed at Portreath on the North Cornish coast in 1837, and another was to be used at Lee Moor, as already mentioned. The wide thoroughfare between Charlestown and Mount Charles, near St Austell, would have allowed for twin tracks, but the line from Plymouth was not built until 1859 and in the meantime the project for the branch line was abandoned .

Pentewan harbour

Charlestown was not convenient for adventurers to the west of St Austell, who met increasing difficulties in negotiating its narrow, crowded and steep streets. Landowners and merchants saw the potential of developing another harbour to the south-west of the town. Three years after Rashleigh's death, Sir Christopher Hawkins, of Trewithen, near Truro, with financial help from London merchants, between 1820 and 1826 rebuilt a small old harbour on his land three miles to the west of Charlestown at Pentewan. A river outlet there had long been used as a beaching place for unloading coal and limestone or loading sand and Pentewan building-stone from a cliff-side quarry, but in the year 1826 Pentewan only handled 100 tons of stone and 20 tons of clay.[25]

A few years later, after a narrow-gauge horse-drawn railway was constructed from Pentewan to St Austell, the port was despatching more clay and stone than Charlestown. Port dues yielded Hawkins a useful income from Spencer Rogers and other clay producers who were paying over £500 per annum. Hawkins, landlord to a number of them, ensured in his leases that they were tied to the use of his port.[26] Yet the very success of china-clay growth in the area proved to be Pentewan's undoing. Waste products from pits upstream were washed downriver into the bay and built up a sand bar that blocked the harbour except at high tide, causing delays which were particularly harmful during the trade depression of the 1830s, when orders, once lost, could not easily be regained.[27] This problem was only partly solved by selling sand from about 1900.

Meanwhile, Austen Treffry had considered improving the harbour at Fowey, but another member of the Rashleigh family, William of Fowey, rejected his proposals to build a tramway across Rashleigh land, and this persuaded him to go ahead with the construction of his own port. In a typically bold move, he built another harbour near Par between Charlestown and Fowey, rejecting the advice of no less an expert than Isambard Kingdom Brunel that such a project was impractical.

The port of Par
Originally Par was a small rivermouth where a few ships were beached for unloading and loading, but Treffry built a completely new port here between 1829 and 1840.

Again, the china-clay trade played only a minor part in Treffry's calculations. His main interests were to ship out copper ore from his mines, Fowey Consols and Par Consols, import limestone for his lime-kilns, operate a smelter for lead and silver ores, run a brick works and a factory making candles for use in his mines. He cut a three-mile canal inland from Par to Ponts Mill in 1835 to serve his high-quality granite quarries in the Luxulyan valley. Then, as demand for granite escalated for dockworks at Plymouth, Portsmouth, Southampton and London, he built his spectacular granite viaduct-cum-aqueduct in 1839, the first structure of its kind in Cornwall,

Par Harbour, built in the 1830s, where clay became the major export.

across the Luxulyan valley to allow rail access to his quarries. For a brief spell in the late 1840s, Par shipped more granite than any other port in England.[28]

In a cleverly integrated process, the aqueduct carried some water across the valley to drive a 30 feet diameter water wheel working an inclined railway. This incline was used to lower wagons containing quarried granite and raise empty wagons. It connected with the canal to Par harbour, which was replaced by a railway in 1855. This water could also go to Treffry's Fowey Consols mine, although some by-passed the aqueduct and went to ponds at Charlestown to maintain water levels and for occasional flushing of the inner harbour. Later, when his mine closed in the 1860s, the water ran china-stone grinding mills built at Ponts Mill near the former canal basin.

In 1840 a scheme was proposed for a harbour at Porthpean, a beach to the west of Charlestown, where a lease of 1835 tells of a malt house and lime kiln. Treffry is thought to have been involved in this scheme for a harbour, sea wall and piers, together with an access road. By 1850 Edward Stocker was also involved, perhaps with the West of England Co.'s trade in mind. This project does not seem to have come to fruition, but it may have stimulated developments elsewhere.[29]

Newquay and the West Cornwall ports
The port now called Newquay also owed its renovation to Treffry. In 1836 he had acquired a harbour in a bad state of repair, together with the surrounding manor of Towan Blistra, for £7,500, and spent a further £3,000 on improving the port, which by 1843 enjoyed a thriving trade.[30] Small quantities of china clay and stone were also dispatched from West Cornwall via Porthleven. In 1810 a prospectus was issued for the construction of a harbour in what was then a small cove called 'Porthleaven' as a refuge for ships caught in bad weather and for fishing boats. It was to be capable of containing 200–300 ships of 200 tons and took 15 years to build. In its earlier years it never attracted enough trade to make it viable, and the chief cargoes handled were imports of coal and timber for local mines and exports of granite from Breage, Sithney and Wendron and china clay from Tregonning Hill, Cookworthy's original source. The amount of china clay and china stone handled was small, typically 30 tons of clay and 150 tons of stone a year, but in 1856 Phillip Wheeler & Co. started to ship out clay from the port and the total expanded, as will be discussed in Part Two.[31]

Modest quantities were also despatched from Penzance, Falmouth and Hayle. However, by the mid-1800s, as a result of the enterprise of Rashleigh, Hawkins and Treffry, out of 80,000 tons of clay exported from Cornwall, half was shipped from Charlestown, 18,000 tons from Pentewan and 10,000 from Par, leaving 12,000 to be shared by Mevagissey, Falmouth, Hayle, Porthleven, St Michael's Mount and Penzance to the west and Newquay to the north.[32] To put this traffic into perspective, it was small compared with metal mining: 100,000 tons a year of copper ore from the central mining district was passing through Portreath alone in 1840.[33]

The shipping of ball clay, china clay and stone
The china-clay trade benefited from a period of unparalleled prosperity in waterborne transport after the hazards of the Napoleonic Wars. As coastal and foreign shipping freights from Britain fell by 80 per cent between 1820 and 1860, British exports went up from £30m a year around 1800 to £100m per annum in the early 1850s, and imports rose even faster, from £40m to £150m during the same period.[34] China-clay producers also enjoyed access to many Staffordshire potters and Lancashire textile manufacturers via improved canal networks in the Midlands. By the 1840s the cost of transport from the Cornish clay ports to the Staffordshire potteries had fallen from £2.50 to £0.60 a ton.[35]

In Dorset, most clay was loaded from small quays onto sailing barges that took it from Wareham or from quays constructed by clay merchants to Poole Harbour for transfer to sea-going vessels, although some shallow-draft sailing ships carried clay directly as far as London. In the later 1700s Isaac and Benjamin Lester of Poole were two prominent handlers of clay for shipment to Liverpool.[36] From the 1840s steam tugs operated from Wareham to Poole, towing barges holding about 50 tons of clay. For North Devon the main outlets were Bideford Quay and, for a time, Fremington Quay, west of Barnstaple on the Taw Estuary. In South Devon the main port was Teignmouth, where sailing barges brought clay from Kingsteignton, close to Newton Abbot, and, from the 1840s, from the Earl of Devon's quay at Newton Abbot, where clay was loaded from his newly opened Decoy pits.

In Cornwall, in the early part of the nineteenth century, single-masted vessels called sloops with a capacity of 40 to 50 tons were used, suitable for beaching on the shore. These were also handy for smuggling, an important, albeit under-recorded, part of maritime activity at the time, and there were rumours that smuggled goods were sometimes hidden under cargoes of china stone.[37] In 1807 Dr William Fitton of Northampton, in a detailed description of the china-clay and stone trade, noted that these materials were 'shipped for Plymouth and thence to the potteries'.[38] This suggests that only small coastal craft or barges, also with a capacity of 40 tons, were used to take clay and stone to Plymouth, where they were transferred to larger vessels. This practice is confirmed by details of the illicit trade in clay for sale to flour merchants, discussed in the previous chapter.

In the 1830s, although coastal trade was now open to ocean-going vessels, only a handful of ships were recorded as sailing from St Austell Bay to foreign ports.[39] Then, after the new harbours in the clay district were built, two-masted schooners and ketches largely superseded the sloops.[40] The average tonnage per British ship went up from 106 to 140, but the vessels used in coastal trade were smaller, and even if the average cargo was as much as 100 tons, this meant 800 shipments from Cornwall a year.[41] This is an underestimate of the number of vessels involved, however, because ship owners often took smaller quantities of clay together with other kinds of freight.[42]

The finance of trade

How was the transport of clay and stone financed? As far as carriage by road from pit to port was concerned, clay merchants could make their capital go further by delaying payment to the wagoners, mostly smallholders who queued up on the chance of a load and who were reluctant to press for payment. Their accounts were traditionally settled once a year and clay producers sometimes kept them waiting for several more months. A more important problem was how to raise the large amount of working capital needed to cover the long period between the despatch of goods from the ports and payment by the final customer. This was especially evident in the china clay and stone trades, since start-up costs, discussed later in Chapter Nine, were comparatively low, whereas lead times between production and payment by the customers were especially long because of the distance between Cornwall and the main users of clay and stone.

Clay merchants could avoid the long wait by selling at a discount to the grow-ing number of other clay merchants or dealers in timber and coal in the clay dis-trict or at Plymouth. Another solution was barter between clay merchants and pot-ters. For instance, in 1828 Benjamin Smith of the Lane End potteries in Staffordshire offered to exchange chinaware at manufacturers' prices for amounts of between 50 and 200 tons of good-quality clays.[43] Around the same time William Jago of St Columb traded clay for chinaware which his wife sold in her shop, and John Lovering, clay producer and chinaware dealer, carried on a similar barter sys-tem with a Swansea potter.[44]

In the early 1800s, when English potters and Plymouth merchants were respon-sible for a substantial proportion of china-clay and stone production, some of them integrated the production and distribution of raw materials and finished products. Thomas Minton and fellow members of the New Hall Co. in Staffordshire formed a syndicate to acquire a ship to carry china clay from their own works to their pot-teries, although this was wrecked a few years later.[45] In 1801 the 80-ton sloop *Hendra,* of Fowey, was built and owned by the clay landlord Henry Lambe and the Staffordshire potters Wedgwood, Bentley, Hollins and Warburton.[46] Some local clay adventurers were also involved in shipping, such as George Pinch, publican of the Cornish Arms at St Stephen, who operated as a clay producer and dealt in clay pur-chased from other producers as well as chartering ships for clay transport.[47]

An outstanding example of a two-way operation run from Plymouth was that of John Warrick of St Austell and John Dickins of Plymouth. In 1812 they were listed individually as earthenware dealers, in 1822 they were recorded as partners and china merchants, and by 1827 they were clay producers, accounting for 4,000 out of 11,000 tons of clay extracted in the St Austell area. As Plymouth tableware importers they offered a range of china and earthenware to traders in Cornwall and Devon, including Staffordshire cream and lustre ware and Bristol stoneware. As shipping agents they controlled the two-way traffic of clay, stone and chinaware, without being involved in the complexities of ceramic production.[48]

Their dealings also illustrate the hazards of trade, even in times of peace. They advertised to local chinaware merchants the advantage of buying from them, rather than undergoing the 'frequent delays in canal conveyance from the Staffordshire potteries, the difficulty of procuring shipment from Liverpool to the smaller ports in the west of England and the great uncertainty of the sea passage'.[49] Another Plymouth shopkeeper, John Luke, who owned the 33-ton *St Austle (sic)* Packet, built for him at Fowey in 1826, traded between Plymouth and Charlestown, and by 1840 his family were established in Charlestown as merchants in china clay and stone, rope and tar, as well as shipbuilding.[50]

Direct sales to customers

Even if the West Country had the best china clay, it had to compete with other producers of whiteners and fillers such as alum, plaster of Paris or chalk, as well as feldspathic alternatives to china stone. Some clay merchants therefore engaged in direct sales to the customer. In 1829, when Robert Martin of the clay producing partnership Lovering, Martin & Nicholls, found that demand was lagging, he accompanied a shipload of clay to the Staffordshire potters and sold it to them directly. In 1849 Edward Stocker joined his cousin William Marshall Grose, formerly a St Austell linen draper and fellow adventurer in the West of England Co., in setting up a selling agency at Stoke on Trent. Alexander Dingle Stocker later joined the firm, married Grose's daughter and continued the partnership.[51] Some of the larger companies also acted as sales agents for other producers. The Varcoes sold on commission for William Jago, Lovering and Treffry, and purchased clay on behalf of Herbert Minton of Staffordshire. Richard Yelland acted as the Varcoe agent in Staffordshire and William Varcoe later settled in the Potteries as a clay merchant.

By the mid-1800s Cornish clay producers such as Rebecca Martin, Edward Stocker, Charles Truscott and Elias Martyn also followed Wedgwood's and Spode's example and showed chinaware made from their clay at public displays, including the Great Exhibition in London.[52] The Staffordshire Rogers family, operating in St Stephens on Grenville land, had earlier taken a leaf out of Wedgwood's book by getting Spode to make some fine porcelain for promotion around 1811.[53] A Devon-based producer who displayed there was William Phillips of Lee Moor, and his clay, he claimed, was the only one to achieve special mention from the judges for its high quality.[54]

Developing export markets

Production-oriented clay historians have tended to underestimate the achievements of the clay merchants in creating export markets. They were not able to use trading networks developed by Devon ball-clay merchants because the overseas distribution of ball clay had been prohibited in 1662 to safeguard the pottery producers' interests until the 1850s.[55] Nor could china-clay producers benefit from export links for the dominant copper industry, since copper ore was sold locally and shipped to South Wales for smelting. Yet, already in the 1830s, a substantial trade in direct exports of china clay and stone from Cornwall was building up to France, the Low Countries and the Baltic,[56] William Jago making a sales trip to France in 1839.[57] By the early 1850s clay was going to potters in America and Russia and to textile manufacturers in Egypt and India.[58]

China-clay producers had access to a mercantile framework which had made Britain the greatest trading nation on earth, with links to commission agents and acceptance houses in every port of the world.[59] The logistics of shipping clay and stone involved two-way traffic in other materials, and it is important to recognise that clay and stone, being low value to bulk items, were not popular with ship owners, who might insist on a mixed load containing more valuable freight.[60] A major import to Cornwall was limestone from Plymouth, together with culm, an inferior coal from South Wales needed to burn limestone in the dozen or so lime kilns that operated around St Austell Bay, supplying a product used to neutralise acidic local soils, make lime-mortar for building purposes and in tin and lead smelting.[61] Coal offered the greatest potential for a two-way trade. Before the days of coal-fired kilns and steam engines, it was not greatly used in the clay industry, but it was essential for metalliferous mining, and a complex and mutually beneficial network developed.

Clay shipped to potteries in coal-mining areas in South Wales, North East England and elsewhere brought return cargoes of coal to Mid-Cornwall ports. Tyneside possessed abundant supplies of coal, and local potters imported white clays from the South West and flint from Gravesend as return ballast in the colliers which exported coal to many destinations.[62] Liverpool, directly or indirectly, was a prominent hub of two-way trade involving china clay and other materials. Potters had set up there in the 1750s, only to be ousted after the opening of the Trent to Mersey Canal in 1777, when Staffordshire potters could undercut them. An exception was the Herculaneum Pottery, so called because of the classical style of its products (like Wedgwood's Etruria). Set up in 1796, with 40 craftsmen from Staffordshire, it employed 300 by 1827. Its success was due to its riverside site, unloading coal and clay and loading pottery directly between factory and ship. This reduced its transport costs below those of its Staffordshire rivals, but ironically the excellence of its location proved to be its undoing, because it became more profitable to demolish the pottery and replace it with the Herculaneum Docks to serve Liverpool's rapidly growing trade in cotton.[63]

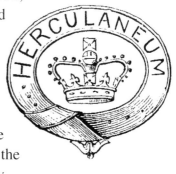

Mark of the Herculaneum Pottery at Liverpool, 1796–1841.

As discussed earlier, coal was transported as a return cargo on clay vessels, but this did not pose a major problem, since coal dust would be burnt off during the firing process for pottery, as indeed would the small particles of lignite sometimes contained in ball clay. However, for china clay used in cotton cloth and paper manufacture, where whiteness was important, coal was not the most compatible of cargoes, and was often unloaded on the west side of a harbour while clay was loaded on the east side. One visitor to Charlestown described the port as 'Little Hades', enveloped in clouds of coal and clay dust, the dockers 'white or black according to which side of the port they worked on'.[64] After unloading coal every part of the ship had to be thoroughly cleaned, including the sails and, for further protection, higher grades of clay were carried in wooden casks.

Summing up, from the end of the Napoleonic Wars to the mid-1800s the cost of marketing and distribution was greatly reduced through the improvement of networks created by, and for, other industries. In the china-clay and stone trade, pit to port transport was often by horse-drawn wagons, but some roads had improved and tramways, waterways and steam locomotives had been developed by entrepreneurs concerned with the export of metal ores and granite and the import of limestone. Eighty-five per cent of china clay and stone production was shipped out in larger, more efficient, vessels through the three new ports that had been built in St Austell Bay. As a result, the cost of delivering china clay fell to about a fifth the level of half a century earlier.[65] In the ball clay trade in North Devon, production became more competitive through the construction of a canal, the primary purpose of which was the import of limestone. In South Devon, two short canals were specifically built for the clay trade, which also benefited from the construction of a quay at Teignmouth for the handling of granite. In Dorset tramways were the main form of pit to ship transport, and the clay merchants themselves built these as well as their own quays.

By the 1850s, marketing and distributing china clay and stone had become a complicated business. A few early customers for china clay, principally English potters, had provided their own shipping, and Plymouth merchants moved into the production and sale of china clay and stone, again chiefly to the potteries. These merchants also purchased chinaware from the potteries and sold it to the public. By now sales of china clay to cotton textile manufacturers and paper makers were of major importance, creating newer and more complex trading networks. However, because ball-clay exports were still prohibited, British potters remained their main customers, although from 1808 sales to the 'sugar colonies' was permitted, probably for use in 'claying' sugar, discussed in Chapter Six.

References
[1] Robert Copeland, *A Short History of Potters' Raw Materials*, Cheddleton, 1972, pp.2–3; L.T.C. Rolt, *The Potters' Field*, Newton Abbot, 1974, pp.140–41.
[2] *WB* 11 October 1867.
[3] Patrick O'Brien, 'Central government and the economy', in Roderick Floud and Deidre McCloskey (eds), *The Economic History of Britain Since 1700*, I, Cambridge, 1994, p.219.
[4] Alison Grant and Peter Christie, *The Book of Bideford*, Buckingham, 1987, p.51; Lois Lamplugh, *Barnstaple*, Chichester, 1983, p.89.
[5] See for example the Agreement between William Morton Pitt and John Calcraft and the carrier Thomas Garland of Norden Farm, with an inventory of horses and wagons, DRO, RWR/E/21. Cecil N. Cullingford, *A History of Dorset*, Chichester, 1984, p.93.
[6] DRO, RWR/E/39, 102, T/72, 75, 293, 306, 433.
[7] Charles Hadfield, *The Canals of South West England*, Newton Abbot, 1967, pp.91–94; Rodney Legg, *Purbeck's Heath*, Sherborne, 1987, pp.70–71.
[8] Hadfield, 1967, pp.120–21.
[9] Alison Grant, *North Devon Pottery*, Exeter, 1983, p.39; Hadfield, 1967, pp.135–41.
[10] Muriel Goaman, *Old Bideford and District*, Bristol, 1968, p.64.
[11] Hadfield, 1967, pp.142–79.
[12] DRO, D/PIT/1/1/6, 9.
[13] Legg, 1987, pp.72–74, 77.
[14] Legg, 1987, p.71.
[15] Grant and Christie, 1987, p.54.
[16] *WB* 18 and 25 January 1833, 17 September 1835, 3 November 1843.
[17] Joan Jones, *Minton*, Shrewsbury, 1993, p.12; Norman Pounds, China Clay, unpublished, n.d., Wheal Martyn Museum, St Austell, chap.4, p.7; this refers to contracts by Kent and Varcoe in 1805; R.M. Barton, *A History of the Cornish China-Clay Industry*, Truro, 1966, pp.32–39; John Penderill-Church, *The Treffry's and Cornish Industry*, n.d., Cornwall Centre Redruth x 00137/69 x.
[18] Peter Stanier, *South West Granite*, St Austell, 1999, pp. 17–19.
[19] Apart from the references in the text, information on the Lee Moor railway comes from R.M.S. Hall, *The Lee Moor Tramway*, Blandford, 1963; Bryan Gibson, *The Lee Moor Tramway*, Plymouth, 1993; Roy E. Taylor, *The Lee Moor Tramway*, Truro, 1999.
[20] William D. Lethbridge, *One Man's Moor*, Tiverton, 2006, pp.174–76.
[21] A trade directory of 1844 described 12,000 tons a year as leaving Fowey, but this was due to a confusion about its designation as a 'registration port' for the area, and the shipping left other harbours. *Pigot & Co.'s Directory*, London, 1844.
[22] Richard and Bridget Larn, *Charlestown*, St Austell, 1994.
[23] Larn, 1994, pp.108–09.
[24] Larn, 1994, pp. 108–09.
[25] William Jory Henwood, 'On the Preparation of China Clay', *JRIC*, 1839, pp.54–56; Henry De la Beche, *The Geology of Cornwall, Devon and West Somerset*, London, 1839, pp.493–512.
[26] He later hoped to develop it for exporting iron ore but this was not successful, although one quay is still known as Iron Ore Quay.
[27] R.E. Evans and G.W. Prettyman, *Pentewan*, St Austell, 1986.
[28] Stanier, pp.6–7 and 36.
[29] Barton, 1966, pp.77, 86, 93.
[30] John Keast, *The King of Mid-Cornwall*, Exeter, 1982.
[31] Stuart N. Pascoe, *The Early History of Porthleven*, Redruth, 1989, pp.8–24.
[32] H.M. Stocker, 'An Essay on the China Stone and China Clay of Cornwall', *RRCPS*, 1852, p.87.
[33] Richard Pearse, *The Ports and Harbours of Cornwall*, St Austell, 1963.
[34] Deirdre McCloskey, '1780-1860', in Floud and McCloskey (eds), 2000, I, p.249; B.R. Mitchell, *British Historical Statistics*, Cambridge, 1988, p.526. Over 1,000 steam ships were registered, although they played little or no part in china-

clay transport. Many steam vessels at this time were used as tugs to assist sailing vessels in and out of harbour.

[35] An example is the ex-works costs of the Hendra pit once run by Minton and partners and later taken over by Treffry; Penderill-Church, n.d.

[36] DRO, D/LEG/F/8, 10.

[37] Private communication from Helen Doe.

[38] William Fitton, 'On the Porcelain Earth of Cornwall', *Annals of Philosophy*, March 1814, pp.181–84.

[39] Inspector's Report, Custom House, Fowey, 1830, quoted in John Keast, *The History of Fowey*, Exeter, 1950, p.100.

[40] C. Bainbridge, *The Wooden Ships and the Iron Men*, Truro, 1980.

[41] These figures are only an estimate because a great variety of boats were used. Average ball-clay shipments from Teignmouth in 1854 were 92 tons: Rolt, 1974, p.130. Shipments of granite from Penryn ranged from 60 to 120 tons in 1868: Peter Stanier, 1999, pp.108–09; Port Navas took ships of 140–150 tons in 1856, *WB* 8 August 1856; vessels carrying limestone from Plymouth quarries to Cornish ports only averaged 40 tons, *WB* 12 May 1843.

[42] H.J. Trump, *West Country Harbour*, Teignmouth, 1975, p.101.

[43] *WB* 18 March 1828.

[44] E. Morton Nance, *The Pottery and Porcelain of Swansea*, London, 1943, pp.473–74; CRO, AD 194. We are indebted to John Tonkin for this reference.

[45] Jones, 1993.

[46] It was transferred to St Ives in 1806. *CRO*, MSR Fowey Register.

[47] CRO, J/403. His son, also called George Pinch, carried on a similar business.

[48] *Universal British Directory*, 1798; *The Picture of Plymouth*, 1812; N. Taperell, *The Plymouth Docks*, 1822; *The Tourists' Companion*, 1823; *The Plymouth Stonehouse and Devonport Directory*, 1822, 1830; *Pigot & Co.'s Directory of Devonshire*, 1830–31; *Thomas' Directory of Plymouth*, 1836.

[49] *WB* 1 August 1828.

[50] Larn and Larn, 1994, pp.118.

[51] Edmund Vale, *China Clay History*, 1954, Wheal Martyn Museum, p.65; Denis Stuart (ed.), *People of the Potteries*, Univ. of Keele, 1985. We are indebted to John Tonkin for the latter reference.

[52] *WB* 9 May 1851.

[53] Copeland, 1972. A fine 'Regency bowl' made by Spode for William Roger is displayed in the British Museum (MLA Accession Catalogue XIV 8). Rogers was a founder member of the 'Cornwall Typographical and Caledonian Society' in 1813. We are indebted for this reference to Jim Lewis; Barton, 1966, p.80.

[54] Crispin Gill (ed.), *Dartmoor*, Newton Abbott, 1970, p.134.

[55] Rolt, 1974, p.130.

[56] Barton, 1966, p.80

[57] CRO, AD/194. We are indebted to John Tonkin for this reference.

[58] Stocker, 1852, p.94.

[59] Stanley Chapman, *Merchant Enterprise in Britain*, Cambridge, 1992.

[60] Rolt,1974, p.102.

[61] Ken Isham, *Lime Kilns and Limeburners in Cornwall*, St Austell, 2000, pp.108–16.

[62] R.C. Bell and M.A.V. Gill, *The Potteries of Tyneside*, Newcastle, 1973, pp.5–25.

[63] Joseph Mayer, *On The Art of Pottery*, Liverpool, 1871, p.41.

[64] Quoted by Cyril Noall, *The Story of Cornwall's Ports and Harbours*, Truro, 1970, p.30.

[65] Henwood, 1839, p.54; Stocker, 1852, pp.77–96.

[66] Trump, 1975, p.68.

Chapter Eight

West Country Landlords and English Potters

T his chapter examines the finance and control of the clay industry in Dorset, Devon and Cornwall. Until after the 1914 War, much of the clay-bearing land was in the possession of great 'claylords' as they were sometimes called, such as the Earl of Eldon in Dorset, the Duke of Devonshire and the Duke of Cornwall. Since many of them also owned vast estates in other counties, they relied heavily on their local stewards to deal with their clay interests, as we shall see. The fixed capital needed to start up a clayworks was relatively small and the same cost-book system of financial control was used as in mining and ship-operating. Profits were shared out at meetings held regularly, often once a month, while if losses were incurred the shareholders had the option of putting more money in or closing the business down. Although often criticised for encouraging short-term attitudes and not building up reserves for expansion or future development, this system worked well when only small, local groups of shareholders were involved, who kept a close watch on proceedings, which was the case with most companies up until 1914.

Clay landlords and clay merchants of Dorset

In Dorset, the main owners of clay-bearing land in the eighteenth century were the Calcrafts and the Pitts.[1] John Calcraft had acquired land in 1757 and 1768, apparently to secure a 'pocket borough' to elect his own Member of Parliament. In 1766 he leased land to Thomas Hyde, to be mentioned later, and in 1771 he entered into an agreement with Hyde to supply clay to Wedgwood. Calcraft died a year later

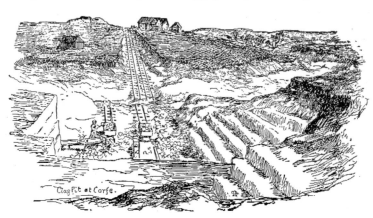

Open ball-clay workings near Corfe, 1882.

and his son became a partner with William Morton Pitt to develop clay land.[2] In 1768 and 1769 Calcraft and Hyde were involved in Parliamentary skulduggery, distributing cash and favours to electors, but were not convicted of corruption. Pitt, who owned the important Norden clayworks near Corfe Castle, also founded craft industries to provide local employment and invested heavily in transforming the nearby fishing village of Swanage into a 'marine retirement' resort.

To finance his enterprises Pitt sold his Encombe Estate in 1807 to John Scott. The son of a Newcastle-upon-Tyne coal merchant, Scott followed a legal and parliamentary career that led him to the position of Lord Chief Justice. By 1792 he had been created Lord Eldon, the title relating to an estate of nearly 12,000 acres in County Durham that he had acquired in that year, and in 1821 he became Viscount Encombe and Earl of Eldon. Legal documents show Pitt progressively raising bigger and bigger mortgages on his properties, and after his death in 1836 these were sold to the Earl of Eldon.[3] When the Earl died two years later his properties, which also included large estates in Gloucestershire as well as 6,900 acres in Dorset, were inherited by his grandson. Throughout the period up to the 1914 War the Earls of Eldon, all of them called John, made Encombe House, near Wareham, their principal residence and retained ownership of much of the clay-land of Purbeck.

Turning to the Dorset clay merchants, land on Arne Heath east of Wareham, was leased for most of the eighteenth century by the Hyde family of Poole, the best known of whom was Thomas Hyde, born in 1731. Hyde farmed an unpromising terrain of scrub and bog, improved tracks to his quarries and clay pits, built a small pier to the west of the Arne Peninsula and carried on a general trade with Newfoundland. He dominated local political life for a couple of decades and by 1790 he owned or controlled over half the Peninsula. However, a bad harvest and a trade recession in 1792 led to his financial collapse.

South of Wareham, the Brown family of that town had been shipping clay from their own land from the 1660s onwards. A century later Giles Brown was a prominent local businessman and two members of the family were partners in the Dorsetshire Bank, formed in 1789, while Thomas Brown carried on the clayworks until his death in 1818. It was then leased to Watts, Hatherley & Burn of Newton Abbot in Devon, who controlled it until, as we shall see, the Pike family bought the land.

Interaction between Dorset and Devon clay merchants

The clay trade of Dorset and Devon was developed by shifting coalitions of dealers from the two counties.[4] William Crawford from Poole, as mentioned in Chapter One, had leased land in the Bovey Basin in the 1720s, while the Pike family from Chudleigh, east of Bovey Tracey, moved in the opposite direction. Joseph Pike was originally recorded as opening clay stores at Hackney on the Devon estate of Lord Clifford, who owned land on the east side of the Bovey Basin between the

River Teign and Kingsteignton. Joseph Pike's sons, however, acquired property south of Wareham in 1760.

One of the sons, referred to as 'William Pike, gentleman of Chudleigh', acted as agent for a syndicate of Staffordshire potters, supplying them with clay from Wareham and also from the Bovey Basin, where, in 1789, he leased land from the Clifford Estate of James Templer of Stover Lodge, mentioned in Chapter Seven.[5] William's father, Joseph, while following his sons to Dorset, also maintained a link with Devon, shipping clay from Teignmouth.

A rival to the Pikes in Dorset was the London merchant Benjamin Fayle, who arrived in 1795 to buy a working pit near Corfe Castle, south of Wareham. The tramways built by these two families were mentioned in Chapter Seven. Both firms prospered, but while the Fayles remained in London and appointed a manager to run their pits, the Pikes were influential in the commercial life of Wareham, supplying coal to the townspeople from their warehouse on the quay. Walter Pike entered into partnership with Henry Hatherley, who was probably related to John Hatherley, partner of Agnes Watts, to be mentioned later.

Although the clay families of Dorset and Devon were linked, Dorset became the focus of the Pikes for almost 200 years, while Devon was the territory of the Watts. The Watts family became involved in ball clay in the eighteenth century based on their ownership of 88 acres in the Preston Manor area of Kingsteignton and flourished throughout the eighteenth century, the head of the family for several generations being called Nicholas. Then, in 1809, Agnes Watts, widow of one of the Nicholas Watts, formed a partnership with Samuel Whiteway, a merchant, sloop owner John Hatherley of Kingsteignton, near Newton Abbot, and William Martin of Wareham. The firm was named Whiteway, Watts & Co., implying that Whiteway was the senior partner. A Dorset connection existed through a lease of 1817 for 21 years for clay pits in Studland.[6]

In 1822 the eldest son of Agnes, the Revd Nicholas Watts, took over her interest in the partnership which, in 1845, expanded production by leasing all the claylands of the Clifford Estate near Kingsteignton. The agreement was to extract 5,000–6,000 tons a year, paying landlord's dues of 3s.6d. (17½p) a ton for 'best clays' and 6d (2½p) for poorer 'saggar clays' used to make containers to protect pottery when fired in a kiln. In 1849 the Revd Nicholas Watts died and his partnership share was taken over by his son, W.J. Watts, JP and High Sheriff of Devon. In 1853 the output was increased by a further 5,000 tons a year of saggar clay. However, about this time relationships between the partners seemed to have deteriorated, because in 1859 Watts surprised his colleagues by granting a new lease on the land he owned to his son, also named W.J. Watts and also a JP, and to two others, Lewis Bearne, County Councillor, and Charles Davey Blake, who formed a new partnership, Watts, Blake, Bearne & Co.

Watts' decision resulted in a Court of Chancery action and the break-up of the old firm of Whiteway, Watts & Co, the other members of which formed a separate enterprise. Although the reason for these events is not clear, it may have something to do with the appearance of a rival South Devon firm in 1848, owned by Frank J. Davey and Edward Blake, whose son, Charles Davey Blake, was the new partner in Watts' firm. Charles was a young man who had been the sales representative for Blake and Davey in Stoke on Trent, and his middle name also suggests that his father had a long association with or might have been related to Frank Davey.

The partnership of Blake and Davey appears to have come about in a curious way. In 1848 Frank J. Davey leased a pit on the Courtenay Estate of the Earl of Devon to produce 1,000 to 1,500 tons of clay a year, paying dues of 4s.6d. (22$\frac{1}{2}$p) a ton. Then Edward Blake of Stoke on Trent, a supplier to the Staffordshire potters, leased a much larger area on the Estate, paying £1,000 a year to raise 14,000 tons per annum, in other words only 1s.5d. (7p) a ton. He was also to enjoy free use of the Earl's wharves. These very uneven arrangements soon ended when the two clay merchants formed Blake, Davey & Co., changing the name in 1856 to the Devon & Courtenay Clay Co. Whatever the possible personal or business connections were that led to these events, however, the outcome was that from the 1860s onwards three front runners had emerged in the South Devon ball-clay industry: Watts, Blake, Bearne & Co., Devon & Courtenay and Whiteway & Co. Their progress will be discussed in Chapter Eleven.

In North Devon the ball-clay deposits at Peters Marland, south of Great Torrington, were probably dug by a number of small local landowners in the early 1600s, but later in the century the Greening family from Bideford consolidated these holdings and ran the clayworks. The Greenings were merchants and mariners, a dissenting family who were said to have come to Bideford from Gloucester to escape religious persecution around 1630. They carried tobacco pipe clay in their own ships to Gloucester and 'sugar bakers' clay', discussed in Chapter Six, for sugar refiners in Bristol. They

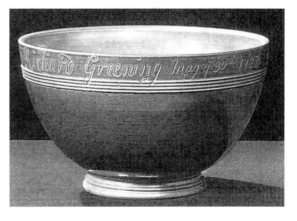

A Bideford ware salt-glazed bowl made for a member of the Greening family involved with North Devon ball clay.

also shipped local earthenware to these ports and to Dublin. By the time of his death in 1692, John Greening appeared to have a monopoly of the ball clay from Peters Marland, as well as owning property in Bideford, and had entered the ranks of the gentry. Later, Josiah Wedgwood and other potters shipped ball clay direct from Bideford to Liverpool.[7]

Cornwall and West Devon clay landlords and ladies

In the china-clay bearing lands of Cornwall and West Devon, the nobility and gentry maintained their ownership of much of the ground in which clay and stone was worked until the 1914 War. They numbered, among others, the Duke of Cornwall, the Earls of Morley, Mount Edgcumbe and Falmouth, Sir Colman Rashleigh of Prideaux, near St Blazey, and Sir William Sarjeant of St Benet's Abbey at Lanivet, as well as clerical gentlemen such as the Revd Lambe and the Revd Stackhouse. Nor should we underestimate the importance of such figures as Lady Grenville of Boconnoc and Ann Maria Agar of Lanhydrock.

A few aristocratic landowners, like Rashleigh, Sarjeant and Lord Morley, took shares in clayworks, but most of them preferred arrangements where they took a part of any profits that were made but ran no risk of possible losses, licensing the use of land to adventurers for periods ranging from as little as a year for prospecting purposes, to 21 years or longer. In their dealings with clay merchants, the clay landlords depended, as mentioned earlier, upon their stewards. For example, in 1815 William Jenkin, the Agar agent, asked Nicholas Cole, the Agar under-steward at Lanhydrock, to enquire from Ann Maria Agar if she wished to have 'a sample of clay', as he called it, from the Lizard sent to her, 'for taking her friend Wedgwood's advice on it'. At the same time he suggested that she might licence her land for 21 years for a minimum production of 50 tons a year at £1 per ton, rising to a charge of £2 per ton for any additional clay extracted. This arrangement, he said, would ensure that if the market price rose, and the producers could sell to other potters at a profit, she would have some protection.[8]

The landowner's share varied according to the expected quality of the clay, the ease of working and the negotiating strength and business acumen of the adventurers and the owners. Occasionally the landowners shared part of the risk, in the sense that their revenue depended entirely upon the volume of sales. Lovering, Nicholls and Martyn paid two-ninths of the sales revenue to the Duke of Cornwall and another two-ninths to John Trevanion,[9] excessive by the standards of metalliferous mining, where landlords' dues ranged from one-sixth of the revenue for shallow and easily worked mines down to one-twenty-fourth for more risky projects that involved adventurers in heavy capital investment. The Agar-Robartes took one-eighth of the revenue from Tincroft Mine.[10]

A common practice was to charge an annual rent merging into payment of dues per ton 'raised and made merchantable'. If the dues exceeded the rent, then the larger sum was paid to the mineral lord. Over and above that, a clause in the licence might specify a payment per acre to compensate for land spoiled by extraction or tipping, varying according to the initial value of the land. Contracts for payment of dues depended upon the wishes of the landlord. To maximise his short-term income, he reduced the dues per ton after a certain minimum was reached. If, on the other hand, he wanted to discourage the clay merchant from taking out large quantities when the price was high and stopping operations when the price fell, he would increase the dues per ton after a fixed amount per annum was produced.

In a typical early example of 1770 at Predannack Downs, near Mullion, the landowner George Hunt of Lanhydrock received a minimum rent of £10.50 per annum for a 21-year licence to dig soapstone, merging into dues of £1.05 per ton if the amount exceeded ten tons a year. In a nearby licence of 1775, a Worcester pottery simply paid £1 for every ton of soapstone, which sold at £20 a ton.[11] Landlords' dues for china stone and china clay were considerably lower than this, typically around £0.20 a ton for china clay, and half this rate for china stone,[12] which was less costly to produce and sold at a lower price.

Poor rates

A major bone of contention between landowners and adventurers was the responsibility for paying poor rates to support destitute local families. Whoever paid them took the risk that if labourers flooded into a parish to work mines or clay pits, they might become a burden on the rates if the works suddenly closed down and if they could not return the unemployed men to their original parish. In mining communities most were paid by the landlord, the shareholders of Tincroft mine, for example, preferring to pay higher dues on production to the landlord rather than run this risk.[13]

Clay merchants complained it was always they who were saddled with the poor rates, which, during a slump in the 1850s, amounted to £0.15 a ton, or a quarter of the ex-works price of 'common' clays. Henry Higman maintained that he had never heard of any metal miner paying such a high rate, and Edward Stocker asserted that granite quarrymen did not pay them at all.[14] Paternalistic producers kept their workers in employment during a slump to remove overburden, or topsoil, so as to be ready to step up production in the next boom, and claimed they sold below the cost of production to keep their labour force together.[15]

The arrival of the potters

As discussed earlier, the English potters had been buying ball clay from Dorset and North and South Devon before they came to Cornwall, but they had not thought it necessary to lease or acquire their own pits there. Why did they do so in Cornwall? When they started to buy ball clay their businesses were small and supplies of ball clay were plentiful, but by the time they became interested in china clay and stone, employment in the Staffordshire potteries alone had risen from a few hundreds to

15,000 and they were among the largest entrepreneurs in Britain. Competition to produce the very whitest chinaware was fierce, and they wanted to control their own sources of the limited supply then available.

As early as 1759 small quantities of china clay had found their way to Staffordshire, where experiments were conducted with samples from Breage, near Tregonning Hill, and also from St Austell.[16] While Wedgwood publicly denounced Champion for holding on to his monopoly of supplies from Cornwall, he had begun secret negotiations in 1775 to create a monopoly of his own from an American supplier, sending his agent, Thomas Griffiths, to Cherokee country, in what is now North Carolina, to bring back five tons of china clay.[17] It was rumoured that clay had also been smuggled out of Cornwall in pilchard barrels by Wedgwood and others. Although an impassioned advocate of free trade and an enemy of monopolies and patents, Wedgwood had to employ underhand practices in order to succeed. He coded his ceramic recipes so that they could not be copied and upset Liverpool potters by poaching their best engraver.

As we saw in Chapter Six, a clandestine approach was advisable in Cornwall for another reason: potters were unwelcome in Cornish mining communities because chinaware was replacing pewter ware largely made of tin. However, the large chinaware manufacturers were too powerful to be denied, and their search for china clay and china stone in Cornwall became even more intensive after possible American sources of supply were cut off during the War of Independence from 1775–83, and blockaded during the Napoleonic Wars. The hazards of war also forced the East India Co., which had a monopoly of porcelain imports from China, to cease trading in them in the 1790s.[18]

This disruption of trade from the Far East and the continent produced a hothouse atmosphere for home producers, who in 1799 were further protected when duty on imported porcelain was raised to over 100 per cent. Nevertheless, in 1814 a Swansea potter applied for a government grant on the grounds that, while his products equalled anything that Sèvres could produce, French goods, when obtainable, sold for three times as much.[19] Protection of the home market, moreover, did not help the big chinaware manufacturers, who sold over half their output abroad. When this outlet was cut off by wars, the loss

Josiah Wedgwood, potter, entrepreneur and scientist.
(IMAGE COURTESY WEDGWOOD MUSEUM TRUST)

of trade resulted in severe unemployment in pottery areas, food riots broke out in the Staffordshire Potteries and many small men went out of business.[20]

The reign of the English Potters in Cornwall

For half a century, from the 1780s onwards, and despite the disruptions of war, English potters developed a substantial part of Cornish china-clay and stone production. Wedgwood had been one of the first on the scene in 1775, licensing a sett from John Trethewey of St Stephen in partnership with his fellow Staffordshire potter, John Turner. Four years later, Thomas Phillips from Lane End in Staffordshire took a sett on the estate of John Carthew, a St Austell attorney, at St Stephen-in-Brannel. Three years after this Wedgwood entered into partnership with Carthew on the same estate, later taking over complete control. Soon over a dozen clay works were active. Wedgwood, Thomas Minton and some other members of the New Hall Co. had bought a licence on land at Hendra Common from William Kent of St Austell, and others looking for setts included Rose, Blakeways & Co. of Coalport and Michael Kean of the Derby China Works.

Basic Derby china mark from c.1820–40.

In 1814 another clay producing consortium included Josiah Spode and Thomas Wolfe of Staffordshire and George Pinch, landlord of the Cornish Arms at St Stephen, who also operated as a clay merchant and shipping agent. Both Pinch and his son were important local adventurers, leasing several properties of different kinds from Sir Christopher Hawkins.[21] Ten years later this syndicate split up, but another Staffordshire firm of potters, the Rogers of Burslem, took over several licences and also opened a new pit in 1828, run by Spencer Rogers. About this date Thomas Broad from Staffordshire acquired leases near Cookworthy's original Tregonning Hill site.

Around 1830 Spode and Wolfe took over the licence of the New Hall Co. for Champion's old sett, and at the same time Minton licensed land near St Dennis.[22] To manage their pits, the out-adventurers either employed or entered into partnerships with local men with experience in the industry, such as John Kent and Edward Slade Varcoe. This pair earned wages of £300 a year each, five times the rate for a labourer.[23]

The Potters withdraw

These acquisitions were the high-water mark of the potters' involvement in Cornish china-clay production, and by the mid-1840s they had gone. Most histories of the clay industry record their departure without comment, but we might wonder why they left. One reason may simply have been a loss of interest, as in the case of Josiah Wedgwood II. He took over the business after his father died in 1797, but drew large sums from it to finance an extravagant lifestyle as a country gentleman. By 1828 he was forced to sell off his London showrooms as well as a large part of his stock-in-trade and irreplaceable models and moulds to pay his debts.[24] To make matters worse, he and his fellow potters found themselves in the midst of a severe economic recession that caused widespread distress in manufacturing districts of Britain,[25] worsened by fears of a plague that disrupted trade over vast areas from Glasgow to Norfolk.[26] Potters were forced to reduce output, cut the length of the working week and close wholesale and retail depots, and some went out of business.

In 1837 what was sometimes known as 'The Great Strike' of pottery workers protesting against their loss of wages lasted 20 weeks and cost the strikers an estimated £50,000 in lost earnings. Many were forced to borrow from Sheffield and London traders but were unable to pay them back, which blackened their reputation.[27] As a result some masters and craft workers left Staffordshire and set up business in Newcastle upon Tyne, Glasgow and other centres.[28] Josiah Spode II was in arrears with the rent on his Hendra Downs sett and, according to Penderill-Church, the landowner's agent Edyvean of Bodmin asked Josiah Wedgwood II to help. But, as we have seen, he was in no position to assist.[29] So severe was the potters' plight that the government imposed an export duty of £0.50 per ton on china clay to discourage sales of clay to foreign pottery competitors. While this may have been of some relief to the potters, it only made things worse for the china-clay producers.

By the time these short-term problems receded the potters no longer felt it necessary to control their own supplies. Economic logic had led them to license pits when rivals were scrambling to control their own sources.[30] But while in earlier days local producers posed a threat by seeking to combine and fix prices, by now their efforts to stabilise prices were defensive rather than aggressive, aimed at preventing newcomers from flooding the market rather than inflating their profits. Moreover, new sources of clay were being identified on Bodmin Moor and in Devon, rumoured to be as great as those of the St Austell district.[31] As far as the potters were concerned, the china-clay trade was a buyers' market. It was also becoming more complicated to organise and control from afar, since different qualities of clay were beginning to be sold to potters, the cotton industry and paper makers.

Developing overseas markets for pottery and clay
When the English potters withdrew from Cornwall and turned their full attention to developing worldwide markets, they indirectly helped Cornish clay merchants

to identify overseas trading opportunities. Generally speaking, potters in Liverpool, Bristol, Worcester and Swansea looked to America for sales, whereas firms on the east coast focused upon Europe. Liverpool potters had already built up a considerable export trade in chinaware by the mid-1700s, much of it in blue and white ware to the colonies, before the American War of Independence put paid to it. Enoch Wood & Sons of Burslem in Staffordshire was equally successful in exporting blue-glazed earthenware for the United States market, decorated with pictures of American mountains, Niagara Falls and early locomotives of the Baltimore and Ohio Railroad.[33]

Enoch Wood, Burslem, c.1814–16.

On the East Coast, Tyneside, possessing abundant supplies of coal, was less well endowed with clays and other materials, but imported white clays from the South West and flint from Gravesend as return ballast in colliers that exported coal to many destinations. Tyneside chinaware, while not equalling the finest work from Staffordshire, Worcester or London, found a ready market among the bourgeoisie at home and on the continent. The Sheriff Hill Pottery, founded at Gateshead in 1773, chiefly produced white-ware, mainly for the Norwegian market, and the Tyne Main

Sheriff Hill Pottery produced whiteware for the Norwegian market.

Pottery, established in 1831 at Gateshead, also made white, printed and lustreware largely for sale in Norway. The Ouseburn Pottery, formed at Newcastle in 1815, exported most of its production to Holland.[34]

Plymouth merchants, Sunderland adventurers and Dartmoor claypits

Apart from the English potters, two of the most important early china-clay producers were John Dickins and John Warrick, whose activities as china merchants and shippers are described in the previous chapter. They leased a sett from the Earl of Falmouth and in 1825 took over leases of several clay pits from their partner Phillip Ball, a Mevagissey banker with farming and mining interests, who had become insolvent.[35] Through these acquisitions they controlled approaching 40 per cent of Cornish china-clay output, which was then running at about 11,000 tons a year.[36] After Warrick retired, the partnership disposed of its clay pits to Staffordshire potter Spencer Rogers, and in 1828 Dickins explored Dartmoor for other sources of clay. In 1830 he entered into partnership with John Cawley, a Plymouth naval officer, to lease a clay sett on Dartmoor from the Earl of Morley, paying dues of 2s. (£0.10) for every ton of clay raised and bringing in clay workers formerly employed in his Cornish pits to work it.

The site was called Lee Moor and three years later Dickins retired from clay production for reasons of old age and Cawley brought in a new partner, Edward Scott, a Plymouth brewer.[37] As neither was experienced in clay working they sublet the pit for £75 a year to William Phillips, an ambitious lustreware and glassware manufacturer from Sunderland, who soon acquired the lease for £250 per annum in his own name. Phillips developed the site on a more advanced scale than most Cornish works of its time: indeed, according to the clay historian Penderill-Church, it was 'a model for all of its kind thereafter'.[38]

Phillips constructed a new leat from the River Plym to supply water and installed a waterwheel to pump the clay from the pit into settling tanks on the surface. He later built sand and mica drags, discussed in Chapter Ten, and cut a major adit which entered the pit 120 feet below the original surface and half a mile long to drain the clay wash from the bottom of the pit. The residual waste water, containing some clay residues, silted up another leat downstream which supplied water to Loughtor Mill, owned by George Strode, who sued Phillips and won his case for compensation. Strode was the landowner who refused permission to build a railway line to Lee Moor, and this incident may help to explain his refusal.

In addition, Phillips started tin streaming operations in a nearby brook and built a coal-fired pan kiln and brickworks, which will be referred to in Chapter Ten. From 1784, bricks had been taxed at 2s.6d. (12½p) per 1,000, and this rate was doubled from 1805. The tax was introduced as a short-term measure to help pay for the war in America but, as so often is the case with such 'temporary' measures, it continued for many years and was not abolished until 1850. It probably inhibited the development of brick making from clay residues in Cornwall, but Phillips may have begun producing bricks before 1850 because he saw the potential for development if the tax was removed. His trade in bricks prospered, selling locally and as far away as Chile, Bolivia and Australia. Such was the demand for Lee Moor bricks that Phillips had to build a bigger works, converting the old one into cottages for his workers.

The scale of this expenditure alarmed Lord Morley, who had lent Phillips £1,250 and who contemplated taking him to court to get his money back, but was persuaded not to do so. To raise funds Phillips took on a partner, another Sunderland glassmaker called Walker Featherstonehaugh, and when even more money was needed he formed the Lee Moor Porcelain Clay Co. with a capital of £100,000. The other partners, with an equal number of shares each, were the Earl of Morley and four Plymouth bankers, all of them close friends of Cookworthy's nephew, Dr Joseph Collier Cookworthy. Sadly, in 1861, having completed the bulk of his great plan of development, Phillips died, and the story of how the works were managed for a time by his son and then passed into the hands of the Martin family of St Austell will be continued in Part Two.

Another Cornish family who, from 1855, had successfully leased setts on Dartmoor, at Hemerdon and Broomage, was that of John Nicholls,[39] an industrial chemist from St Austell, who owned a naphtha works at Charlestown where tar was refined and dealt in explosives made at Kennal Vale, near Ponsanooth. These pits were profitable for many decades under his ownership and that of his son, also called John, who will be considered further in Chapter Twelve. Less fortunate were Robert Stephens and John Bray, two Cornishmen who had worked at Lee Moor for John Dickins. They opened a pit near Hemerdon but unfortunately a lower than expected clay content, combined with heavy rainfall which impeded extraction and high transport costs from this remote area, led to closure in 1858.

Meanwhile, the important Devon ball-clay partners, Watts, Blake, Bearne & Co., were making their first joint appearance as china clay adventurers by opening the Wigford Down Clay Works at Brisworthy to the north. They had been purchasing china clay from Lovering and selling it to Staffordshire potters and to Pochin, the Lancashire chemical manufacturer but, becoming dissatisfied with deliveries, and suspecting Philippa Lovering of poaching their customers, they went into production on their own. By now most of the high-quality clay deposits in the Lee Moor area had been leased and newcomers had to look elsewhere.[40] To put these developments into context, however, china clay output in Devon was only running at about 5,000 tons a year at this time, compared with 60,000 tons in Cornwall.

References
[1] Apart from the references cited in the text, the sources for the section on Dorset are *The Clay Mines of Dorset*, London, 1960; Rodney Legg, *Purbeck's Heath*, Sherborne, 1987; Terence Davis, *Wareham: Gateway to Purbeck*, Wincanton, n.d; Barbara Kerr, *Bound to the Soil*, London, 1968; Kenneth Hudson, *Industrial Archaeology of Southern England*, Dawlish, 1965; *VCHD*, 1908.
[2] DRO, RWR/E/21, 28.
[3] DRO, D/SEN/3/3/2/15, 9, 2/14/4, 1/8/2, 1/9/71, mortgages of 1786, 1798, 1802, Chancery case re purchase by Lord Eldon.
[4] The sources for the sections on Devon are L.T.C. Rolt, *The Potters' Field: A History of the South Devon Ball Clay Industry*, Newton Abbot, 1974; Ball Clay Heritage Society, *The Ball Clays of Devon and Dorset*, Newton Abbot, 2003; Watts, Blake and Bearne, *Devon Ball Clays and China Clays*, Newton Abbot, 1954; M.J. Messenger, *North Devon Clay*, Truro, 1982; Rod Garner, *The Torrington and Marland Light Railway*, Southampton, 2006.
[5] DRO, D/PIT/E/107.
[6] DRO, D/PIT/E/93.
[7] Alison Grant, *North Devon Pottery*, Exeter, 1983, pp.39–40, 111.
[8] Letters from John Jenkin of 13 October and 4 November 1817 and 21 December 1821. We are indebted to Jim Lewis for this information.
[9] CRO, DDTF/3436,3444/5.
[10] J.B. Lewis, 'A Look at Cornish Copper Mining 1795–1830', *JTS*, 2004.
[11] E.G.Harvey, *Mullyon, Its History, Scenery and Antiquities*, Truro, 1875, pp.41–42.
[12] R.M. Barton, *A History of the Cornish China Clay Industry*, Truro, 1966, pp.32–39.
[13] J.B. Lewis, 'Captain William Davey of Redruth and the reopening of the Consolidated Mines', *JTS*, 2004.
[14] Evidence given by Henry Higman and Edward Stocker at a Parliamentary Enquiry, *Parliamentary Paper*, 1857, XI, pp.246 et seq.
[15] H.M. Stocker, 'An Essay on the China-Stone and China-Clay of Cornwall', *RRCPS*, 1852, pp.77–96.
[16] J. Allen Howe, *A Handbook to the Collection of Kaolin, China Clay and China Stone*, HMSO, 1914, p.8; A.C. Todd and Peter Laws, *The Industrial Archaeology of Cornwall*, Newton Abbott, 1972, pp.103–26.
[17] Todd and Laws, 1972, pp.103–26.

[18] Norman Pounds, *China Clay*, unpublished manuscript, n.d., Wheal Martyn Museum, St Austell; Lynn Miller, 'Wedgwood in Cornwall', *JTS*, (15), 1988, pp.27–35; Geoffrey A. Godden, *Encyclopaedia of British Porcelain Manufacturers*, London, 1988, p.10.

[19] E. Morton Nance, *The Pottery and Porcelain of Swansea*, London, 1943, pp.473–74.

[20] Barton, 1966, p.30.

[21] CRO, J/403.

[22] John Penderill-Church, *The China Clay Industry*, n.d., p.4, Wheal Martyn Museum, St Austell; Barton, 1966, pp.33–34; Joan Jones, *Minton*, Shrewsbury, 1993.

[23] Penderill-Church, n.d., pp.4–5; Barton, 1966, pp.35–37; 'Goonvean China Clay Pit', *MQE*, November 1958, p.477.

[24] Brian Dolan, *Josiah Wedgwood*, London, 2004, pp.385–88.

[25] Nick Harley, 'Foreign trade', in Floud and McCloskey (eds), London, 2000, I, p.312.

[26] Arthur A. Eaglestone and T.A. Lockett, *The Rockingham Pottery*, 1973.

[27] William Evans, 'Art and History of the Pottery Business', Shelton, 1846 in *JCH*, 3, 1970, p.23.

[2] H. Coghill, 'The Ceramic Manufactures of Staffordshire', in E.S. Timmins (ed.), *The Resources, Products and Industrial History of Birmingham*, London, 1866, pp.145–46. We are indebted to Jim Lewis for this reference.

[29] Penderill-Church, n.d., pp.4–5.

[30] Jones, 1993, p.12.

[31] Peter Joseph, 'Durfold China Clay Works, Bodmin', *JTS*, 2001, 28, pp.37–67; Helen Harris, *The Industrial Archaeology of Dartmoor*, Newton Abbott, 1972; Crispin Gill (ed.), *Dartmoor*, Newton Abbott, 1970.

[32] Joseph Mayer, *On The Art of Pottery*, Liverpool, 1871, p.41.

[33] Examples of these may be seen in the Brooklyn Museum of Art, New York.

[34] R.C. Bell and M.A.V. Gill, *The Potteries of Tyneside*, Newcastle, 1973, pp.5–25.

[35] *WB* 8 February 1822, 7 May 1825.

[36] Barton, 1966, p.55.

[37] *Thomas' Directory of Plymouth*, Plymouth, 1836.

[38] John Penderill-Church, *China Clay on Dartmoor*, Wheal Martyn Archives, 1983, p.2. The subsequent account is based upon Penderill-Church, 1983, pp.3–7; William Lethbridge, *One Man's Moor*, Tiverton, 2006, pp.176–213; Crispin Gill, *Plymouth, A New History*, Devon, 1966, repr. 1995, p.219; R.N. Worth, *History of Plymouth*, Plymouth, 1872, repr. 1890, p.339.

[39] John V.N. Silverlock, 'Hemerdon Clayworks', *CCHSN*, 3 April 2003.

[40] Harris, 1972; L.T.C. Rolt, 1974; Wade, 1982, pp.12, 19, 81.

Chapter Nine
Mining Magnates, Welsh Connections and Cornish Clay Merchants

Mining of copper and tin dominated the Cornish industrial scene throughout the eighteenth and nineteenth centuries. However, the 1860s saw a rapid decline in the tonnage of copper produced as large overseas operations, particularly in the Americas and Australia, were able to exploit large deposits at a lower cost per ton. Tin mines reached a peak of production but also fell thereafter into a long period of decline. This led to rising unemployment and large-scale migration.

Graphs show the tonnage of copper and tin metal produced from Cornwall, including some Devon production. An outsider looking at the increasing tonnages of china clay being produced, as shown in Chapter Five, might ask whether metal mining adventurers might have been interested in the mining of non-metallic minerals such as ball clay and china clay.

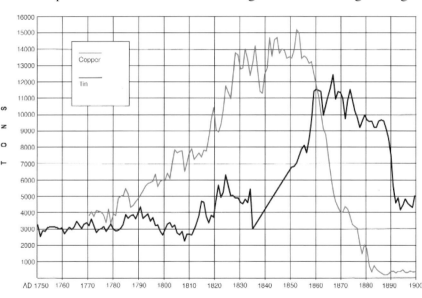

Tin and copper metal output 1750–1890 based on Geological Memoir for Falmouth and Camborne, 1906.

In the previous chapter we described how Dorset and Devon families established a thriving trade in ball clay, with some outside involvement from the London merchant Fayle in Dorset and the Staffordshire merchant Blake in Devon. In the Cornwall and West Devon china-clay industry, in contrast, it was the English potters and Plymouthian merchants who made most of the early running and then withdrew. Now we discuss some hitherto neglected issues concerned with the control of the china-clay industry. Why did Cornish mining adventurers, with strong links to South Wales collieries, not add value to china-clay and stone extraction by combining it with pottery manufacture? And why were London stone merchants, who took over a large part of Cornish granite quarrying, and Plymouth traders and professional men who acquired the great Delabole slate works in North Cornwall, not interested in china stone?

The South Wales connection

In the mid-1850s, Cornwall's wealthy and powerful mining magnates had the opportunity to develop a vertically integrated industrial complex, shipping china clay and stone to South Wales and converting it into pottery. The transport and manufacturing industry was already in place. Cornwall was Britain's main source of china clay but had no coal, whereas in South Wales easily worked seams of coal ran down to the coast just opposite Cornwall. Since several tons of coal were needed to smelt one ton of copper ore, it made economic sense to send Cornish copper ore to South Wales for smelting and metal manufacture and bring back a return freight of coal and timber for the mines. In contrast, smelting tin ore required considerably less fuel and continued to be carried out in Cornwall. Major Cornish copper producers or merchants – the Vivians, Williams, Foxes, Pascoe Grenfell and R.A. Daniell – set up or bought existing smelters and foundries in Glamorgan and during the first half of the nineteenth century, trade between Cornwall and South Wales quadrupled.[1] As London, Bristol and Birmingham brass and metal manufacturers withdrew from the Welsh smelting trade, the Cornish and their associates took it over and also played an important role in local commercial and political affairs as JPs, MPs, High Sheriffs and Lord Mayors.[2]

A similar logic seemed to apply to clay and pottery, where several tons of coal were needed to fire one ton of pottery. In the early days of china clay extraction a small but increasing trade was developing between china clay producers and Welsh potters. One of John Lovering's first customers as a clay merchant in 1815 was the Cambrian Pottery in Swansea. He exchanged china clay for Swansea ware and by 1850 this pottery was importing 300 tons of china clay a year.[3] John Henry Vivian and his wife were important local collectors of the products of the Cambrian pottery, which produced high-quality ware, especially that decorated by the gifted Thomas Baxter, son of an artist at the Worcester pottery and himself employed there before moving to Swansea. Vivian, according to his descendants, may have taken a financial interest in the Swansea works.[4] Another copper magnate who was usually ready to engage in any venture that looked promising was John Williams of Scorrier, and in 1817 he applied for a licence to extract china clay from Burngullow, near St Austell, for export to Swansea and also to Bristol and Liverpool.[5]

The Quaker Fox family were also well aware of the progress of the clay trade.

Connected by marriage with Cookworthy of Plymouth and Richard Champion of Bristol, they were partners in the Polgooth tin mine near St Austell, described around 1800 as 'the most considerable tin mine in the world', said to have a workforce of 1,000,[6] in which Cookworthy had earlier taken shares. Barclay Fox worked closely with Treffry to promote a Cornish railway at a time when Treffry was entering the clay trade, but Fox's diaries contain no evidence that he himself took any interest in clay.[7] The Foxes, prime movers of the Royal Cornwall Polytechnic Society at Falmouth, invited H.M. Stocker to present a paper on clay working, which he did, but then complained of their indifference to the industry.[8]

An influential Polytechnic member was the Norwich-born John Taylor, who dominated Cornish tin and copper mining for a time. A restless man of wide industrial interests, he had earlier been involved in canal construction in Devon, followed by chemical manufacture in Essex and further mining experience in North Wales, Staffordshire, Yorkshire and Devon. He would have been well aware of the potential of the china-clay and stone trade, having run mines at Holmbush, near St Austell, and a tin-smelting works at Charlestown, where he built a house for one of his daughters. Yet, like the Foxes, he did not get involved in the china-clay industry,[9] although, as we shall see in Chapter Seventeen, his son Richard advocated the integration of clay and pottery manufacture in Cornwall.

John Taylor, whose many mining interests included Polgooth, Pembroke, Crinnis and Charlestown United mines.

Why did such rich and enterprising men not diversify into clay production and ceramics? Probably they were used to playing for much higher stakes and, like the English potters, had plenty of more profitable irons in the fire. To start a clayworks, only a modest capital was needed. In 1782 Wedgwood and the St Austell attorney Carthew formed a china-clay company with a capital of £800, and even in the mid-1800s an outlay of £1,000 might be sufficient.[10] In comparison, the Foxes spent £34,000 in 1802 to reopen Lord de Dunstanville's old Dolcoath mine, and it cost £80,000 to get United Mines, near St Day, running in 1809.

Clay profits were low compared with those of metalliferous mining. In 1828 Lovering's pit made £80 in its first year,[11] and in the mid-1800s a larger pit recorded a revenue of £204 in its first year against expenditure of £532. In the following year it halved costs to £262 and nearly doubled revenue to £396, enabling it to make its first net return on capital invested in the third year.[12] Since the average output per pit at that time was around 730 tons a year, this suggests a profit of between £0.15 and £0.25 a ton, which means that a few larger pits, with an annual production of 2,500 tons a year in the 1850s might have made a profit of £400–£600 a year.

In contrast, in the 15 years or so up to 1837 the Consolidated Mines of Gwennap paid out nearly £250,000 in dividends, as well as paying back the initial investment of £66,000, and the East Wheal Rose lead mine made an average of £22,500 a year from 1842 to 1854.[13] Such enormous gains were, of course, sometimes matched by huge losses. In the early 1800s one copper mine reported profits of £81,000 over a decade while another declared losses of £80,000. From 1811 to 1816 the Crinnis Copper Mine near St Austell Bay produced a total profit of £180,000, yet in the sudden slump of 1816 many mines in Cornwall were running at a loss.[14]

In South Wales, too, the expenditure and opportunities of copper-related ventures dwarfed those of possible china-clay and pottery projects. Pascoe Grenfell & Sons had invested £40,000–£50,000 in producing Muntz metal, named after a Birmingham metal manufacturer, an alloy composed of 60 per cent copper and 40 per cent zinc used to protect the hulls of ships from wood-boring worms. Pascoe Grenfell, moreover, had no direct interests in Cornwall other than purchasing copper ore for his smelting works.[15] Perhaps he, like the other Cornishmen with Welsh connections, shared the view of the great South Wales iron master William Crawshay about the coal industry in South Wales. At that time, like clay, it offered low entry costs but modest profits and was, he remarked, 'always a lean one, best left to smaller men'.[16]

Austen Treffry, the 'King of Mid Cornwall', with his finger in every commercial pie, seemed an obvious candidate to make an early entrance into the clay industry. However, his biographer, John Keast, argued that as an outsider, Plymouth-born and Fowey-based, Treffry chose to keep out of the industry and confine himself to transport-

Left: Austen Treffry of Fowey who had interests in mining, china clay, quarrying, ports and railways.

ing and shipping clay for fear of falling foul of the clay families . But was this not out of character for such a bold opportunist as Treffry? Running clay works would have been a mere bagatelle compared with his many other activities. Perhaps he had too many irons in the fire, and certainly he was often financially over-stretched – when he died he was found to be £119,000 in debt – yet this had not prevented him from pressing on with other ambitious schemes. He was reputed to be the largest employer in the whole of the West Country, with some 5,000 men working in his many enterprises.

He always maintained that profit was not his main motivation and that he spent his 'whole life in devising schemes for the employment of the labouring classes',[18] but he had fallen out with the clay merchants and landlords after he thwarted their plans to build railways and turnpike roads. He was also out of favour with some of the local gentry because of his radical political activities in Fowey when he was a young man.[19] Although he campaigned together with Lord Falmouth for rail links to Cornwall, his application for one licence, according to Penderill-Church, was turned down by Lord Falmouth in favour of a 'local man'.[20]

However when he did enter the clay trade in the mid-1840s the massive difference in scale between metal mining and clay working would have been only too clear to him. He only made around £400 a year from his clay works, whereas one mine alone, his Fowey Consols, was producing ore worth from £34,000 to £50,000 per annum.[21] Nevertheless, at the time of his death at the age of 68 in 1850, Treffry finally entered the exclusive circle of the leading clay families and was about to become a director of the large West of England China Stone & China Clay Co. It is intriguing to speculate what his impact upon the industry might have been had he lived longer.

Contrasting control in granite, slate and clay extraction

During the middle decades of the nineteenth century, the pattern of control in Cornwall's granite and slate quarries and clay works diverged. In china clay, as we have seen, outsiders who had made most of the early running withdrew, leaving the industry largely in the hands of local clay merchants. In quarrying, the opposite took place, as outsiders gained control. Although an important granite producer, Treffry sold most of his output to the Freemans, the London stone merchants. William and John were the sons of Sarah Freeman, a redoubtable widow in the mould of Rebecca Martin and Philippa Lovering, who kept a tight grip on the lucrative business of supplying paving stones for London streets around 1840. This, however, was part of a general movement by London stone merchants to take over the South-West granite industry. By 1858 the Freemans accounted for nearly half of the Cornish output of granite, running quarries from Penzance to Launceston.[22]

The working of Cornwall's dominant slate quarry at Delabole also passed out of local control, but to Plymouthians rather than to Londoners. Unlike the Cornish clay families or the London granite masters, however, the new men in charge of slate production had no prior knowledge of the industry. The Plymouth link began in the 1830s, when the sister of a partner in a Cornish slate works married a Plymouth grocer, who moved to Delabole to take over the lease of a quarry. To provide more capital he took other Plymouth merchants as partners, including a Mayor of Plymouth, William Hole Evens. Obviously a persuasive man, Evens enlisted all of the Freemen of the Borough of Plymouth during his mayoralty as shareholders or directors of a company he formed in 1841, the Old Delabole Slate Co. The first chairman was a Plymouth doctor and a later chairman had interests in a large paper mill at Ivybridge, in Devon. Evens moved to Delabole to oversee the works, and the first two quarry managers were also Plymouthians, although later Welsh managers were brought in. By 1864 the company employed 400 workers.[23]

How did Cornish china-stone producers manage to preserve their independence when local granite and slate quarrymen did not? Possibly because granite and slate quarrying was a simpler operation to understand than china-stone production. The marketing and distribution of china stone was closely integrated with that of china clay. the world of ceramics, and involved destinations different from that of granite and slate. Moreover, the technological leaders of the granite and slate industries were sited in Aberdeen and North Wales, and the Londoners and Plymouthians who ran the Cornish quarries could call upon expertise from those places. Cornish china-stone producers, in contrast, were the only significant suppliers and had a better knowledge of how to extract, grind and market them than most others.

St Austell and the china clay families

Trade directories of the time reflect a growing recognition of St Austell as the centre of clay production, and dozens of traders and professional men dabbled in the industry as a sideline. One example was Adam Thomson, a Lostwithiel merchant associated for a time with Treffry, whose other interests included agriculture, shipping, tramways and the mining of manganese and iron.[24] Here we shall concentrate on the families who were to dominate the industry in years to come.

Perhaps the most venerable of the long-running clay families were the Varcoes, associated with the trade since the days of Cookworthy and Pitt in the later 1770s. At that time Jacob Slade had leased a sett from Pitt and passed it to his nephew, Edward Slade Varcoe, who also managed pits leased by Staffordshire potters including Wedgwood, Minton and the New Hall Co. Varcoe's son, William, served for many years as an agent for the Staffordshire Rogers family, as well as participating in leases of other clayworks. One of the first significant entirely Cornish-owned pits was Goonvean, near Trethosa, taken over by four members of the Varcoe family, together with another well-known clay name, Richard Yelland.[25]

In 1823 St Austell was simply described as a town that was expanding due to the presence of the mighty Polgooth tin mine nearby. At the same time the list of local traders who became clay merchants included the druggist John Higman senr, the draper John Higman junr, the grocer, maltster and draper John Lovering and yet another draper, Elias Martyn. In 1830 a cluster of four copper and tin mines was mentioned, but the only reference to the china-clay industry was a description of Charlestown as a port for shipping out copper and clay. However, the traders listed in 1823 were still recorded, joined by the saddler Edward Stocker, and six 'agents for china clay' were mentioned for the first time, including Elias Martyn, John Higman junr and John Martin. At St Columb, Thomas Broad of Staffordshire also appeared as a 'china clay merchant'. The other three St Austell clay traders were Ephraim Close, Thomas Thriscott and R.R. Geach.[26]

In 1844 china clay was acknowledged alongside copper, tin and pilchards as a mainstay of the St Austell economy, with 'immense quantities of china stone and china clay exported annually to the Staffordshire Potteries, the North of England and France'. The number of china-clay merchants remained at six, but Close, Geach & Martyn had been replaced by the grocer Phillip Wheeler, the Charlestown shipbuilders and rope makers Anthony Luke & Sons and the agent Joseph Morcom.[27]

By this time Higman and Wheeler had been associated as clay producers for some years, after Higman took shares in the leases of Wheeler, his brother-in-law. Elias Martyn's links with clay production went back further, when his father bought land in 1790 which Elias junr worked in 1820 as Wheal Martyn pit (near the china-clay museum), producing 400 tons of clay annually by 1827. Although not listed in directories as a clay producer, John Lovering had been involved at least as early as 1828 as a partner in a clayworks. However, while holding 70 per cent of the shares, he took little part in running the business and his obituary in 1834 recorded him as a maltster.[28]

By the mid-1800s the mounting involvement in clay production of these families was shown by the number of licences they took on land owned by the Hawkins family, the Fortescues, Lord Falmouth and others.[29] More than 80 clayworks were operating, with a total output of 30,000 tons a year. Rebecca Martin, John Lovering, Elias Martyn, Philip Wheeler and the Varcoes controlled half of it, while Edward Stocker had just entered the trade.[30]

Adding value to clay

As we saw in Chapter Four, the exploitation of profitable sources of china clay and stone on a significant scale was associated with the pioneering ceramic experiments of Cookworthy in Plymouth. From the very first, others seized upon the possibility of creating additional local wealth and employment by manufacturing pottery and other products from these raw materials. In 1758 William Borlase hopefully remarked that: 'For making porcelain, as well as preparing ochres and other painting-earths for the artist, a great many clays and mineral-earths may be found in Cornwall.' He foresaw no great production or distribution difficulties: 'Water-mills may easily be procured, fuel is cheap, and water-carriage to London and Bristol so convenient on either side of the county, that a sufficient undertaker might at least find as many encouraging circumstances to set up such manufactures in Cornwall as anywhere in England.'[31] Borlase did not realise how much fuel would be needed. Later, Wedgwood was said to have contemplated opening a pottery at Charlestown but had decided against it because of fuel costs, although the reluctance of Rashleigh to allow strangers into his port may have been a stumbling block.

Despite Borlase's confident prediction, Cornwall, for reasons that will be explained in Chapter Eighteen, never saw a significant ceramics industry of the kind later established in North and South Devon and the Wareham district of Dorset. As discussed in Chapter Four, Cookworthy's difficulties in firing kilns with wood or mixtures of coal and wood led to the transfer to Bristol, where coal was cheaper. Nevertheless a number of Staffordshire potters settled in the Bovey Tracey area and attempted to reduce fuel costs by using local lignite. This, however, proved unsatisfactory because of it low burning temperature and dense fumes that discoloured chinaware.

Early Bovey Tracey potters
After visiting a Bovey works in 1775, Wedgwood famously dismissed their Queensware, boasting that he could carry their clay to Staffordshire, make it into pottery and 'send it back to their own doors, better and cheaper than they can make it'.[32] His dismissive appraisal seems to have coloured opinions about Bovey pottery for many years, but recent studies have suggested that it played a significant role in commercial and decorative production between 1750 and 1836. The first reference to the possibility of establishing potteries in the area appears to be in Richard Pococke's *Travels through England* in 1750. Early potteries were established by John Hammersley and his sons from Staffordshire and by Edmond Carthew of St Austell and Robert Edyvean of Bodmin.[33]

The two most important ceramics producers, however, were the Indeo Pottery, set up by William Ellis of Bovey Tracey and his partners around 1766, and the Folly Pottery, started around 1801 by a local woolcomber and a blacksmith but soon taken over by John Honeychurch of Bovey Tracey and his partners. Both concerns imported craftsmen from Staffordshire. The Indeo works produced earth-

Billhead for the Folly Pottery, c.1830. Some of the buildings situated on the left-hand side have survived and house the Bovey Tracey Pottery Museum.

enware and printed and enamelled tableware said to be of good quality, and the Folly firm manufactured blue and white earthenware and black ware. This black ware imitated Wedgwood's, but its glaze was said to crack after use. The Folly Pottery was apparently so called because of the number of creditors who lost money in it, but both firms had an unstable financial history and in 1837 both closed down. The Folly Pottery was successfully reopened six years later, however, by naval officer Thomas Wentworth Buller and his partner, John Divett, and its subsequent history will be outlined in Chapter Eighteen.[34]

Later Plymouth potters

In the years that followed Cookworthy's transfer from Plymouth to Bristol, other potters remained in that port. Most contented themselves with coarse brown and yellow earthenware, although a potter from Staffordshire came to the city to make white, blue-painted earthenware and a few produced cream-coloured Queensware that was claimed to rival the best that Staffordshire could turn out. By the mid-1800s, Plymouth potters were trying to capture the lucrative West Indies trade, where prosperous British plantation owners were good customers for high-quality tableware, but again the cost of imported coal was said to make them uncompetitive . Moreover, as we shall see in Chapter Seventeen, the second half of the century proved to be more rewarding for West-Country potters, in some places at least.

To sum up, the withdrawal of the English potters, the departure of Plymouthians, the death of Treffry, the lack of interest of Taylor as well as other Cornish mining magnates and the London stone merchants, left the china-clay merchants of the St Austell area in control. In Devon and Dorset, too, most of the families who dominated the clay industry for the next half century were in place, ready to rival and then take over from the copper and tin magnates as the prime movers in the

Cornish extractive industry. However, the optimistic expectations that exploiting china clay and stone would stimulate large-scale manufacture of porcelain and other products in Cornwall were not fulfilled. In Part Two we shall see how growth of this kind was achieved in Devon and Dorset but not in Cornwall.

References

[1] Even in 1800 Cornish mining used 60,000 tons of coal, and by 1837 a sixth of the coal shipped from Neath, Swansea and Llanelly was used by Cornish mines. (A.H. John, *The Industrial Development of South Wales*, 1950, repr. 1995, pp.114–15.) As the rapidly expanding iron and steel industries of South Wales sought new sources of supply, quantities of Cornish iron ore were also shipped to Welsh smelters, for instance in 1838 and 1839 some 10,000 tons of iron ore from Ruthern Bridge, near Bodmin, were transported to South Wales. (RCG 1 April 1842.)

[2] J.R. Harris, *The Copper King*, Liverpool, 1964; Ronald Rees, *South Wales and the Copper Trade*, Cardiff, 2000.

[3] E. Morton Nance, *The Pottery and Porcelain of Swansea*, London, 1943, p.112.

[4] John O. Wilstead and Bernard Morris, *Thomas Baxter: the Swansea Years*, Swansea, 1997, pp.13 et seq.

[5] RIC, Courtney Library, HJ/1/13/1. In a letter of November 1, William Jenkin, steward to the great Robartes estates, advised that 'John Williams of Scorrier House hath lately applied to me for a grant of taking clay at a spot of waste land called Hallybeggan Moor', possibly Halviggan Moor. See also letters of 13 October, 4 November 1817. We are indebted to Jim Lewis and Penny Smith for this information.

[6] *The Universal Directory of Trade, Commerce and Manufacturing*, London, 1798, p.363.

[7] R.L. Brett, *Barclay Fox's Journal*, 1865, pub. London, 1979.

[8] H.M. Stocker, *RRGSC*, 1852, p.160. Did the impracticality of the drying machine, powered by wind or water, a model of which Stocker demonstrated to the members, and also the doubtful nature of the statistics of employment and production which he presented to them, discussed in Part Two, contribute to their apparent lack of enthusiasm?

[9] Roger Burt, *John Taylor*, Buxton, 1977; R. and B. Larn, *Charlestown*, St Austell, 1994, p.83.

[10] H.M. Stocker, 'An Essay on the China-Stone and China-Clay of Cornwall', *RRCPS*, 1852, pp.77–96.

[11] Cornwall Centre Redruth, 622. 361094237.

[12] CRO, DDTF/3436, 344/5.

[13] Jim Lewis, 'Captain William Davey of Redruth and the Reopening of the Consolidated Mines', *JTS*, 2004; H.L. Douch, *East Wheal Rose*, Truro, 1964, p.98.

[14] Lewis, 2004.

[15] Grenfell complained to Matthew Boulton in November 1804, 'The high price of copper can only be advantageous to the miners: to me who am a smelter and manufacturer without interest in any copper mine, it is a serious evil'. While John Rowe states that Pascoe had considerable interests as a mine adventurer in western Cornwall, according to Penny Watts-Russell, he in fact limited his interest to buying copper for his Swansea smelter and Flintshire manufactury; John Rowe, *Cornwall in the Age of the Industrial Revolution*, St Austell, 2006, p.121; Penny Watts-Russell, 'A copper-bottomed life', and 'Cutting the connection', *ABK*, Gorran Haven, May and August 2003.

[16] John, 1950, repr. 1995, p.40.

[17] John Keast, *The King of Mid Cornwall*, Redruth, 1982, p.133.

[18] CRO, TF 939, letter of 9 April 1847. We are indebted to Jim Lewis for this reference.

[19] Jim Lewis, Treffry Papers, RIC Courtney Library.

[20] John Penderill-Church, *The Treffrys and Cornish Industry*, Cornwall Centre, Redruth, X001 37169X.

[21] Keast, 1982, p.133.

[22] Peter Stanier, *South West Granite*, St Austell, 1999, pp.145–48.

[23] Catherine Lorigan, *Delabole*, Reading, 2007, pp.9–17.

[24] Jim Lewis, 'Adam Thomson', *JCALH*, 51, 2006, pp.27–31.

[25] Philip Varcoe, *The China Clay Industry: the Early Years*, St Austell, 1978; it was held by the Varcoe family until 1931.

[26] *Pigot and Co's London and Provincial Commercial Directory*, 1823, 1830.

[27] *Pigot*, 1844.

[28] Michael Barefoot, *My Great Grandmother was Cornish*, Harberton, 1989; Richard Pearse, 'One of the Pioneers', *ECC Review*, 1957.

[29] CRO, J402, 403 and 429, Lives Standing on the Manors of Trevethelick, St Stephens and Treveneague.

[30] R.M. Barton, *A History of the Cornish China-Clay Industry*, Truro, 1966, pp.87–88.

[31] Wm Borlase, *The Natural History of Cornwall*, 1758, p.320.

[32] Quoted in L.T.C. Rolt, *The Potters' Field*, Newton Abbot, 1974, pp.28–29.

[33] Brian Adams and Anthony Thomas, *A Potwork in Devonshire*, Bovey Tracey, 1996, pp.8–13.

[34] Adams and Thomas, 1996, pp.15–44.

[35] Llewellyn Jewitt, 'History of the Plymouth China Works', *Art Journal*, 1873, pp.630–31; C.W. Bracken, *A History of Plymouth*, Plymouth, 1931, p.239.

Chapter Ten
Riots, Strikes, Clay Work and Clay Pay

In the final chapter of Part One we focus upon the working conditions and wages of the men, women and children who produced clay and stone, and their relationship with their employers. To put these matters into context, the open-air life of many china-clay workers was, unlike ball-clay working, free from the dangers of underground mining and neither was affected by the harsh disciplines of factory work in the increasingly overcrowded and unsanitary towns of industrial Britain. This does not mean that their life was always untroubled, for the development of the clay trade occurred against a background of political, economic and social turbulence. The French Revolution of 1789 instilled fear of a peasant rebellion among the propertied classes and as a result the riots caused by food shortages during the enemy blockades of the Napoleonic Wars and later by bad harvests were brutally repressed.

Potatoes and bread were the staple diet of the labouring classes and in the famine year of 1816, the 'Bideford Potato Riots' were triggered off by the spectacle of farmers loading potatoes onto ships to sell them for a higher price up-country, and the North Devon Yeomanry Cavalry were called out to apprehend the ringleaders and take them to Exeter goal.[1] In the same year, in the Dorset town of Bridport, several hundred rioters attacked bakers who were believed to be hoarding flour and again the leaders were arrested and sentenced to hard labour.[2] Cornwall, according to Philip Payton, was 'especially notorious for its propensity to riot', while David Mudd claimed that it had 'a record of protest and violence as colourful and ferocious as that of any of the traditional cauldrons of unrest in other parts of Britain'.[3]

In 1830–31 a wave of rural violence spread into Dorset from Buckinghamshire, Berkshire, Hampshire and Wiltshire, where farmhands broke threshing machines, seen as causing unemployment, at a time of economic crisis in the countryside and political storms at Westminster.[4] Hayricks were burned, machines were smashed and Dorset landowners and magistrates, besieged in their own homes, were protected by the local militia. Disturbances were again frequent during the 'hungry forties', when conditions were sometimes as bad as in the more widely chronicled Irish potato famine. In 1847 the Cornish potato harvest was said to amount to 'not one hundredth part of a crop'.[5]

Early in that year 200–300 china claymen marched through St Austell on their way to the docks at Pentewan to try to stop a ship leaving the harbour with corn destined for one of the big cities, where it would fetch a higher price. There they were met by the respected figure of John Tremayne of Heligan, the County Magistrate, who promised he would do all he could to limit outward shipments of grain while food was scarce. He also took the precaution of sending for coast-guards and swearing in special constables, and the men peacefully withdrew.[6] A few months later mobs of claymen and miners ransacked the barn of a Grampound farmer suspected of hoarding grain, and also marched on to Wadebridge where, it was rumoured, corn was being sent off by ship.[7]

The following month, in another food riot, men from mines around Charlestown downed tools and toured the clay pits, persuading claymen to join their march to St Austell. All the inns and beer houses in the town were closed and shopkeepers, fearing they would be plundered, also shut down. Under Sheriff Thomas Coode and Magistrate Sir Arthur Graves Sawle alerted the Sheriff of Cornwall, who sent for the militia and swore in 90 special constables. This show of force prevented any further disturbance, but the ringleaders were arrested.[8]

Industrial disputes

As for trade union activity, the Combination Acts of 1799 and 1800 made them illegal. Despite this, associations of journeymen in the skilled trades were tolerated provided they carried out their business in a peaceful fashion, a West Country example being the hand paper-makers, who were able to restrict entry to their craft and combine to secure relatively high levels of pay.[9] Ordinary labourers, on the other hand, enjoyed no such privileges, and any attempt at collective action on their part was vigorously suppressed. The most notorious case was that of a group of farm workers from the Dorset village of Tolpuddle, 20 or so miles north-west of the ball-clay area. They received lower wages than in neighbouring districts but when, in 1833, they asked for a rise, the farmers responded by cutting their pay and threatening to reduce it still further if they gave any trouble. However, trade unions were now legal and the workers retaliated by forming 'The Friendly Society of Agricultural Labourers' with about 40 members, whereupon James Frampton, a Dorsetshire magistrate whose house had been besieged by rioters three years earlier, had them arrested, tried and sentenced to deportation to the colonies for life.[10]

The harsh treatment meted out to the 'Tolpuddle Martyrs', as they became known, caused a national outcry, but it certainly dampened enthusiasm for industrial action among rural workers for several decades. Industrial strikes were not unknown, however. In Cornwall, in 1831, copper miners at Fowey Consols combined to try to stop piecework practices that encouraged competition rather than solidarity between workers. The militia were called in, the miners' leaders were incarcerated in Bodmin goal and the movement collapsed.[11] A brief and inconclusive miners' strike also occurred at Gwennap, near Redruth, in 1842.[12]

In the granite quarries, where, again, the piecework system was unpopular, a few disputes took place. Moreover, the quarrymen's work was low skilled, producing rough blocks that were finished off by skilled masons at the final destination, and quarrymen could easily be replaced. Nevertheless, a dispute broke out during a recession in 1834 when employers at Haytor in the east of Dartmoor, tried to cut piece rates by 40 per cent, and four years later a nationally organised Stonemason's Union formed a branch at Constantine, soon followed by a dozen others, from Gunnislake in the east of Cornwall to Lamorna in the west. In 1842 a strike that started in London spread to Cornwall when London stone merchants advertised for Cornish quarrymen to come to London to break a strike there. The Cornishmen refused and machinery was damaged at Carnsew quarry, near Penryn.[13] A rare agricultural strike took place in 1850 when Cornish farmers tried to cut wages during another trade recession.[14]

Pay and working conditions

The strikes we have described so far involved large mines, or quarries influenced by UK-wide labour unions. In contrast, claymen had no national affiliation nor any tradition of striking in sympathy with other groups of workers. Their working practices were loose, informal and small scale, and the labour force fluctuated in size from season to season, even from day to day, depending upon the state of the weather or the demands from other local activities. China-clay work ceased in times of drought or icy conditions when clay could not be washed out of the pits and water-wheels and pumps could not operate.

Many claymen, moreover, were unwilling to conform to the rigid disciplines of the factory or workshop, and some preferred the freedom to take time off to work for local farmers or fishermen when the need arose, or look after one or two pigs, a cow, a few chickens and a patch of potatoes. If labour disputes occurred, they were usually settled on the spot, although in a rare case of collective action in 1845 a mass meeting of clayworkers to demand an increase of 3d. per day in pay was reported in St Austell. The employers, taken by surprise, asked for 'a day or two' to think matters over, and the men went home to await their response. Since no further report appeared, it seems that their reply was favourable.[15]

The organisation of china-clay work in the 1850s

Over the nineteenth century, as china-clay pits grew bigger and deeper and refining and drying methods became more complex, the larger clay merchants did not supervise the works themselves but employed pit foremen known as 'clay captains'. When clay merchants operated a number of pits their visits became less frequent, and the clay captains hired workers and made adjustments and improvements to working practices within the framework of a production target set by the merchants. Specialised types of work evolved and one of the toughest jobs was that of the 'burtheners' or 'burdenmen', who shifted the 'overburden' or layer of topsoil and stained clay before the good clay could be worked. This layer could vary from a few feet to as many as 50 feet deep. This was a task for the young and the fit, and if they worked on piece rates they could make 50 per cent more than the ordinary labourer, or about as much as a skilled tradesman.

Essential to china clay production was a good supply of water to wash the clay from the pits and carry it to the refining and drying areas. Fortunately the topography of the clay country was such that water could usually be obtained locally, but if it was not available on the sett it had to be bought from other landowners. The 'washers' or 'breakers' in the early years worked in or beside a stream of water that was diverted over the clay face to break down the decomposed granite, and they wore tall leather boots of the kind used by tin-streamers, since they were often working knee deep in water, clay and sand 'in the stream' or breaking clay ground 'in the strake' or working area.

The resultant mixture of clay and sand then ran down to the lowest part of the pit where the coarser sand settled, and in very early workings it was shovelled out by 'sand men' in one or two stages to one side of the shallow pits. As these deepened, wood-walled structures were built to contain the sand, and two or more of these pits were developed so that while one was filling another could be emptied. The sandmen then shovelled sand onto 'trams', or wagons, mounted on rails, which were discharged after being hauled up an inclined railway and 'trammed' to the edge of finger-shaped tips, which are a distinctive feature of Ordnance Survey maps of the period.

From the mid-1800s onwards, as space became scarce, these inclines were extended upwards to form what became known as 'sky tips' using special self-tipping wagons such as the 'tumblejack'. The work of the sandmen was as arduous as that of the burdenmen and was paid at a similar high rate. A major and increasing problem was the disposal of waste material for, by now, five tons or more of debris had to be cleared away for every ton of clay produced.[17]

In the earliest days, clay and fine sand remained in suspension in the water that ran through wood-sided channels called 'launders' to pits where the fine sand settled out. When pits increased in depth it was sometimes possible to transfer the sand and clay in suspension via a tunnel or adit to lower ground, where refining and drying took place. Alternatively, the slurry was pumped out by plunger-pumps through pipes in a shaft to one side of the pit. This shaft was dug by 'shaftsmen', who usually had mining experience, and was connected to the bottom of the pit by a horizontal 'level' and a short vertical 'rise'. The shaftsmen were paid at a higher rate than clay labourers.

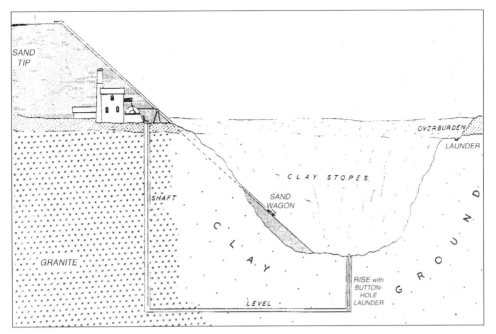

Section of a larger claypit, c.1820, showing pumping shaft, level and rise. The rise was timbered but had a series of holes sealed by timber battens, called buttons, which were removed one by one as the pit deepened.

Plunger-pumps were first worked by water-wheels and later by Cornish beam engines to bring clay, water and fine sand to the surface.

Early micas

By the mid-eighteenth century the single pits seen by Fitton, used to remove fine quartz sand, were called 'catch pits', and the micaceous residue pits were replaced by sets of shallow channels called 'micas'. Both devices aimed to have an even, slow flow of feed slurry which encouraged the natural tendency of coarser particles to settle faster than fine particles.

These channels were probably based on the 'strips' used in tin dressing from the early-eighteenth century for separating fine tin-bearing material from finer 'slimes'. The strips were often similar to mica channels in their dimensions and usually held transverse partitions to give some control of the flow rate of particles. The tin-bearing particles sank because they were denser than the non-tin-bearing minerals.

In the clay industry a bar of timber could be placed across the settling channels to encourage steady conditions. Over the years the series of trays was extended to form long sets of channels to cope with increasing tonnage and quality requirements. The drags were monitored, often by full-time workers, who generally discharged residues into local rivers.

In china-clay refining the separation was dependent on the size of the coarse and

Early set of micas with fine sand removal (nearest camera) followed by micaceous sand removal.

fine particles; there was no difference in density, as seen in tin dressing. The possibility that the micas were inspired by strips does not seem to have led to any recorded comments. The world of metallic ores, where metal content was the all-important factor, stood apart rom the non-metallic clay sector, where the physical properties of minerals determined their worth. The micas were more efficient than the original settling pits but nevertheless a precise separation of coarse china clay and mica particles of a similar size was still difficult. The residues from the micas, which continued to contain some china clay, were discharged into local 'white rivers', where a low-grade china clay containing some mica, called 'mica clay', could be recovered by working the rivers downstream. As china-clay operations grew rapidly in the later 1800s, the clay content caused pollution problems which will be discussed later. From the mid-1800s, when coal-fired kilns were introduced, 'dry men' were employed for firing and handling clay before and after drying, as discussed in Chapter Fourteen.

For many years clay was dried naturally in the open air. In the initial stage the clay slurry thickened in shallow tanks called 'sun pans' constructed of 'stent', or waste stone from clay pits, with a thin layer of sand at the bottom. 'Moss boys' were paid small sums to collect moss from the moors to plug the cracks between the stones. As the clay settled, clear water was run off until the thickened clay could be cut out in cubic blocks. The wet blocks were then carried away by boys and women to be stacked in the open to dry, protected by 'reeders' or reed hurdles. Blocks

Fine sand from mica drags was discharged via drains to local rivers.

made in the spring normally took about four months to dry, whereas those made in the summer were left through the winter and lifted in the next spring. When dry, they were scraped by women to remove any sand and green algae. After this the blocks were sometimes stored in a roofed building, and eventually loaded on horse-drawn wagons for transport to the ports or to rail sidings.

The size of the workforce

Estimating the number of workers involved in particular industries in the earlier years of clay production is a task that is fraught with difficulties. The Census of Population only began in 1801, and in the early versions of this exercise workers were enumerated on occupational rather than industrial lines. For instance, the 1851 Census of Population only recorded around 1,000 male china-clay labourers in the clay parishes of Mid Cornwall, while a later estimate put the total at 1,700.[18] The latter figure seems nearer to the mark, since the Census under-recorded employment in the industry. China-stone workers may have been enumerated as quarrymen, and those who ground china stone as 'mill men'. To the total enumerated workforce we must also add clerks and clay captains, as well as tradesmen who, in increasing numbers, were working full time for clay firms and may have been listed separately as carpenters, blacksmiths, boot and shoemakers and coopers. This last category became of greater importance in the china-clay trade when paper makers, who demanded high standards of purity, took a larger share of output. 'Slack coopers' made casks holding a half or quarter ton of clay which did not have to be water-tight since the clay was protected by a paper lining.[19]

Women played a significant part in the extractive industries of Cornwall in the nineteenth century. At the Delabole slate quarry, women carried slate up to the storage yards and looked after the donkeys that carried the loads, and later worked as 'splitters', the most skilled and best-paid job.[20] Women and children were also involved as part of family groups in brick making. The Childrens' Employment Commission of the 1860s was particularly concerned with the poor working conditions of boys and girls at Bealeswood Brickworks, near Gunnislake.[21] The Revd Doctor Fitton, mentioned in Chapter Two, seems to have been the first to report on the employment of women and children in Cornish china-clay works in 1808.[22] Another account of the same year by the Revd Richard Warner described them as engaged in soap-rock extraction on the Lizard.[23] As early as 1736 the Revd William Borlase was complaining of the difficulty of obtaining female servants because they were better paid in mining, although domestic work remained a more respectable occupation.[24]

In metal mining, 'balmaidens', as female workers were called, possibly made up around a third of the entire workforce in Cornwall and Devon by the mid-1800s, and according to the 1851 Census, over half the women at work in the mining heartland of Illogan and Gwennap were employed at the mines. Because they did not work underground they accounted for a much larger proportion of the surface labour. We might therefore have expected women and children to play at least as important a role in china-clay production, since virtually all the work took place above ground, as opposed to ball-clay working, where some was mined underground. However, the 1861 Census of Population, which possibly marked the peak of female participation, recorded fewer than 150 balmaidens in the Mid-Cornwall china-clay parishes, only one or two per cent of the available female labour force.[25]

Why were women apparently so thin on the ground in the clayworks? Lynne Mayers identified a considerable under-recording of their numbers in the Census, where enumerators were instructed to exclude part-time or casual workers. While mining balmaidens worked on a fairly regular basis, few if any female clayworkers were employed from October to January, according to Mayers, and many of them only worked during the twin peaks of activity from April to June, at the end of the winter drying season, and from August to September, at the end of the summer drying season. As a result perhaps only a fifth to a tenth of the females who were recorded on the payrolls of some pits were actually enumerated as clayworkers in the 1851 Census.[26] If this proportion held good for all pits, then possibly as many as 700 or 800 women were employed during the peak periods of early summer and the autumn. In addition to these, a few women wagoners carried china clay and stone from the pits, sometimes working as part of a husband and wife team, and a 70-year-old woman was recorded as a 'sand settler', possibly looking after the mica drags, in the 1841 Census. Early hand-pumps were also said to have been worked by women.[27] Much later, during a labour shortage in the 1914–18 War, women were to be employed as 'washers'.[28]

A surprising lapse by china-clay historians
Here we need to comment on a miscalculation by Rita Barton, Kenneth Hudson and A.L. Rowse, who accepted without question another estimate of clay and stone employment made in 1852 by H.M. Stocker, of no fewer than 7,200 men, women and children.[29] Yet this was more than the entire male and female population, including children, of the main clay parishes of St Stephen, Roche and Luxulyan, and at least half of the men in those parishes were miners or farmers. Of course some labourers travelled greater distances to the pits from Bodmin, Lostwithiel or Grampound, but hardly in significant numbers.[30] Although any estimate of the size of the clay workforce is bound to be problematic because of the part-time and casual nature of much of the work, it seems likely that in Cornwall it would have amounted to something of the order of 2,000 men, perhaps rising to a total employment of 2,500–3,000 during peak periods when women and children were engaged.

Although the tonnage of ball clay produced in Devon and Dorset in the mid-1800s was roughly the same as that of china clay and stone in Cornwall and West Devon, we cannot assume that the ball-clay industry employed nearly as many workers. Ball-clay operations at this time were far less complex: they consisted of digging the clay, loading it onto wagons or trucks and transporting it a few miles before unloading it onto a barge or boat (in North Devon the journey was somewhat longer). Unlike the china-clay industry, there were no 'washers' to wash down the clay, 'sand men' to empty the pits, 'stackers' to store it for drying in the open air, coopers to make barrels to keep it clean, nor indeed any workers to clean the coal dust from the barges or boats that carried it to sea. In the absence of any figures for ball-clay employment, all that perhaps can be said is that it was probably not more than 1,000 in total in North and South Devon and Dorset.

Hours of work and pay

As mentioned earlier, hours of work were flexible, since operations were greatly dependent upon the state of the weather, fluctuations in the demand for clay and the needs of other local activities such as farming. Night work might be required when trade was busy. Pay rates for the earlier nineteenth century are poorly documented, but the china-clay industry grew up alongside local metal mining, where many underground miners were virtually sub-contractors, bargaining with their employers on payment for specific tasks performed. Many china-clay works followed a similar pattern, and this kind of piece work, which was known as 'tut' work in the mines, was called 'contract' or 'task' work in the clay pits. A basic rate of pay applied to all workers, known as the 'owner's account' or colloquially as 'onner's count'. Each individual clay company set its own rate, but they tended to be broadly similar.

Workers performing straightforward tasks such as washers and mica drag men received the basic hourly rate, but those employed on tasks where more effort was required and where output could be measured were paid contract rates. These covered jobs such as shaft sinking and tunnelling, the removal of overburden, which was measured in cubic fathoms, and the loading of skips by shovelling sand. China-stone quarrymen were paid according to the grades and the tonnage of stone raised. In the earlier days of the industry, when clay was dried by the sun and wind, the handling of clay blocks by men, women and children was paid at separate rates, based largely upon the thickness of the blocks and the distance they were transported. Women employed to scrape sand and lichen off the blocks also received a separate rate. 'Casking', or packing 'best' clay in barrels, was another task paid at a contract rate.

The hours worked per day also varied according to the task performed. Those employed on tonnage rates would work until the job was done, which might only take four or five hours. Such men would then either go to a second job or tend their smallholdings or bargain for another task, such as unloading a wagon of coal. For labourers on straightforward tasks, however, the morning china-clay shift commonly lasted from around 7a.m. to 3p.m. or so, with half an hour's break at 9.30, that is to say a seven and a half or eight hour working day, shorter than the ten or twelve hours customary in many other industries. The normal working week on the morning shift ran from Monday to midday on Saturday. Annual holidays included Christmas Day, Good Friday, the Parish Saint's Day and Midsummer Fair Day. In some pits the anniversary of the opening of the clayworks would be marked by a day off, when a procession of workers would proceed to the homes of the clay landlords and clay merchants, where they would be rewarded with small sums of money. Alternatively, clay merchants might give a dinner for their workforce at Midsummer, and one account of 1835[31] shows what healthy appetites they had. Fifteen men, nine boys and five females sat down to a dinner at which they consumed 90lbs of beef and pork and large quantities of old potatoes, cabbage and household bread, together with 14oz of bacon and 300 pints of beer.

Comparative pay rates
How did clayworkers' pay compare with labourers' wages in other industries? Wage rates fluctuated widely during the early decades of the nineteenth century, falling sharply at the end of the Napoleonic Wars when demobilised soldiers flooded the labour market and again in the 1830s when a general depression reduced the demand for clays and stone. The going rate for farm labourers in the Dorset clay area in the early 1830s was 10s. (50p) a week and, as we saw earlier, the Tolpuddle men had only received 9s. a week and the farmers had threatened to reduce their wages to 6s. (30p) a week. A Dorset ball clayworker told a visitor to Corfe Castle in 1854 that he received only 2s. (10p) a day and nothing at all in rainy weather.[32] Although he complained about his low pay, he seemed to earn more than Cornish claymen. In Cornwall in the early 1830s the Ninestones pit, run by Lovering & Martin, was paying clay washers 8s.9d. (43p) a week[33], and 20 years later, when the clay trade was prospering and male wages were rising,[34] clay labourers pressed members of the China Clay Producers' Association for a rise in pay to £0.50 a week, based upon an eight-hour day. The employers' answer was to offer the same daily pay as before, but for a seven and a half hour instead of an eight-hour day, giving men the chance to make 2s. a day, or 60p a week, if they worked a ten-hour day which, as the employers pointed out, was the standard length of the working day in mining and other manual industries.[35]

This wage compared not unfavourably with that of a surface labourer at the Fowey Consols mine, who might average 50p a week. For older boys in the clay works the weekly wage might be 4s. (20p), rather less than at a mine, where they might expect 25p, or more if the mine was very profitable. China-clay workers who were burtheners and sandmen on piece rates did better than this, making 65p a week, the same as underground miners in Fowey Consols, and mining wages at this time were generally rather higher for surface workers in the St Austell area than in West Cornwall, probably because competition for labour for the clay pits helped to push them up.[36] Young 'moss boys' earned 3s.6d. (17½p) a week. Long-serving china-clay labourers who could no longer keep up the expected pace of work were paid a lesser sum, the general rule being that as long as they could manage to walk to work they would be paid. Women were paid at hourly rates for cleaning clay blocks, but sometimes on tonnage rates for carrying or stacking them, working in groups organised by a sub-contractor.[37] They earned 15p a week on hourly rates and up to 20p or more on piece rates, compared with up to 30p at the mines.[38] Clay captains might receive a salary of £1.50–£2 a month depending on the size of the pit.

For men, therefore, clay work offered about the same financial rewards as surface work at the mines, and more than in agriculture, although with benefits such as cheap housing and food, farm labourers may have been better off. Underground mining was a better short-term financial proposition, and there was always a chance of a windfall from striking a rich lode. But a miner could be incapacitated by his 40s, whereas claymen could and did continue working until their 70s. Moreover, by the 1850s, while the output of clay was increasing and daily rates of pay were rising, Mid-Cornwall mines were showing signs of failure. After decades of rich dividends, mining adventurers, instead of taking money out, had to put it back in again to keep the mines going.

A steady job in the clay trade was thus not without its advantages, but in Cornwall it did not enjoy the esteem of metal mining, and for an out-of-work miner, 'going to clay' was seen as a step down in status. On the other hand, a diligent and reliable clay labourer could progress from job to job to become a clay captain, whereby he would receive somewhat higher pay, often a superior house and, perhaps most important of all, great respect in his community, responsible as he was for hiring and promoting workers in the dominant local activity – truly a big fish in his small local pool.

References
[1] Duncan Fielder, *A History of Bideford*, Southampton, 1985, p.55.
[2] Cecil N. Cullingford, *A History of Dorset*, Chichester, 1984, pp.98–101.
[3] John Rule, *The Labouring Classes in Early Industrial England*, 1986, p.348; Philip Payton, *The Making of Modern Cornwall*, Redruth, 1992, p.141; David Mudd, *Cornwall in Uproar*, Bodmin, 1983, p.3.
[4] Paul Carter, 'Agricultural Disturbances in Southern England', *LHR*, 65, 2000, pp.90–92.
[5] *RCG* 11 January 1847.
[6] *WB* 22, 29 January 1847.
[7] *WB* 28 May 1847.
[8] *WB* 18, 25 June 1847.
[9] D.C. Coleman, *The British Paper Industry 1495–1860*, Oxford, 1958.

[10] Walter M. Citrine, 'The Martyrs of Tolpuddle' in Trades Union Congress, *The Book of the Martyrs of Tolpuddle*, London, 1934, pp.1–15. These sentences were later quashed.

[11] John Rowe, *Cornwall in the Age of the Industrial Revolution*, St Austell, 1953 repr. 1993, p.143; Payton, 1992, p.142.

[12] Bernard Deacon, 'Attempts at Unionism by Cornish Metal Miners in 1866', *Cornish Studies*, 10, 1982, pp.27–36.

[13] Peter Stanier, *South West Granite*, St Austell, 2000, pp.164, 169–70.

[14] Rowe, 1993, p.252.

[15] *RCG* and *WB* 3 October 1845.

[16] Unless otherwise stated, the information in this section comes from Charles Thurlow, *China Clay*, Truro, 1996.

[17] Henry Higman, *Parliamentary Papers*, 1857, XI, pp.246 et seq.

[18] *WB* 24 June 1857.

[19] John Seymour, *The Forgotten Arts*, London, 1984, p.94. Casks were expensive but much easier to manhandle than cheaper wooden boxes, see John Tonkin, 'How Deep was the Great Depression?', *JPMMC*, 2001, pp.31–32.

[20] Catherine Lorigan, 'Thomas Rowlandson and the Delabole Slate Quarry', *JRIC*, 2002, pp.30–49.

[21] John Ferguson and Charles Thurlow, *Cornish Brick-Making and Brick Buildings*, St Austell, 2005.

[22] William Fitton, 'On the Porcelain Earth of Cornwall', *Annals of Philosophy*, 2 November 1813, pp.348–51. This account is based on a visit of 1808.

[23] Reverend Richard Warner, *A Tour through Cornwall in the Autumn of 1808*, Bath, 1809, p.224.

[24] Quoted in Rowe, 1953, repr. St Austell, 1993, p.8.

[25] Lynne Mayers, *Balmaidens*, Penzance, 2004, p.97.

[26] Private communication from Lynne Mayers, 2005.

[27] Joseph M. Coon, 'The China Clay Industry', *RRCPS*, 1927, p.56.

[28] Mayers, 2004, p.97.

[29] H.M. Stocker, 'An Essay on the China-Stone and China-Clay of Cornwall', *RRCPS*, 1852, p.88.

[30] Even the most cursory glance at Stocker's calculations showed their inaccuracy. To take another example, J.H. Collins pointed out that 'presumably by inadvertence', Stocker valued the output of clay and stone at £240,500 a year, which assumed a selling price about 50 per cent above Stocker's own figures. J.H. Collins, *The Hensbarrow Granite District*, Truro, 1878, p.14.

[31] *RCG* 18 July 1835.

[32] Walter White, *A Londoner's Walk to the Land's End*, London, 1879, p.17.

[33] John Penderill-Church, *The Claypits of Cornwall and Devon*, Wheal Martyn Archives, n.d., p.20.

[34] Higman, 1857.

[35] Leaflet 'To China Clay Labourers', from 'An Employer' of 25 February 1853, Cornwall Centre, Redruth.

[36] J.B. Lewis, *A Richly Yielding Piece of Ground*, St Austell, 1997.

[37] Mayers, 2004, pp.28–29.

[38] John Taylor, 'On the Economy of Mining', *Quarterly Review*, 1837; R.N. Worth, *Historical Notes Concerning the Progress of Mining Skills in Devon and Cornwall*, Falmouth, 1872.

Chapter Eleven
The Growth of the Clay Family Companies, 1860–1914

The history of ownership and control of the clay industry in the South West in the later Victorian and Edwardian period is largely that of partnerships among local families, combined in some cases with newcomers who settled in the region. The main Dorset producers continued to be the Pikes, originally from Devon, and the descendants of the London merchant Fayle. These two families acquired or leased more ball-clay-bearing land south of Wareham, developing further railway networks and quays which will be discussed in Chapter Thirteen. Another important merchant was Edward Pinney of Bucknowle, who leased land from the Earl of Eldon in 1884 for £2,400 yearly to dig 9,600 tons, reduced in 1903 to £860 for 8,000 tons.[1]

By the mid-1800s the Pikes also exercised a considerable political influence. In the 1857 Parliamentary Election for Wareham they helped to end the monopoly of Erle Drax, who had occupied the seat since the Parliamentary Reforms of 1832. As their champion the Pikes backed John Calcraft, lord of the manor, from whom they leased clay-bearing land. While their candidate was a former Tory who had become a Liberal, his opponent, Drax, was once a radical but had become a Tory, causing some confusion among the electorate. Calcraft won with a majority of just three votes, which was upheld after a Select Committee of the House of Commons had investigated allegations of bribery. The villain of the piece was said to be William John Pike, who promised electors reduced rents, or reduction of rental arrears, and threatened them with loss of employment. On the other hand, a Drax agent paid £5 to a known Calcraft supporter to stay away on election day.[2] The Pikes continued their involvement in local affairs, L.W. Pike being a prime mover in securing better water supplies for Wareham in 1898.[3]

Watts, Blake, Bearne and other South Devon clay merchants

In South Devon the scene was set for the growth of the ball-clay industry by the formation of the three partnerships described in Chapter Eight: Watts, Blake, Bearne & Co., the Devon & Courtenay firm of Frank Davey and Edward Blake, and Whiteway & Co. The driving force in the first, and most prominent, of these partnerships was the young Charles Davey Blake, who had been the sales representative for his father Frank in Staffordshire. C.D. Blake opened up new markets abroad, to be described in Chapter Thirteen, after the ban on exports of ball clay was lifted. To expand production his company took out more leases on land owned by the Watts, the Diocese of Salisbury and Lord Clifford, and also diversified into other related industries, to be discussed in Chapter Eighteen. In 1914 the partnership was transformed into a limited liability company with Blake as chairman and largest individual shareholder, with a third of the £90,000 capital. Two members of the Watts family and three Bearnes held the other shares.[4]

By 1902 the Devon and Courtenay Clay Company had passed into the hands of the Somerset bankers, solicitors and mill owners Fox, Fowler & Co., who merged it with another firm, Candy & Co., of Bovey Tracey, that they had acquired from Frank Candy, who had founded it in the early 1870s. Candy supplied some ball clay to potters, as well being a manufacturer of pipes and terracotta, which will be discussed in Chapter Eighteen. Whiteway & Co., the third of the original South Devon partnerships mentioned earlier, leased or owned clay land near Kingsteignton but lost leases to Watts, Blake, Bearne & Co. After the last of the male line died in 1881, his widow passed on the firm to her nephew, Dr W.H. Wilkinson of London. He changed his name to Whiteway-Wilkinson and under this title the partnership became a limited liability company in 1905, with the doctor as chairman and his three sons as directors.

Another ball-clay firm, formed around the 1890s, was Hexter Humpherson & Co., pottery merchants whose main activity was the production of bricks and pipes. From Teignmouth they shipped out clay dug from their pits in Lord

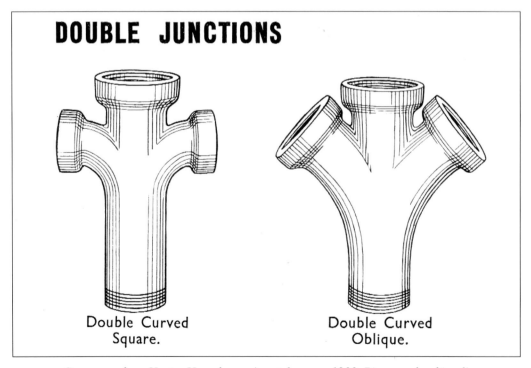

Stoneware from Hexter Humpherson's catalogue, c.1900. Pipes made of 'real' Devonshire Stoneware Clay.

Clifford's land. In 1903, with C.D. Blake as chairman, the firm was turned into a limited liability company, but its clay business was later acquired by the partnership of Hexter & Budge. Mrs Budge was the housekeeper and later the wife of C.D. Blake, and in 1912 she purchased land from the Hexter brothers and, rather like Rebecca Martin and Phillippa Lovering half a century earlier in Cornwall, she took an active part in the trade. Finally, mention should be made of Browne, Goddard & Co., another firm that shipped clay from the Clifford Estate from 1872 onwards, which was acquired in 1919 by a new firm, Newton Abbot Clays.

The Wrens and Holwills of North Devon

Among his many other interests, the enterprising C.D. Blake lent money to the Marland Clay Co. in North Devon, for whom his firm acted as sales agent. By this time the influence of the Greening family who, as described in Chapters Seven and Eight, pioneered the ball-clay industry south of Bideford, had faded. One of the Greenings seems to have married into the Wrens, the family of another Bideford merchant, and 'Greening's Moor', as it had been called, was recorded in 1845 as belonging to Robert, Josias and Thomas Wren, together with Robert Greening. Five years later the firm traded as Robert Wren and Richard Greening.[5]

George Braginton, son of the steward to Lord Rolle's Estate, was operating a pottery and brickworks on the same moor as the Wrens, and Lord Clinton was also extracting clay nearby. To transport his products, Braginton had leased Lord Rolle's canal and by 1862 was shipping a range of sanitary ware and drainage pipes down the canal to Annery, a few miles south of Bideford, where he had another pottery. However, in 1865 he became bankrupt, by which time the usefulness of the canal was drawing to an end because of the construction of a railway, to be discussed in Chapter Thirteen. The prime mover of the railway project was William Alderly Barton Wren, JP, who had opened a brick and tile works at Marland and who now seemed to be the sole lessee of the Marland Clayworks. In 1879 Wren formed a limited liability company, the Marland Brick & Clay Works, with a capital of £25,000, and leased the brick and tile works for a lump sum of £9,200, plus a rent of £100 a year and royalties of 2s. (10p) for every 1,000 bricks, 9d. (4p) per ton of clay sold, and $2^{1}/_{2}$ per cent of the revenue from glazed terracotta bricks.

In 1880 Wren acquired Braginton's pottery, which was now producing nearly 15,000 tons a year, from its new owners for about £1,700. Two years later he took on the lease of the clay-bearing land that Lord Clinton had been working and then leased his entire clayworks, and also his railway (to be discussed in Chapter Thirteen), to Henry and Eustace Holwill for a payment of £1,500 plus a rent of £650 a year.[6] They were sons of Frederick Holwill, a manager of Wren's pits and a man of substance, Alderman, High Bailiff to the County Court, ship owner and coal, lime and timber merchant. Later in 1882 Wren died, leaving the estate to his two sisters. The Holwills then took over the firm, naming it the North Devon Clay Co. Ltd, and soon bought out the lease of the brickworks.

In 1875 this works had seemed a thriving concern, employing 75 men and selling 25,000 paving and fire bricks a day produced by a circular continuous kiln, as well as decorative terracotta, which it supplied to Paignton, Brighton, Chelsea and elsewhere. However, the brickworks appeared to be losing money and in 1891 was leased to a Manchester brick manufacturer, who immediately sub-leased it to another group of Manchester merchants. It still did not pay and the lessees fell behind with their rent and gave up the lease. The Holwills invested heavily in new equipment and production climbed to a peak of 2,800,000 bricks in 1901, but then declined. On the other hand, clay sales to Europe and America increased steadily, as will be discussed in Chapter Thirteen.

Finally, there appear to be a number of intriguing links with ball-clay production by the North Devon Clay Co. and clay merchants in South Devon and Cornwall. Watts, Blake, Bearne & Co. lent it money and acted as sales agents; J.W. Ludham, who for a time managed the works for Wren, became manager of Candy & Co., and Wren's bankers were Fox Fowler & Co, who had acquired Candy & Co. and the Devon & Courtenay Clay Co. in South Devon. Clay deposits identified in the early 1900s by Henry Holwill at Meeth, south of Marland, were acquired by Eustace Holwill. However when a company was later formed to exploit these clays, with Eustace as managing director, it was largely financed by Cornish clay merchants, including E.J. Hancock and John Lovering, who forced Eustace to resign.

The Varcoes, Higmans, Wheelers and Martyns of Cornwall[6]
A list of Cornish clay holdings in the later 1840s contained the names of over 20 producers, and a decade later this figure had doubled. The number of clay workings identified also doubled over the same period, from over 40 to 85, plus four in Devon. In 1875 the number of Cornish clayworks peaked at about 120, plus eight in Devon, but the 'Great Depression' that followed reduced the Cornish total to 94.[7]

The Varcoes were among the longest lasting of all the Cornish clay families. From the earliest times they held shares in other clay firms,[8] including one managed by Robert Varcoe for just 2s. (10p) a day, only a third more than the wages of a day labourer. Richard Varcoe acted as agent for two more clayworks and, by the mid-1800s the family were directly responsible for four or five per cent of Cornish clay output. By this time John Varcoe was listed in the *St Dennis Post Office Directory* under 'gentry', while Robert Varcoe was simply recorded as a clay merchant. Although they were never in the very top echelon of clay produc-

Martyn advertisement of 1873.

ers, over the next half a century the family played an important part as clay producers and agents for other clay firms. In the twentieth century they were to become millers of stone from china stone quarries and many other non-metallic minerals using buildings originally erected for Treffry's lead-smelting works on Par Harbour.

A common form of dividing responsibilities within family partnerships was for one member to take charge of production, another to focus on sales and another on bookkeeping. The original partners of Watts, Blake, Bearne & Co., the largest Devon ball-clay company, conformed to this division of labour, and china-clay examples were the Higmans and Wheelers, who, by the mid-1800s, operated several pits around St Austell. A few years later H.W. Higman joined John Tresidder of Helston in a clay venture at Tregonning Hill, site of Cookworthy's early interest. By now they produced six or seven per cent of total Cornish clay output, and H.W. Higman was entered in the *St Austell Directory* under 'gentry'.

By no means all the families who led the way in the clay trade lasted the distance. In the 1840s Elias Martyn controlled five pits contributing over ten per cent of total production and later acquired a stake in the West of England Clay & Stone Co., the largest clay firm of its day. Yet the expansion of output in his pits, although continuing to rise, did not keep up with the growth of the trade as a whole, and by 1858 his share of overall production had fallen to five per cent. Perhaps he was hampered by a lack of capital or a failure of judgment, because he suffered a series of losses in toll roads and railways and his projects seemed to come on stream just as a recession was hitting the clay trade. This pattern was continued to a disastrous degree by his son, Richard Uriah Martyn, after the death of Elias in 1872. In one of the most severe slumps of the Victorian era, Richard lost money on a railway venture and had to sell two pits. Another, Wheal Martyn, went into liquidation but was acquired and revived by John Lovering junr, who, after Richard Martyn's sudden death in 1887, bought the sole remaining family sett.

Parkyns, Sellecks, Nicholls and Olvers

By the 1870s most top-quality clay in the St Austell district was already being worked, and adventurers seeking to expand had to look further afield, notably to Bodmin Moor. Although china clay had been identified there earlier, it was not until 1860 that it was exploited by Henry Phillips of St Austell and later by Andrew Cundy and John Truscott on behalf of Frank Parkyn, who acquired Phillips' lease in 1870.[9] Parkyn, born in 1850 in Lerryn on the Fowey River, was the son of one of the last of the 'merchant princes' of Mid Cornwall, a dealer in wool, corn, sand, fertiliser and timber. Parkyn senr. also operated a farm, lime kilns, a flour mill and a saw mill, where he cut staves to make casks for china clay, yet another of his products.

Frank Parkyn junr. hardly seemed a likely successor to control such a diverse business empire, since his health was so delicate that he was sent to live on Bodmin Moor, where it was thought the fresh air would do him good. It appeared to do so, since he lived to be 90. His searches for china clay on Bodmin Moor met with success and he opened a clay works at Greenbarrow, near Temple. Then he entered into partnership with Woodman Peters, who managed the clay side of the operation while Parkyn looked after administration and sales. Taking over an unpromising sett at Burngullow that lacked water supplies, they raised its output to 40,000 tons a year.[10] They acquired other pits as well as dries near Par, and became important clay producers in the late-Victorian and Edwardian period. Around 1900 Parkyn bought a fine mansion and estate at Penquite, near his home village of Golant, and adopted the life of a landed gentleman, patron of the church and supporter of local charities and community activities.[11]

On Dartmoor, Captain Christopher R.H. Selleck reopened the Wotter Pit, about half a mile from Lee Moor, which had been unsuccessfully worked for a time by the Martins. Selleck had worked for the Martins at Lee Moor before leaving and going to America, but returned to become under-manager of the works in 1891. Ten years later he took the risky step of restarting Wotter, helped by his nine children and half a dozen casual labourers. He had an uphill struggle to keep it going in the face of strong local opposition. Essential supplies of water for all the pits in the area came from the upper reaches of the same stream, which the Martins had diverted to the Lee Moor works, and Selleck had to make do by constructing ponds to collect water from his own limited watershed.

He was also unpopular with local clay merchants and farmers, who claimed that his traction engine, drawing two ten-ton trucks, was ruining their lanes. The Martins forbade him from using their roads, although on one occasion, when their works manager, Captain Bray, tried to stop him, he enveloped Bray in a cloud of steam from his traction engine and drove on. To appease other road users and the District Council he carted many loads of rock to make good the damage to four or five miles of road. His perseverance was rewarded in 1906, when three speculators in Malayan rubber shares injected capital into his enterprise which enabled him to build a pipeline down to Marsh Mills, a descent of some 800ft, in 1910.[12] Such was his success that he also developed the hamlet of Wotter to house his workers. His son, Captain William Selleck, also purchased a disused pit on Dartmoor, and other Sellecks, including a younger son, Claude Selleck, joined by ties of marriage and business with the Nicholls, played an important part in the development of china clay in the twentieth century.

Thomas Nicholls was, for a short time, a partner with Lovering and Martin in a china-clay venture of 1827–28 at the Lower Ninestones Pit, and four generations of Nicholls later worked there, Captain Sam Nicholls retiring in 1957. A Nicholls was a partner in Carluddon Moor Works and Carbis Pit nearby in the 1860s and '70s and, at that time, John Nicholls of St Austell began extracting clay around Balleswidden, near St Just in West Cornwall. He was almost certainly the John Nicholls referred to in Chapter Three, a manufacturing chemist and owner of a naphtha works at Charlestown, who operated the pit on Dartmoor in 1855. Nicholls & Co. were also working North Goonbarrow by 1874 and were still in control in 1927, when they joined English China Clays, of which Hart Nicholls became a director. Hart Nicholls, with his father, organised the pumping of clay six miles from their works to Par harbour in 1911,[13] and he was also known for selling as many of the by-products of stone, sand and mica-clay from his pits as possible.

Another clay family that prospered on Dartmoor were the Olvers of St Austell, who ran the Smallhanger China Clay Works on land that had been prospected in 1852 by the ball-clay merchant Edwin Blake. He passed on the lease to a Staffordshire man, Frederic Bishop, who in turn sold it in 1869 to John Olver. The clay was not of such good quality as that of Lee Moor, but the Olvers operated there for over 80 years, employing some 40 men, and kept in touch with C.D. Blake of Watts, Blake, Bearne, the largest ball-clay partnership, who were also on the lookout for china-clay ground. In 1872 they leased ground in the Cornwood area belonging to the Parkers, who had earlier sold land near Lee Moor to Lord Morley. These became Headon China Clay Works, the flagship of the Watts, Blake, Bearne & Co. china-clay operations, but some other pits that they developed in the area led them into disputes with the Martins about water supplies similar to that which had troubled C.R.H. Selleck.[14]

Truscott, Gaved, Luke and Veale

Charles Truscott, recorded in the 1820s as a boot and shoe maker and clay merchant, was by the mid-1800s producing about five per cent of Cornish china clay from his three pits, and also took part in clayland railway ventures, but by the 1880s his pits were being sold off.[15] John Gaved, active in clayworking in the 1830s, produced six or seven per cent of total clay output in his five pits in the mid-1800s, some in partnership with William Varcoe, but again they were disposed of from the 1880s, one of them to Parkyn and Peters.

In 1849 William Luke joined a cartel with other leading clay producers to ration output in order to keep up clay prices and at that time his allocation exceeded those of Martin, Treffry and Truscott. He later had shares in a small chinastone works with Stocker and Martin, but in the 1880s his leases were sold, one to John Wheeler Higman. John and Benjamin Veale, listed as wool merchants and drapers in the 1820s, formed two clay partnerships. The Veale and Phillips clayworks were taken over by Edward Stocker in 1877, while Veale and Olver's pit, recorded solely under the name of A.S. Veale, a Quaker family from 1874, was acquired by Frank Parkyn four years later. The name of Veale was associated from 1886 with electrical generation and contracting in St Austell.

Thriscutt, Bale, the Cocks and the Brownes

A number of other family holdings are worthy of mention. In the 1830s Thomas Thriscutt (or Thriscott) worked a Caudledown pit and also one at Sheldon (or Shilton or Shilliton) with Thomas Giddy. By the later 1840s Thriscutt was producing about 1,000 tons a year, or some three per cent of total Cornish output. His son-in-law and heir, Henry Adolphus Bale, followed him into the clay holdings.[16]

William Cock opened a clayworks around 1828, and some 30 years later his daughter Angelina married William Marshall, who took a considerable share in the company.[17] Meanwhile, William Cock's son David had joined him in the enterprise. David Cock's well-known treatise of 1880 on the clay industry has already been referred to in previous chapters, but he had a chequered career in the trade. After managing the Mid Cornwall Mines Co., which became bankrupt in 1874, he speculated heavily by acquiring tin, chinastone and china-clay companies during the trade depression of the later 1870s, when share prices were low, only to run deeply into debt and depart for New York. However, he returned to acquire new clayworks and became associated with William John North, a Wolverhampton mining and clay trader.[18] North's interests will be discussed in Chapter Nineteen, as will those of other important outsiders who came to exercise an increasing influence on the clay trade in the later nineteenth century.

In 1851 the Duke of Leeds granted a lease on Tregonning Hill, where Cookworthy had made his original discoveries a century earlier, to William Browne of St Austell and Robert Dunn, a Redruth draper. By 1858 Browne's pits were producing 1,800 tons a year, approaching three per cent of total Cornish output. Four years later he acquired, in partnership with Martin Rickard of Roche, a tin and clay works at St Dennis, but his holdings did not prosper. In 1879 his sons passed on to Pochin (who will be referred to in Chapter Nineteen) a pit at Hensbarrow that they had briefly worked, and in 1884 their West Cornwall output had dwindled to nothing.[19]

Dyer, Perry and Gilbert

Three other men who worked their way up the ladder were Jack Gilbert, Samuel Dyer and James Perry. Dyer began life with very little schooling, starting work at eight years of age as an errand boy earning just 3d. (about 1p) a week. He was soon walking six miles to tend sheep, starting out at four o'clock in the morning. At the age of 21, earning £0.50 a week, he married a balmaiden, and his diligence impressed one farmer for whom he worked, who bequeathed him a small clay pit. He was steward of the Lambe land-owning family, managed Rosevear China Clay Works for 40 years, operated as a cooper and auctioneer, became sole manager and shareholder in two other pits and was co-manager and shareholder with James Perry in two more. An outspoken man, Samuel Dyer never hesitated to voice his thoughts on all manner of subjects. He upset claymen by asserting that they could manage very well on the wages he paid them if they did not squander them on drinking or gambling, and his condemnation of trade-union organisers at a chapel ceremony during the china-clay strike of 1913 made him unpopular with workers and Nonconformists alike.[20]

James Perry, another self-made man, was born in Helston and educated at Truro, moving to St Austell as a grocer's assistant around 1873 at the age of 18. After working as a commercial traveller he acquired what became Perry's Temperance Hotel in East Hill, St Austell, and then entered into his first china-clay venture as partner to Samuel Dyer. As well as his involvement with Dyer in the merchandising firm of Dyer, Daley & Co., which he ran himself, he also acquired other clay companies, which his son Archie operated as secretary and manager,[21] and later became a prominent member of the China Clay Association.

Jack Gilbert, from a Stenalees family, was a big man both physically and commercially. A prize-winning Cornish-style wrestler and a successful claywork manager, he owned considerable interests in clay companies, as well as actively participating in community affairs. In the early 1900s he invested in, or managed, or both, pits near Bugle, prospected near Carne and opened a clayworks to be called Prideaux.[22]

The Martins of St Austell and Lee Moor[23]

The families discussed so far were all significant players in the clay business, but giants of the later Victorian period were the Martins. The founder of the Martin clay dynasty was John Martin of Higher Blowing House, St Austell, who by 1835 was leasing Goonbarrow Pit from Henry Lambe. In 1843 he joined with Elias Martyn and others to lease Lady Grenville's clay and stone setts in St Stephen. In the same year, with Elias Martyn and others, he leased Pentewan Harbour and the St Austell to Pentewan Railway from the landowner, John Hawkin, for seven years at a rent of £305 per annum. He died shortly afterwards.[24]

William Martin inherited clay setts from his father John and, after his father's death, benefited from advice given by his mother, Rebecca. A formidable figure in her own right, she guided the family fortunes during a period of rapid growth, proving a match for any clay merchant. She was a prime mover in attempts to regulate prices and production in the slump of 1848 and in the acquisition in 1862 of two setts on Lee Moor in Devon, a particularly shrewd investment since she acquired the use of a business with a capital of £100,000 for a rent of just £800, rising to £1,200, a year, or landlord's dues, whichever was the larger. Her acumen is all the more remarkable in that the Census of Population of 1861 recorded her as blind.[25]

In Chapter Eight we saw how William Phillips had spent a fortune on the works, with two-thirds of the capital provided by Lord Morley and three Plymouth bankers. After his death, these men sought advice on how to run the pit from William Pease, former right-hand man of Joseph Austen Treffry. Lord Morley's brother-in-law, the Hon. William Clements, visited the West of England Co.'s clay

---

MARTIN BROTHERS,
CHINA CLAY & CHINA STONE
MERCHANTS,
ST. AUSTELL, CORNWALL;
AND MANUFACTURERS OF
LEE MOOR PORCELAIN FIRE BRICKS,
LEE MOOR, PLYMPTON, DEVON.

Offices:—HIGHER BLOWING HOUSE,
ST. AUSTELL.

Martin advertisement of 1873, from Kelly's Directory

and stone pits in Cornwall, reckoned to be some of the most progressive works in the industry, and was much impressed. Pease then inspected the Lee Moor works, which were being managed by Phillips' son, John.

Pease commented unfavourably on the 'enormous sum of money' that William Phillips had spent on the works, and the way that his son was running it. When Pease was asked to take over from John Phillips and become a director, he refused, although he was willing to visit once a month. He recommended two clay captains from St Stephen, William Harris and John Arthur, to run the pit, which they did, but resigned after a month for reasons which were not made clear. During their stay they only managed to raise about 10 per cent clay from the clay matrix, whereas John Phillips had claimed a 25 per cent yield. John Phillips left and the directors decided to sell and Pease was authorised to ask Stocker, Lovering & Martin to tender for the lease. They made very similar offers: £800 a year rent, and dues of £0.125 a ton from Stocker and £0.2 from Lovering and Martin. Rebecca Martin, however, also proposed a rise in rents after seven years and better dues on the bricks and drainpipes produced.

Although her competitors, the Stockers and Loverings, had turned down the proposal to acquire the lease of Lee Moor on the grounds that the terms made it unprofitable, it soon became so lucrative, and her acquisition seemed such a bargain, that it aroused suspicions that the former leaseholders had been hoodwinked by Cornishmen into letting Lee Moor go at an unrealistically low price. After Rebecca's death, her son, William Langdon Martin, ran the Devon operations. Within a decade or so he had taken over Whitehill Yeo, a small existing pit producing high-quality pottery clay.[26] He soon raised the output to over 25,000 tons a year, compared with the 2,000 that Pease had erroneously stated was the output in 1861. Had the Cornish advisers deliberately misled the Devon men? Scams involving false valuations were by no means unknown in West Country metal mining, and Stocker's company had earlier been accused of understating output in Cornwall to reduce the dues paid to the landowner, Lady Fortescue. On the other hand, according to Penderill-Church, Edward Stocker had believed Pease's critical appraisal of Lee Moor.[27]

William Langdon Martin was said to be well liked by his workers, building more rows of cottages for them, as well as a Methodist Chapel and a primary school, but not a public house.[28] Perhaps this helps to explain why, as we shall see in Chapter Eighteen, they did not join their St Austell counterparts in the clay strikes of 1875–76. Nevertheless, he had no sympathy with the St Austell strikers and condemned those magistrates who refused to call in the militia to quell the unrest and called for their dismissal. William Langdon Martin died in 1885, after which his brother Edward ran the family business, while the younger brother, Thomas, moved to Lee Moor. However, Edward died in the following year and

Thomas became managing director. He was said to be the brother who inherited most of his mother's fiery spirit. Known as 'old ginger whiskers', he was a stickler for maintaining the cleanliness of china-clay ships.[29]

The Martins also played an active role in developing the transport infrastructure of the Mid Cornwall clay district, including the St Austell to Pentewan and the Newquay Junction Railways, and were one of the earlier investors in steam engines for pumping. Thomas was also, from 1885, a partner in Charlestown Foundry. While the Lovering family operated more pits, those leased by the Martin family were larger, making them the leading producers of china clay in Cornwall and Devon. The key to their success was their acquisition of Lee Moor. Before this event, their annual production of clay had been of the order of 8,000 tons a year. By 1905 their combined output from Devon and Cornwall had risen to 80,000 tons, of which Lee Moor and Cholwichtown contributed 73 per cent. What is more, the Dartmoor pits accounted for 84 per cent of total profits, suggesting that they were producing a higher proportion of best clays, and in 1908 the managing director reported that their 'prospects were far beyond their utmost expectations'.[30] Brickmaking, discussed further in Chapter Twelve, was also of particular importance at Lee Moor.

In 1891 the Martin brothers' partnership was transformed into a limited liability company, all the subscribed capital of £176,000 being held by members of the family. By this time their head office had shifted to Plymouth, reflecting the greater importance of the Dartmoor operation. When Thomas Martin died in 1913, his sons, Reginald and Claude, took over as joint managers. Claude was killed in the 1914–18 War, but Reginald, who was in charge at Lee Moor, became one of the three joint managing directors when a china-clay consortium was formed in 1919.

The Loverings

The Loverings originated from the Padstow area and operated a malting business at St Columb, but in the early 1800s John Lovering set up in business at St Austell as a shopkeeper and maltster. He engaged in the clay trade from at least as early as 1815 as a purchaser or agent and in 1828 became a partner in a clayworks. Although he held 70 per cent of the shares, he took little active part in the enterprise,[31] and his obituary in 1834 simply recorded him as a maltster, not a clayman.[32] His widow, Phillippa, like Rebecca Martin a skilful businesswoman, survived him by 38 years and successfully carried on his trade with the help of four of her sons, of whom John was the most prominent. His father's holdings had given him about ten per cent of clay output and, advised by his mother, he acquired other profitable pits as they came on the market, as well as opening new setts. By the 1850s he had converted the Great Carclaze Pit, which had been an open pit for tin working, the china clay being discarded, to a china-clay works with clay

Lovering advertisment of 1910, from Kelly's Directory.

refining and drying operations. Sometimes he worked in partnership with the Martins, and by 1858 the two families together controlled 25–30 per cent of Cornish production.

At that date both John Lovering junr and Rebecca Martin were listed under 'gentry'.[33] Lovering took a less public role than William Martin in railway ventures and in the 1876 strike, although this did not mean that his influence in such events was not felt. He was keenly interested in the economics of raising clay, and his technical innovations will be discussed in Chapter Fourteen.

The Stockers

Along with the Martins and the Loverings, the third leading family to appear were the Stockers. Edward Stocker, son of Francis Stocker, a Customs Officer at Fowey,[34] was, according to family tradition, involved in the distillation of peat to produce naphtha and other products at a works on Dartmoor.[35] When Prince Albert, who took a great interest in the welfare of Princetown Gaol on Dartmoor, visited that place in 1846, he was shown around a nearby naphtha works by the manager, 'a Mr Stocker',[36] and if this is the same person, it suggests that he had some scientific knowledge. In any case an Edward Stocker was recorded as supplying clay producers in the 1830s with products from his St Austell shop, and in the 1840s was listed as a St Austell plumber, brazier, tinplater, ironmonger and saddle-maker. Clearly a versatile man, Stocker acquired shares in the late 1840s in the large West of England China Clay Co. along with Elias Martyn, Treffry and Stocker's cousin, William Marshall Grose, a St Austell linen draper. With Grose he set up an agency in the Staffordshire Potteries and soon established himself as the prime mover in the company, as well as taking leases on other setts. A decade after entering the clay industry he effectively controlled over ten per cent of the entire Cornish output.

West of England China Stone & Clay Co. advertisement listing
depots at home and abroad. From Kelly's Directory of 1910.

While the Martins were the only one of the top three families to operate clay-works outside Cornwall on a large scale, in 1905 the West of England Co., run by Edward Stocker's grandson, Thomas Medland Stocker, bought a half share in the Devon clay pits of John Nicholls, the industrial chemist from St Austell mentioned earlier. Edward Stocker had, moreover, started tests to investigate the possible distillation of naphtha from peat at Ridding Down above Lee Moor, although it is possible that his real motive was to keep an eye on Martin's operations.[37]

A man of strong but volatile opinions, Edward Stocker at first rejected the use of coal-fired clay drying kilns discussed in Chapter Fourteen because of unevenness in drying and, as we have seen, turned down an offer to acquire the profitable Lee Moor pits. He also declined a project to extend a railway to Newquay, later joined it but let down his partners by failing to meet his financial obligations. On the other hand he took shares to extend the Pentewan – St Austell railway and was quick to seize the opportunity to lease water supplies from Treffry's Luxulyan viaduct and aqueduct to power machinery to grind china stone at Ponts Mill in the valley nearby. He took a militant role on the employers' side in the 1876 clay strike, to be discussed in Chapter Sixteen, and it was largely his decision to dismiss union men that triggered off the strike. Continuing to acquire leases for clay setts, by the time of his death in 1880 he had an interest in 15 or more pits, and his share of Cornish clay production was second only to Martin's. His three sons followed him into the clay business and his grandson, Medland Stocker, as we shall see in Chapter Eighteen, proved to be one of the enlightened leaders of the clay trade and a prominent figure on the employers' side in the 1913 clay strike, but in a much more conciliatory role than that of Edward Stocker in 1876.

The clay families we have discussed so far, although they produced the bulk of the output, were surrounded by dozens of smaller producers. Many of these came and went without leaving a permanent mark on the china-clay story, and the remains of old pits are still being discovered. An increasingly important part was also being played by newcomers to the industry from outside the region, who will be discussed in Chapter Nineteen. This coexistence of a few large and many small firms was common in other British industries, such as coal mining, iron working, textiles and manufacturing in general.[38]

The origins of the clay families

Summing up the origins and careers of the families who dominated the clay trade up until the 1914 War, only a small proportion came from the upper classes or the lower orders. Few landowners apart from Lord Morley in Devon took much of an interest in the industry, and although 'rags to riches' was not entirely a myth, most of the clay merchants rose from within the middle class, often from a lower stratum of traders to a higher level of wealthy, or at least well-to-do, industrialists. Many came from the *petit bourgeoisie*, some entering the clay industry almost casually, as a sideline.

Few possessed knowledge of the technicalities involved, although the Dorset clay merchant Benjamin Fayle and the Devon producer Edward Blake had knowledge of the needs of the potters. Edward Stocker may have had some chemical experience and William Langdon Martin was a keen amateur mineralogist. However, the lack of a technical background was in no way exceptional, either in the West Country or in England as a whole. For example, in Redruth William Davey had run the prolifically profitable Consolidated Mines at Gwennap while trading in gunpowder and leather hides, acting as a partner in a brewery and renting out properties.[39] The lessees of the slate quarry at Delabole were shopkeepers and doctors. On Dartmoor John White, a spirit merchant, had combined with Joseph Edgcumbe, a chemist, to operate a granite quarry[40]. Coal and tin-plate firms in South Wales, clothing factories in East Anglia and tobacco manufacturers in Glasgow had been developed by brewers, maltsters, farmers, estate agents, bleachers and dyestuff and hosiery manufacturers.[41]

Although fiercely competitive, the clay merchants bought and sold leases from one another, traded with each other to provide the range of grades which the various markets demanded and worked together in shifting coalitions as partners in clay, stone and other ventures. The talents that they possessed in common were a gift for organisation, skills in buying and selling and an interest in quality control, both in the processes they operated and the products they dealt in. Some also acquired a reputation for holding their cards very close to their chests but, as we shall see when discussing their operations in Chapter Fourteen, they often had sound commercial reasons for their reticence.

References

[1] DRO, D/SEN/1/10/32,4.

[2] Richard Ryder, *The Calcrafts of Rempstone*, Tiverton, 2005, p.100.

[3] The information for this section from comes from Rodney Legg, *Purbeck's Heath*, Sherborne, 1987, p.77; Terence Davis, *Wareham: Gateway to Purbeck*, Wincanton, n.d., p.58.

[4] The sources of information in this section are L.T.C. Rolt, *The Potters' Field*, Newton Abbot, 1974; The Ball Clay Heritage Society, *The Ball Clays of Devon and Dorset*, St Austell, 2003.

[5] The information in this section comes from M.J. Messenger, *North Devon Clay*, Truro, 1982, pp.14–22; Rod Garner, *The Torrington and Marland Light Railway*, Southampton, 2006, p.22.

[6] In addition to the references given in the text, this section on the china-clay families uses information from Michael Barefoot, *My Great Grandmother was Cornish*, Harberton, 1989; Richard Pearse, 'One of the Pioneers', *ECC Review*, Christmas 1957, on the Loverings; Philip Varcoe, *The China Clay Industry: the Early Years*, St Austell, 1978; *WB* 20 June 1889 supplement on Cornwall County Council, for the Varcoes; R.M. Barton, *A History of the Cornish China Clay Industry*, Truro, 1966; *BCW*, 1894–1914; *St Austell Star* 1889–1915; *China Clay Trade Review*, 1919–20; *Slaters' Royal National and Commercial Directory*, 1852–53; *Kelly's Directories*, throughout the period.

[7] We are indebted to John Tonkin for this information.

[8] CRO, J403 leases of William, James and John Varcoe from Hawkins.

[9] Peter Joseph, 'Durfold China Clay Works, Bodmin', *JTS*, 28, 2001, pp.37–67.

[10] *SAS* 11 March 1897.

[11] Andrew Foot, *A History of St Veep Church and Parish including Lerryn*, St Veep, 1989; *RCG* 18 September 1940.

[12] A.D. Selleck, *The Selleck Family*, St Austell, 1970.

[13] Kenneth Hudson, *The History of English China Clays*, Newton Abbot, n.d. p.55.

[14] William D. Lethbridge, *One Man's Moor*, Tiverton, 2006, pp.207–20.

[15] CRO, J403 leases by Margery, George, Lewis, William and John Truscott from Hawkins and others.

[16] Barton, 1966, pp.62, 78, 88.

[17] The Marshalls were still involved with it in the 1930s.

[18] Barton, 1966, pp.59, 128, 175–76.

[19] Barton, 1966, pp.97–98, 125–26, 142.

[20] *SAS*, 26 October 1894, 28 January 1909, 20 October 1911. We are also indebted to John Tonkin for information on Samuel Dyer.

[21] *CG* 26 April 1912; *The Chemical Age*, 18 April 1925. We are indebted to John Tonkin for these references.

[22] In addition to the references given in the text, information in this section on Lee Moor comes from E.A. Wade, *The Redlake Tramway and China Clay Works*, Truro, 1982; John Penderill-Church, *China Clay on Dartmoor*, 1983, Wheal Martyn Archives, pp.8-11; Lethbridge, 2006, pp.176–213.

[23] M.J.T. Lewis, *The Pentewan Railway*, Truro, 1960, pp.28–29. A detailed account of the Martins is given in Derek Giles, 'Some less known china clay firms: Martin Brothers', *China Clay History Society*, 2000, 14, pp.2–5.

[24] We are indebted to Lynne Mayers for this information.

[25] John Penderill-Church, 'The claypits of Cornwall and Devon', Wheal Martyn Archives, n.d., p.27.

[26] Penderill-Church, 1983, p.14.

[27] Penderill-Church, 1983, pp.12–13.

[28] Penderill-Church, 1983, p.13.

[29] Giles, 2006, p.4.

[30] CRO, DDTF/3436, 344/5.

[31] Barefoot, 1989, pp.48–50.

[32] *St Austell Post Office Directory*.

[33] James Whetter, *The History of Gorran Haven*, Gorran, 1991, p.44.

[34] Edmund Vale, *China Clay History*, unpublished manuscript, Wheal Martyn Museum Archives, 1954, pp.23–24.

[35] 'Crossing's Hundred Years of Dartmoor', *WMN*, 1901, repr. Newton Abbot, 1967, p.120.

[36] Penderill-Church, 1983, p.12.

[37] Maxine Berg, 'Factories, workshops and industrial organisation', in Roderick Floud and Deidrie McCloskey, *The Economic History of Britain Since 1700*, Cambridge, 2000, I, pp.123–50.

[38] Jim Lewis, 'Captain William Davey', *JTS*, 2004.

[39] Peter Stanier, *South West Granite*, St Austell, 1999, p.62.

[40] A.H. John, *The Industrial Development of South Wales*, Cardiff, 1995, pp.38–39. Berg, 2000, I, pp.123–50.

Chapter Twelve
New and Changing Industrial Uses

Between the mid-1800s and the First World War, potters remained the main customers for ball clay and china stone, and it was not until the mid-twentieth century that non-ceramic applications were developed for ball clay in such fields as coating fertiliser pellets, diluting animal feedstuffs and as a filler in rubber and linoleum. The usage of china clay changed significantly, however. Already, in 1878, the well-informed geologist J.H. Collins suggested that larger quantities were going to paper makers than to textile manufacturers and to potters[1] and by the Edwardian era, more than three-quarters of the entire output of china clay was consumed by the paper industry. Yet, although the share of china clay taken by potters fell, the actual amount they used continued to increase. The following table shows the enormous rise in production.

Output of china clay and stone (000 tons)			
	Cornish china clay	Devon china clay	Cornish china stone
1858	66	5	22
1912	763	78	73

Adapted from J. Allen Howe, *A Handbook to the Collection of Kaolin, China Clay and China Stone*, HMSO, 1914.

In 1858 china-stone production stood at 22,000 tons a year, and since potters used a roughly similar amount of china clay, this suggests that the ceramics industry probably took about a third of total clay output, which had reached 66,000 tons a year. By the early 1900s china-stone production had risen to 73,000 tons a year and, assuming that china clay consumption by the ceramics industry still kept in line with that of china stone, potters' demand for the two products had at least trebled over the period.

The steady growth of the china-stone industry

The earliest china-stone quarries were mainly worked side by side with china-clay pits but as the tonnage of china stone increased they became separate units. The hardness of stone increased as excavations became deeper, and initially there was some resistance from potters about the suitability of harder stone. This may have led H.M. Stocker, in his 1852 paper, to speak about a need to limit stone produc-

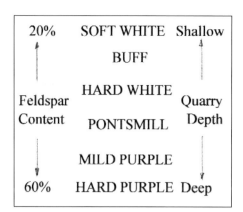

20%	SOFT WHITE	Shallow
	BUFF	
Feldspar	HARD WHITE	Quarry
Content	PONTSMILL	Depth
	MILD PURPLE	
60%	HARD PURPLE	Deep

Table listing hand selected grades of china stone, range of feldspar content and relative depth of working.

tion to 18,000 tons a year, at which rate, he claimed, all the local china stone would be removed in 50 years. He was proved wrong, and in his prediction he may have limited his estimates to a quarry depth of 20–30ft, where most of the softer stone occurred. In fact, the harder stone had a higher feldspar content and potters needed to reduce the amounts used in their recipes. To ensure a uniform quality of stone with a low iron content, chipping axes and hammers were used to remove impurities from the hand-selected stone for potters.

Producers in St Dennis and St Stephens, who enjoyed a monopoly of china stone, in 1860 formed the Cornwall China Stone Co. and were allocated a portion of sales by an arbitrator, the St Austell magistrate R.G. Lakes. After some years of amicable and prosperous trading the monopoly was broken up by some of the landlords.[2] As quarries developed an increasing amount of lithionite granite, the inferior 'shell stone' had to be quarried and stock-piled to one side. Some was used in Cornish buildings, including Caerhays Castle in 1805, the front of St Mary's Chapel in Truro in 1830 and Glendorgal, near Newquay, in 1850–60. The best-known example is the buff-coloured stone from 'Cathedral Quarry' at Nanpean, used as ashlar for the interior walls of Truro Cathedral. In the latter part of the nineteenth century a small area of china stone was found around Stenalees and worked by William Varcoe & Sons. This quarry also provided some stone for buildings in St Austell such as the public rooms, designed by Silvanus Trevail, and the Bible Christian Chapel.

Staffordshire potters bought 'lump' china stone and preferred to grind it themselves. It was probably not until the mid-nineteenth century that grinding mills were set up in Cornwall. The earliest mills may have been located at Tregargus, near where Cookworthy had briefly established his first mill. Over the years these Cornish mills ground stone for potteries at home and overseas. As many as 12 mills contained about 70 grinding pans in total. The ground stone was dried on heated floors and handled and shipped using the same facilities as china clay. Lump stone was often shipped from Charlestown, where an open storage area, now a recreation ground, was set aside for it.

Traditional uses of clay in ceramics
Large-scale economic and social changes transformed demand for pottery clay in the traditional ceramics sector, as well as for new products such as Parian Ware,

sanitary ware and art pottery. In the traditional sector, technological improvements reduced costs through such innovations as lithographic printing and the use of aerographics, or spraying of colours. Demand spread from aristocratic patrons to the lower middle classes, and new products were introduced such as porcelain doorknobs, nameplates, candle extinguishers and spill holders, although these last-named objects gave way in their turn to porcelain matchbox holders.

Styles proliferated and decoration became more lavish, embracing classical Greek and Roman, Persian and Indian, Chinese, Dutch, French, Italian and German Renaissance, English arts and crafts and art nouveau. Victorian bourgeois taste favoured increases in size if not in elegance. Gigantic porcelain statuary and ornamental pillars became fashionable, together with enormous jars, vases and bowls which, to Georgian eyes, might have appeared vulgar and ostentatious.[3]

High-class Parian statuary and cheap Goss seaside souvenirs

However much the middle classes wished to emulate the tastes of those of a higher rank, they could not afford marble statuary and ornaments, and had to be content with cheaper reproductions in chalky white porcelain biscuitware, until enterprising potters in the 1840s developed a new material which was less costly but looked as smooth and glossy as marble. They called this Parian Ware, named after the Greek island Paros, where the stone for the Elgin Marbles was quarried, evoking comparisons with classical antiquities. It was usually hollow, and differed from porcelain in containing glass and a high proportion of china clay and feldspar, giving it a translucent, silky, ivory-coloured appearance. China stone was sometimes used in the recipes but Norwegian feldspar was often preferred. A further advantage was that it could be skilfully touched up to remove blemishes, which meant that fewer pieces had to be rejected after firing. Sales were boosted when an Art Union was formed which ran a lottery in which the winners chose Parian Ware statues of famous people or copies of classical ornaments. Among the first to make it were the Staffordshire potters T. & R. Boote, Minton and the Copeland-Garrett partnership, and it was also produced by Worcester manufacturers and at Belleek in Northern Ireland, which became famous for its open-work basketry ware,[4] and which will be discussed in Chapter Eighteen.

After three or four decades, large-scale Parian Ware statuary and ornaments became less fashionable, but a new demand developed lower down the social scale, as thousands of working-class railway trippers purchased cheap holiday souvenirs made of the material. William Henry Goss, a former apprentice of the Copeland factory at Stoke on Trent, opened a pottery nearby to make small items selling from 6d. (2.5p) each. They ranged from models of famous buildings such as Shakespeare's home and the Longships Lighthouse off Land's End to mugs, jars and pots decorated with the heraldic crests of holiday resorts or local views. Queen

Victoria herself was an enthusiastic collector and the firm founded the League of Goss Collectors, encouraging its members to buy series of items such as small reproductions of the shoes of famous people. The success of Goss led other potters to follow suit, and during the 1914–18 War they made mementoes of battleships, tanks, bombs and aeroplanes, producing their copies immediately news of them appeared in newspaper reports or photographs.[5]

Sanitary ware and 'art pottery'

Meanwhile the advance of knowledge in public health, hygiene and medicine led to a remarkable growth in demand for ceramic products for a surprising variety of uses. One minor loss was that sales of 'leech jars', made of porcelain or chinaware, died out, since these creatures were no longer used to suck out patients' blood. On the other hand, porcelain false teeth became fashionable, especially in America.[6] Much more important, the passing of Public Health Acts led to the installation of proper drainage systems on a vast scale, accompanied by a large increase in demand for bathroom wall and floor tiles and sanitary ware. Britain led the world in the production of tiles and drainpipes, usually made of non-white fireclay, often salt glazed to give a smoother finish, and some glazes used china clay as a component. Twyfords began making WC pans around 1870 with complicated moulds using plaster of Paris and wood, and after this date fire clay was gradually replaced by china clays and ball clays with a transparent glaze.

A key figure in these new developments was Henry (later Sir Henry) Doulton in London. Son of the founder of a local pottery factory, he foresaw the growth in demand for sanitary ware and supplied half of the market for drainpipes in London and a fifth of the requirement throughout the country. His drainpipe factory at Lambeth in London was said to be the largest of its kind in the world, and he also ran two big works in Staffordshire and at Paisley in Scotland. Doulton also became the largest producer of 'art pottery' worldwide.[7] His 'Lambeth faience', as it was called (a fine white stone ware decorated with under-glazed painting), created a new range of products both in body and decoration, receiving the Royal Warrant in the early 1900s.[8] Other local producers of sanitary ware and terracotta will be discussed in Chapter Eighteen, when we deal with the issue of adding value to the clay industry.

Terracotta and bricks

Brickworks in the Tamar Valley produced small amounts of terracotta from stained, fine-grained granite, and in 1893 a clay 'too rich for brick making' was discovered at Tregullon, near Bodmin. This may be the same sett of 'red dust of great value' with 'good stones of tin and lead' offered for lease in 1844 on the Tretoil Estate, two miles from Bodmin.[9] The clay identified in 1893 was used experimentally by Henry Dennis, a local man who had made a fortune in mining and quarrying in North Wales. Lord Robartes granted him a lease to exploit the clay, but nothing came of it. His main source of terracotta was his works at Ruabon, North Wales.[10]

Although, as discussed in Chapter Four, the Lee Moor pit in Devon had begun to make bricks out of waste material on a significant scale in the 1840s,[11] it was not until later in the nineteenth century that brick manufacture became important in Cornwall. John Lovering acquired the Wheal Grey China Clay, Brickmaking & Tin Co. east of Germoe, while Edward Stocker managed brickworks at Carloggas, near Nanpean. Other clay district brickworks included those at Carbis, near Roche, and at Par Harbour, and four – Chytane, Burthy, Gavrigan and Wheal Remfry – near Indian Queens. Coal-fired clay-drying kilns, described later in Chapter Fourteen, with their tiled floors supported by brick-built flues, used substantial quantities of bricks and tiles from these works. At Carloggas, edging bricks for individual plots in graveyards were made. At Carbis, bricks were used for fire-places in homes and Great Western Railway waiting-rooms. The latter were probably inspired by those produced by Candy & Co. in Devon.[12]

Cheap cotton cloths for the Empire

Sales of china clay to cotton manufacturers at home and overseas continued an upward trend, although they took second place to demand from paper makers. The cotton trade was subject to massive fluctuations, however. Large-scale overproduction and strong competition from abroad led to depression in the later 1850s and early 1860s, followed by a sharp recovery in the early 1870s as rival producers in France and Germany were handicapped by the Franco–Prussian War. This was followed, in the later 1870s, by the most severe slump in trade of the Victorian period.[13] By the later nineteenth century, the British textile industry had lost its technological lead to foreign industrialists, who turned out a superior, finer woven material using more scientific and capital-intensive techniques.

The uses of china clay in the manufacture of cotton goods was not without criticism. In 1872 the weavers of Todmorden, in Yorkshire, made representations to the Privy Council, with the support of a Dr Buchanan,[14] on aspects of china clay used for sizing cotton. They pointed out that clay usage, in sizing mixtures, had increased from 1854, when the Crimean War reduced flour supplies. In 1862, when the American Civil War had reduced supplies of cotton to the United Kingdom, the proportion of china clay in size mixtures rose again, from 20 per cent up to 100 per cent in some cases.

Dr Buchanan claimed that size with a high clay content was 'adulteration' of the size and could encourage experiments with other materials in the size that might be poisonous in combination. He even wondered whether arsenic might have been

Cotton spinning in the mid-nineteenth century. Note the scavenger cleaning under the machinery

tried for size, but thought this was unlikely. Dr Buchanan also supported the suggestion that china clay dust might cause chest disease in cotton workers, a belief encouraged by the white, dusty haze that was often found inside cotton mill buildings This was some years before the disease caused by fne cotton fibres, byssinosis, had been identified.

The *Journal of the Royal Society of Arts* for 1877 contains an article by W. Thomson[15] in which he says it suited sizers to use china clay because they were paid according to the weight of size added, and china clay was heavier than flour – three times heavier, in fact. He also said that weavers used china clay as a finishing agent because it not only added weight to the cloth, it helped to fill up the interstices of a loosely woven cloth and gave the appearance of a more closely woven material. Thomson went on to point out that finished goods were not washed before they were sold, but the cotton merchants replied to this by saying that their Asian customers did not wash their clothes, and in any case would always buy the cheapest, which were finished with china clay. Even so, the use of china clay in the cotton industry continued for many years, and china clay was even referred to as 'Cornish cotton' amongst Lancashire manufacturers.

Other countries, moreover, were erecting barriers to protect their own infant manufacturing industries, and so British producers turned to the protected mass markets of the Empire: by 1905 India was taking over 40 per cent of cotton cloth exports.[16] Although some high-quality manufacturers continued in production, the success of overseas trade depended mainly on the low price of their product, which in turn required a high proportion of filler between the interstices of the coarser cloth in order to reduce its threadbare appearance. As much as 70 per cent of the weight of the cheapest cloth exported consisted of china clay,[17] although, since clay was a much denser material than cotton, the proportion of clay by volume was considerably less.

Victorian social reformers were quick to condemn what they saw as an exploitation of native peoples by capitalist entrepreneurs, but David Cock, member of a well-known St Austell clay-producing family, claimed that the poor benefited from the low price of cloth, preferring to pay 7s. (£0.35) a piece for material containing 50 per cent sizing rather than 12s. (£0.60) for pure cotton. 'Heavy sizing is a good thing,' he argued, and no filler was lost in the wash because: 'Indians or Chinese do not wash grey cloth before or after wearing.'[18] Other clay producers remained silent on the subject, probably because, as discussed in Chapter Four, they were aware that public prejudice against the use of clay in non-pottery products lingered on.

High-quality papermaking

Wrapping paper and cardboard, which made up about a fifth of total paper production,[19] did not use china clay, but public mistrust extended to its application to other kinds of paper. According to a technical handbook for china-clay salesmen, it was originally used as an 'extender' to reduce the cost of writing and printing paper. However, it was soon found that it made paper whiter, heavier and capable of taking printer's ink, whether the chief ingredient was rags, esparto grass or wood pulp. Already, by the mid-1800s, supplies of rags could not keep pace with the rapidly rising demand for paper for packaging and postal services, while mass literacy created a vast reading public for books and magazines. The price of rags rose significantly in the 1850s, and rag merchants relied on local collectors, including dustmen, to sort and save them. In Paris an army of 25,000 youngsters, most of them juvenile offenders, searched for rags, and collection was overseen by the prefect of police. Similar schemes were tried in London to recover rags, paper, bones and metal.

As rags became more expensive, the search for alternative materials intensified. The growth of the cotton industry led to an increase in cotton and linen waste, and supplemented rags for use in paper production. But this was not enough, and in 1854 the *Times* of London offered a reward of £1,000 to anyone who could discover a substitute within a year. Two papermakers were appointed to judge the claims but after the year was over they decided that they were unimpressed and no prize was awarded. Over the next 30 years no fewer than 125 applications were made for patents on the use of raw materials for papermaking, and no doubt these proposals were watched by china-clay producers, with possible applications for china clay as a filler in mind.

1000 POUNDS

REWARD.

The Proprietors of a leading Metropolitan Journal OFFER the above REWARD to any person who shall first succeed in

INVENTING OR DISCOVERING

the means of using a

CHEAP SUBSTITUTE

FOR THE

COTTON & LINEN MATERIALS

NOW USED BY

PAPER-MAKERS,

Subject to the following conditions:

1. The material must be practically unlimited in quantity, and be capable of being converted into pulp of a quality equal to that which is at present used in manufacturing the best description of newspaper, and at a cost, ceteris paribus, not less than ten per cent. lower.

2. It must be tested, approved, and adopted by three eminent manufacturers of paper (two of them to be named by the advertiser), whose certificate shall entitle the inventor to the payment of the reward.

3. This offer will be in force only for a period of 12 months from the 20th of May, 1854.

Apply by Letter to A. B., Messrs. SMITH & SONS, 136, STRAND.

Times advertisement of a reward for alternative papermaking materials.

Sugar cane 'bagasse', the trash left after the juice had been extracted, is said to have been used to make paper from as early as 1856 in Baltimore. In 1862 it was claimed it could be shipped to London for half the price of rags. In 1898 an English syndicate set up a mill in Louisiana but it was a complete failure and the use of this material seems to have died out.[20]

Paper adulteration by china clay?

An 1877 article for the Royal Society about the adulteration of cotton goods contains a section on frauds perpetrated by the paper industry. 'Some years ago we heard a good deal about adulterated paper and one Scotch (sic) chemist distinguished himself by pointing out the frauds which paper manufacturers began to practise, viz., of introducing plaster of Paris and other mineral substances into writing and other kinds of paper.' The article went on to criticise the perpetrators of such frauds and praised those representatives of Her Majesty's Government who, in giving their orders, stipulated that this paper should be made from pure vegetable pulp and be free from adulteration. David Cock, a clay merchant, in his Treatise on China Clay, tells that, 'Some years ago we heard a great deal about adulterated paper, but it was soon shown that, with the use of china clay, an excellent paper could be sold at prices at which even passable paper could not be made.' These comments, sharing the same opening words as the RSA article, highlight the range of views betwen the clay suppliers, the users and the government in the 1880s.

Esparto grass

British paper manufacturers, especially those in Scotland, showed some enthusiasm for the use of esparto grass, a cellulose material which was found in North Africa and southern Spain. Before it could be used it had to be sorted and weeds and roots removed – a labour-intensive process – and treated with lyes such as soda or potash. This resulted in a pulp which was beaten and then used to make paper in the same way as rags. A Cornish reference to the use of esparto grass appears in one of Quiller-Couch's stories. The owner of a paper mill in a Cornish creek, alert for a new line in papermaking, receives 'a truss or two of esparto grass'

156

sent to him as a sample by 'some Algerian enterprise'.[21]

British imports of esparto grass escalated from a mere 16 tons in 1861 to 176,000 tons in 1877, by which time only 19 vat mills, utilising rags, remained in Britain.[22] Other European and American papermakers preferred wood as an alternative to rags. As early as 1840 Keller had secured a German patent for a machine that pressed timber logs against a grindstone to separate the fibres in the wood, and his machine was improved by a German paper-mill director, Voelter. By 1851 the method was economically suitable for the conversion of wood, the cellulose of which was white. However, at this time, paper made from wood was considered to be not as good as that made from esparto grass, which in turn was said to be slightly inferior to that made from rags. It is possible that the judicious, perhaps secret, use of china clay as a filler improved the lower-grade papers. The quality of wood-pulp paper began to rise with the advent of chemical treatment, initially by caustic soda and later in the 1880s when sulphate and sulphite methods were adopted. These processes are still in use today.

By this time, wood pulp was taking over from rags and esparto grass as the main ingredient in papermaking in Britain, imports of wood pulp rising from 80,000 tons in 1887 to 978,000 tons by 1913.[23]

As demand for higher-quality print-worthy paper rose, china clay was incorporated in the pulp in increasing proportions, up to 35 per cent in weight. It was not used to reduce costs, as some suspected, but to impart a smoother and more solid surface, especially for the reproduction of halftone photographs, in which fine spots of ink of varying density produced gradation of tone. The basis of vegetable fibres, cellulose, retains hygroscopic moisture and is therefore not so receptive as clay to wetting by printer's ink. Half of the surface of paper made from fibres alone consists of interfibre spaces, giving poor opacity to the final page. The solution was to use a white 'loading' in the pulp to fill the gaps, producing a more even surface and also improving opacity and ink retention. Although other materials had been tried, such as calcium carbonate, gypsum and talc, china clay was preferred at this time.

For the highest quality white paper, only 'best clays' would suffice, and even for cream-coloured or tinted paper, clays of at least medium quality were specified. The use of pictorial illustrations increased, and these were best displayed on the smoothest paper, which meant taking paper already filled with china clay or other material and coating it with the brightest and finest agents available.[24] Second-quality clay could also be made whiter in appearance by the producers adding a small proportion of a blueing agent.[25] This was dripped into the stream of clay, in a discreet fashion, on its way to the settling pits. The provision of massive quantities of china clay of appropriate grades would become the major challenge for china-clay producers in the twentieth century.[26]

The development of wallpaper

Hundreds of years before wallpaper was made, tapestries were hung on the walls of the rich, both to embellish the room and to insulate it, the Bayeux Tapestries being a famous example. From the fifteenth century, painted cloths provided a cheaper wall covering, but in the following century the Chinese started to export wallpaper to Europe and production later began in France and elsewhere. This paper was thicker than that used in more recent times, giving it an insulating quality, and it was stuck directly onto walls or hung from beams or rollers. The paint used was the same as on painted cloth, or a layer of glue was applied and 'flock', or wool clippings, was scattered over it.

By about 1760 a method of hand printing developed using a series of engraved wooden blocks, 18 by 20 inches in size, each imparting a different colour to build up a design. Wallpaper manufacture, whether involving a dusty flock or paints containing lead or arsenic was an unhealthy occupation, and also harmful to those living in the rooms in which it hung.[27] A government enquiry of 1832 condemned the use of green paint containing arsenic as injurious to health, and these compounds were later banned. A West Country example of early hand-painted and coloured wallpaper, made in Paris by Dufour & Leroy about 1820 and depicting Vesuvius and the Bay of Naples, was found in the dining-room of what became the White Hart Inn in St Austell, formerly the residence of Charles Rashleigh, solicitor and entrepreneur. Most of this paper was removed by Victoria & Albert Museum staff to London, where it is now stored.[28]

The first successful machine for printing wallpaper was patented in 1839, and many examples of machine-made paper were on display in the Great Exhibition in London in 1851. By 1860 production had risen nine times and this was the era when china clay was used as a filler, giving a 'loading' of up to 20 per cent by weight of the paper.[29] Some papers were relatively thick, so that they could be embossed to make a premium product. Middle-class households could now afford the product, and nursery wallpaper, some of which involved an educational theme, became fashionable. The late nineteenth and early twentieth centuries saw a massive rise in demand. Art nouveau designs, such as 'Willow Boughs' (c.1887) by William Morris, and floral motifs (c.1900) by Arthur Sanderson, including 'The Cedar Tree', remain popular to this day. By 1900 over 40 major wallpaper firms operated in Britain, and virtually the entire British output was controlled by a monopolistic association, the Wallpaper Manufacturers' Trust.[30]

Alum, paints and ultramarine

Apart from its main uses in pottery, cotton textiles and paper, china clay had long been used for other purposes, including alum and ultramarine, as discussed in Chapter Six. It was the production of alum in which H.D. Pochin became involved.

In 1855 he patented a process for roasting china clay and reacting it with sulphuric acid which, according to one authority, was 'in effect a repetition of Chaptal's methods' of 1789, referred to in Chapter Six.[31] By using china clay he made a low iron alum which became popular for many markets. A rival chemist of the time was Peter Spence who, from 1847, had evolved a similar process using alum shales from the Manchester area. Pochin went on to build several alum production plants in England and acquired china-clay works in Cornwall to supply them. His involvement in the china-clay industry will be discussed in Chapter Eighteen. A brochure issued by H.D. Pochin & Co in about 1915 claimed that Pochin had also patented an alum process in 1870 under the trade name of 'patent aluminous cake', but we have found no record of this and the patent may not have been granted. However, the use of china clay for alum provided a small but steady market for a century or so, amounting to perhaps 10,000 tons of clay a year.

J.H. Collins, in 1878, referred to well-established applications of china clay in the manufacture of alum and ultramarine.[32] In 1880 the cost of synthetic ultramarine was said by the clay merchant David Cock to be a mere four per cent of the cost of the natural material, and was being manufactured in Germany and England as well as France.[33] According to Henry Collins, the son of J.H. Collins, between 12,000 and 15,000 tons of clay a year were used for this purpose in Germany alone in the years before the 1914 War.[34]

By 1893, in the production of cream enamel paint, feldspar and ground Cornish china stone almost superseded the use of lead, according to one authority.[35] However, the presence of china clay in paint raised possible allegations of adulteration. It could legitimately be added to reduce intensity of colour, and it has been claimed that it conferred special properties to mixtures which other fillers did not.[36] However, if it diluted the quantity of lead or zinc below the level stated on the container, it was illegal.[37] China clay was also applied in the clarification of wine in Spain and Portugal, in face powder and increasingly in drugs,[38] and kaolin gradually replaced bread and linseed in poultices for medical use.[39]

Rubber, oil cloth and linoleum

A new demand arose with the rapid rise of the rubber industry, as British imports of rubber rose from a few thousand tons a year to over 20 million tons per annum by the early 1900s.[40] However, the rising demand for rubber pushed up its price, and china clay was introduced as an inert, non-reinforcing filler to lower the cost without excessively impairing its hard-wearing properties.[41]

Rubber was also, for a time, a component in linoleum until it became too expensive. Oil cloth and linoleum were two products that increased enormously in use in Victorian times as covers for tables ('table braids'), chairs, walls, windows and, predominantly, floors.[42] Their production basically involved spreading layers or 'paints' of various ingredients, including pigments and ochres mixed with hot linseed oil, onto a canvas or linen backing. This was a time-consuming operation, since each layer had to dry before the next could be applied, which took days until steam heating greatly reduced the waiting period. Oil cloth, or 'wax cloth', was the dominant product until the 1860s, when Frederick Walter, an India-rubber manufacturer, founded a works at Staines to develop a heavier-weight floor covering which was called linoleum, from the Latin name for linseed oil. This incorporated resin, oak dust and other materials, and once Walter's patent ran out other oil-cloth manufacturers moved into production of linoleum on a large scale. It was softer to walk on than oil cloth and became the main floor covering for public rooms and commercial premises as well as for domestic use.[43] Famous artists, such as William Morris, were commissioned to produce designs.[44]

Before the paints or layers could be applied, the canvas base had to be sized, and china clay may have been used for this purpose, as well as in the composition of the paints themselves. Just how much clay was used, however, seems impossible to say, since we have found no figures in china-clay company records which identify this use. Nor can we gain information from oil-cloth and linoleum firms, which kept their 'receipts', or formulas for their paints, a closely guarded secret.[45] Only one 'receipt' involving china clay has been found, a recipe used by the Storey Co. of Lancaster for the manufacture of oil cloth around 1870. Four layers of paints were applied: the first contained about a quarter by weight of china clay, but in subsequent layers no clay was used, although the final layer contained chalk as a whitener.[46]

On the basis of this one example, china clay did not seem to be a significant component of oil-cloth manufacture. Yet local port records in Lancaster for 1888 show 20 china-clay ships from Fowey and other West Country ports arriving at that harbour, with a combined capacity of some 4,000 tons.[47] At that time this figure represented about one per cent of total West Country clay production, and the possible recipients of these shipments included Williamson's, another Lancaster oil-cloth firm employing 2,500 hands, and James Helme at Halton on the River Lune, three miles upstream. These, however, were only three of several dozen British oil-cloth makers of the period, including Nairn's of Kirkcaldy in Fife, who employed 1,500 hands at about that time, as well as owning factories in America, France and Germany and operating a world-wide sales network.[48]

If all the other oil-cloth firms were using china clay to the same extent as the Lancaster operators, total demand might possibly have amounted to five or ten per cent of West Country clay output, but histories of Nairn and Williamson contain no reference to the application of china clay, only to the use of 'whitening', which may have been some other material. Moreover, 1888, the only year in which we have details of clay shipments to Lancaster, appeared to mark the high point of china-clay usage. From then on, linoleum manufacture took over as the primary product, and by 1914 the Lancaster producers had largely replaced china clay with powdered chalk both as a filler and as an extender for the paints used in surface treatment.[49]

China clay in a Cornish art colony

In this context it is of interest to mention a curious connection between the use of china clay, the linoleum industry in Lancaster and a group of artists who settled in Lamorna Cove, west of Penzance, in the late-nineteenth and early-twentieth centuries. The doyen of the colony was Samuel 'Lamorna' Birch, son of a Birkenhead painter and decorator, whose family moved to Manchester and set him to work in the linoleum warehouse of James Helme. When Samuel began to show symptoms of tuberculosis, the paternalistic Helme gave him a much healthier job in his warehouse on the River Lune, near Lancaster.

The young man loved to paint the local landscape and, noting his talent, Helme commissioned designs from him and bought some of his paintings. By this time the neighbouring linoleum manufacturers and philanthropists, the Storeys, had funded an Institute in Lancaster where Birch developed his artistic skills before seeking fame and fortune in Lamorna, near Newlyn.[50] Among the members of the art colony that Lamorna Birch founded were Laura (later Dame Laura) Knight and Alfred (later Sir Alfred) Munnings. As we shall see in Chapter Fifteen, Lamorna Birch and Laura Knight were among the first to romanticise china-clay working in a painting, possibly following the example of Lamorna Birch's glamorous pictures of granite quarries.

The canvasses that artists used were sealed with a primer to prevent oil paint from eventually degrading the canvas, and a sizing was either applied by the supplier of the canvas or done by the artists themselves. It usually consisted of animal glue and white lead, sometimes mixed with yellow ochre or other pigments.[51] If the artists were doing well, they could, as Laura Knight put it, '... go a-bust on canvasses and paints... I waste dozens of yards of canvasses and paper without counting the cost.'[52] At other times, however, she was reduced to priming her own canvasses with linseed oil and china clay, which seems akin to the process used in the manufacture of oil cloth. She passed this knowledge on to Munnings and, since the artists lived in a close-knit community, it seems likely that others followed her practice.[53]

Low-grade 'mica clay'

The china clays that were sold to most customers were composed almost entirely of kaolinite, the white platy mineral formed when the feldspar in granite is decomposed. Kaolin particles can vary in size from about 100 microns to as little as one micron, which is one millionth of a metre. Another component of decomposed granite is mica, which can be similar in size to the coarsest clay fractions. For over 200 years the industry had to cope with a mixture of these fine minerals which had the same specific gravity and were difficult to separate in the refining process using mica channels called 'micas'.

To some extent the problem of eliminating mica in high-grade clays was overcome by caution in refining, and accepting the loss of some coarse china clay along with similar sized mica particles rather than risking a higher clay recovery with some mica present, which would have been unacceptable to most users. During the progress of waste materials along local rivers, the tumbling action of the stream broke down some larger kaolin particles, making it easier to separate them from mica, so the materials could be reworked downstream to extract additional china clay, usually of low grade, on one or more occasions before they reached the sea.

There were customers at the cheaper end of the market who accepted slightly contaminated clay. This 'mica' clay represented about $2^1/_2$ per cent of total china-clay and china-stone tonnage sold from Cornwall in the early 1900s.[54] Amongst these customers were makers of coarse pottery, fire bricks, terracotta, steam boiler coverings, linoleum, cement and brattice cloth, a coarse tarred curtain coated with mica clay used to control mine ventilation. For such clients mica clays were prepared containing up to ten per cent mica, obtained by secondary refining in mica channels of the residue from the first stage of refining. This sometimes took place at the same location, but more commonly beside rivers because most primary refining residues were discharged into 'white' rivers in the days before strict pollution control. For example, in 1869 Cornelius Varcoe worked mica cleanings from Pochin's clay dries at Gothers and later built his own drying kiln.[55]

The mica clay produced from these secondary operations was dried like china clay in air dries or on coal-fired pan kilns. In many cases the 'dries' used had been built by small companies for china clay drying but were redundant due to company closure or obsolescence.

Scouring powder and edible clay

An early use for decomposed granite was as scouring powder, known in West Cornwall as 'gard'. Knowledge of the abrasive qualities of fine quartz sand in decomposed granite dates back for centuries, and a reference appears in William Bottrell's collection of West Cornwall folk stories published in 1880. In the preparations for St Just in Penwith's Feast Day, some 50 or more years earlier, a sandy substance called 'gard' was employed by a housewife to 'scour the life out of timbran [timber]... the dairy door and all, as well as the benches'.[56] Another reference appears in a story by J.H. Pearce, where an old lady made her living by selling 'scouring gard' and other items from door to door in the St Just in Penwith area.[57]

We conclude this chapter with one application of china clay that entered Cornish mythology. Inevitably it was the most disreputable of uses, namely the adulteration of flour for bread making. In Chapter Four we saw how this malpractice was detected during the Napoleonic Wars, but it probably continued for many more decades, since bread in those days had a grey appearance and could be made to look more attractive with a whitener like chalk or clay. Later in the century the mining expert Robert Hunt published a collection of stories recounted by 'droll tellers' who wandered through the West Country with folk tales and legends, often updating them by casting well-known local personalities as heroes or villains. In one such story a strange mark on a rock called Tolcarne, near Penzance, was said to have been made by the Devil when he stamped his foot in rage after a sinner escaped from his clutches by jumping off the rock. In a variant of this story, the sinner became a miller, whom the Devil claimed as his own because of his dishonesty in mixing clay with flour. An article from the influential *Mining Journal* in 1881 repeated this version, adding that the illegal misuse of clay was prevalent throughout England and the Continent.[58] Edwardian clay merchants were aware that such practice was worldwide, from the West Indies to northern India, northern Russia and Sweden. Germans were said to spread clay on bread, Venezuelans to eat it raw or baked and some addicts consumed as much as six pounds in weight a day.[59] An Edwardian writer on Cornish folklore suggested that it was good for the health, in moderation, of course, and attributed the robust appearance of china-clay workers to the clay dust that they consumed with their pasties.[60]

Food adulteration continued to be a major preoccupation of the British public and, curiously enough, Henry Davis Pochin made it the subject of his maiden speech in Parliament in 1869, when he supported a motion of Lord Cecil urging the government to take action on the 'widespread and reprehensible practices of... adulterous foods, drinks and drugs'.[61] As we have seen, flour adulteration was only one instance of a widespread mistrust of the introduction of china clay in cotton textiles, paper, paints and other non-ceramic uses, but J.H. Collins defended some of these applications, calling china clay 'a most useful ally in many of these forms of adulteration which we still regard as legitimate forms of competition'.[62] Public suspicion, moreover, had no discernible effect upon sales of china clay, which rose at an ever-increasing rate to unprecedented heights.

References
[1] J.H. Collins, *The Hensbarrow Granite District*, Truro, 1878, p.23.
[2] Hichins and Drew, *The History of St Austell*, St Austell, c.1892, p.42.
[3] George Savage, *Pottery through the Ages*, Penguin, 1959, p.35.
[4] Dennis Barker, *Parian Ware*, Princes Risborough, 1998. Examples are sometimes displayed by Copeland descendants at Trelissick, Truro.
[5] Lydia Pine, *Goss and Souvenir Heraldic China*, Princes Risborough, 2003.

[6] *British Clay Worker*, XIII, January 1905, p.324.

[7] *British Clay Worker*, IV, 1895, p.119.

[8] Information from the Boca Raton Museum of Art, Florida, which houses a large collection of Doulton Lambeth ware. Sales of Royal Doulton china were world-wide and important items can be found, for example, in Havana, Cuba.

[9] *Mining Journal*, 10 February 1844.

[10] *British Clay Worker*, IV, July 1893; Ronald Perry, 'Silvanus Trevail and the Development of Modern Tourism in Cornwall', *Journal of the Royal Institution of Cornwall*, 1999, pp. 33–43; Dan Cruikshank, 'Terracotta Revival', *Architects' Journal*, 29 June 1988.

[11] *Western Morning News*, 24 October 1911; Helen Harris, *The Industrial Archaeology of Dartmoor*, Newton Abbot, 1972; Crispin Gill (ed.), *Dartmoor*, Newton Abbot, 1970; John Penderill Church, *The China Clay Industry*, n.d., p.8.

[12] When Wheal Remfry, which supplied specialist bricks for china-clay kilns, shut down in 1972, this brought the Cornish brick-making industry to a close. For a full discussion of this topic see John Ferguson and Charles Thurlow, *Cornish Brick Making and Brick Building*, St Austell, 2005.

[13] Geoffrey Timmins, *Four centuries of Lancashire Cotton*, Preston, 1996, p.40.

[14] Dr Buchanan 'On certain sizing processes used in cotton manufacture', Journal of the Royal Society of Arts, June 14 1872, p.625.

[15] *China Clay Trade Review* Vol. 3, No. 33, June 1922.

[16] *SAS* 1 February 1906.

[17] Stephen Broadberry and Andrew Marrison, 'External Economies of Scale in the Lancashire Cotton Industry', *Economic History Review*, LV, I, 2002, p.55; William Lazonick, 'The Cotton Industry', in Bernard Elbaum and William Lazonick (eds), *The Decline of the British Economy*, Oxford, 1986; Timmins, 1996.

[18] David Cock, *A Treatise, Technical and Practical, on the Nature, Production and Uses of China Clay*, London, 1880, pp.80–90.

[19] Gary Bryan Magee, *Productivity and Performance in the Paper Industry in Britain and America 1860–1914*, CUP, 1997, p.197.

[20] Joel Munsell, *Chronology of the Origins and Progress of Paper*, New York, repr. 1980, pp.177–78; Dard Hunter, *Papermaking*, New York, 1947, pp.575, 579.

[21] Sir Arthur Quiller-Couch, 'Faithful Jane', in *Q Anthology*, London, 1948, p.379. Q may have had a paper mill at Gragon, near Bodinnick, in mind.

[22] D.C. Coleman, *The British Paper Industry*, Oxford, 1958, p.285; Gary Bryan Magee, 1997.

[23] Magee, 1997, p.138.

[24] J. Strachan, 'The Functions of Mineral Fillings in Printing Paper', *British Journal*, 1933.

[25] Ultramarine was available from the late 1830s and blue aniline dyes from the late 1860s.

[26] R.H. Clapperton, *Modern Paper Making*, London, 1952, p.185; J.P. Casey, *Pulp and Paper*, London, 1960, p.985; Edwin Sutermeister, *Chemistry of Pulp and Paper Making*, London, 1929, p.352; F.H. Norris, *Paper and Paper Making*, London, 1952, p.95; Alfred H. Shorter, *Paper Making in the British Isles*, Newton Abbott, 1971, pp.139–41; Henry F. Collins, 'The China Clay Industry of the West of England', *Mining Magazine*, 1920, pp.25–32.

[27] R. Hills, *Papermaking in Britain 1488–1968*, London, 1998.

[28] Charles C. Oman and Jean Hamilton, *Wallpaper*, London, 1982, p.216; R. and B. Larn, *Charlestown*, Charlestown, 1944, p.10.

[29] R. Higham, *Handbook of Papermaking*, London, 1968, p.346.

[30] Examples of these designs are to be found in the Wallpaper Collection of the Whitworth Museum, Manchester.

[31] Charles Singer, *The Earliest Chemical Industry*, London, 1948, p.253.

[32] J.H. Collins, *The Hensbarrow Granite District*, Truro, 1878, repr. St Austell, 1992, p.23.

[33] Cock, 1880, p.90.

[34] H.F. Collins, 1920, pp.25–32.

[35] *British Clay Worker*, II, September 1893, p.107.

[36] Alfred B. Searle, *Clay and What We Get From It*, London, 1925, p.135.

[37] J. Allen Howe, *A Handbook to the Collection of Kaolin, China Clay and China Stone*, HMSO, 1914, p.143.

[38] Howe, 1914, p. 144.

[39] Martindale, *The Extra Pharmacopia*, London, 1941.

[40] B.R. Mitchell, *British Historical Statistics*, Cambridge, 1988.

[41] D.E. Highley, *China Clay*, Mineral Dossier 26, HMSO, 1984, pp.30–31.

[42] M.W. Jones, 'The History and Manufacture of Floorcloths and Linoleum', *Society of the Chemical Industry*, 1918, pp.8–16.

[43] Board of Trade, *Working Party Report on Linoleum and Felt Base*, HMSO, 1947.

[44] Linda Parry, *William Morris Textiles*, London, 1983, p.74.

[45] Searle, 1925, p.98.

[46] Guy Christie, *Storey's of Lancaster*, London, 1964, pp.82–83.

[47] Private communication from George Leven of 2005.

[48] Augustus Muir, *Nairns of Kirkcaldy*, Cambridge, 1956.

[49] British Whiting Federation, *The Story of British Chalk Whiting*, Welwyn, c.1960, p.16.

[50] Austin Wormleighton, *A Painter Laureate*, Bristol, 1995, pp.41–48.

[51] David Bomford *et al*, *Impressionism*, London, 1991, pp.44–50.

[52] Laura Knight, *Oil Paint and Grease Paint*, London, 1936, p.179.

[53] Jean Goodman, *The Life of Alfred Munnings*, London, 1988, pp.104–06.

[54] In 1909, it accounted for 24,000 out of a Cornish total of 894,000 tons, according to J.H. Collins, 'The china clay industry of Cornwall', *Mining Magazine*, 1911, p.452. Mica sales from West Devon amounted to 11,000 tons, a higher proportion relative to china-clay sales from that area, perhaps because the mica associated with Lee Moor clay was largely a silvery form called muscovite. This potassium mica derived its name from its reputed source in *Muskowia* in Russia.

[55] Kenneth Rickard, *St Dennis and Goss Moor*, Tiverton, 2004, p.110.

[56] William Bottrell, *Stories and Folk Lore of West Cornwall*, London, 1880, repr. 1996.

[57] Joseph Henry Pearce, *Cornish Drolls*, 1893–94, repr. Llanerch, 1998, p.84.

[58] Robert Hunt, *Popular Romances of the West of England*, London, 1865, p.43; *Mining Journal*, 9 April 1881, p.425.

[59] John A. Service, 'Potters Clay and how It is Obtained', *American Pottery Gazette*, repr, West of England China Stone & Clay Co., St Austell, c.1910, p.22.

[60] J. Henry Harris, *Cornish Saints and Sinners*, London, 1906.

[61] H.T. Milliken, *The Road to Bodnant*, Manchester, 1975, p.51.

[62] J.H. Collins, 1911, p.452.

Chapter Thirteen
Worldwide Destinations and Competitors

A changing geographic pattern of clay and stone movement in the Victorian and Edwardian eras, while reflecting the important shifts in industrial usage described in the previous chapter, was also affected by radical improvements in land and sea transport, both positively and negatively. It made West Country clays more competitive in some parts of the world but less competitive in others. This chapter sketches in the background to developments in international transport and communications, before examining new sources of china clay and discussing the contrasting fortunes of the clayports.

In expanding their sales worldwide, the clay merchants had the advantage of access to a more comprehensive trade system than producers in any other nation. UK exports more than quadrupled between the mid-1800s and the 1914 War, and international shipping freights dropped by 20 per cent from the 1880s to the early 1900s.[1] Clay was still mainly sent out in sailing ships, and a Merchant Shipping Act of 1854 stimulated the construction of shallow draft vessels which could navigate tidal rivers and creeks. On the eve of the First World War a great increase in china-clay exports to the USA in larger vessels of 5,000–13,000 tons was reported.[2]

However, the most dramatic transformation did not occur in carrying merchandise or people, but rather in the transmission of information. With the coming of the telegraph and the telephone, news which had taken days, weeks or even months to arrive now took only minutes. As will be discussed in Chapter Nineteen, financiers and merchants, some of a shady nature, could more easily speculate in the market for ownership of china-clay firms. At the same time cyclical fluctuations became more violent and spread quickly from one nation to another. What British industrialists of the period called the 'Great Depression' was triggered by financial crashes in New York and Vienna in 1873, and Cornish iron, lead, tin and clay sales fell, while exports of cotton goods, important users of china clay, did not regain their 1872 levels until the cotton textile boom of 1904/05.[3]

The demand for clay was even more volatile than that for consumer products such as pottery, textiles or paper, because manufacturers of these commodities ran down their stocks of clay as trade slackened, only to accelerate orders sharply when business picked up. These fluctuations could be very short term: for instance in January 1894 Thomas Martin reported to his board that he 'had never known trade to be in a worse state than at present, especially in the Pottery district', yet

Cornwall and Devon china clay tonnage, value and price 1858–1912			
	Output (000s)	Value (£000s)	Av. Price per ton(£s)
1858	71	64	0.90
1868	111	102	0.92
1888	412	266	1.64
1902	546	391	0.72
1912	861	597	0.69

Adapted from J. Allen Howe, A Handbook to the Collection of Kaolin, China Clay and China Stone, *HMSO, 1914.*

in April of the same year he was able to announce that 'there was such a brisk demand for clay' from his Virginia pit near St Stephen, that 'no stocks were left'. The sale of bricks was similarly spasmodic: in 1895 the kiln at Lee Moor was closed down because 2,000,000 unsold bricks were in stock. In 1902 trade was again at a low ebb and the Lee Moor workers were put on half time, yet four years later a short-lived boom in tin mining led to a shortage of labour.[4]

One example of the fluctuations in demand from the paper industry occurred in the early 1900s, when demand from American paper-makers exercised the greatest influence. A yearly reduction in clay sales occurred in late spring and early summer, when supplies of water to American paper-mill owners diminished. Just before spring 1908, however, a financial crisis in America led to a downturn in trade, leaving American mill owners with large stocks of unused clay. They cut their orders from Cornwall even further, and as a result nearly 100 Cornish coopers were out of work, since the USA was the chief customer for clay in casks. In circumstances like this the larger clay companies would, in their turn, try to unload their stocks of 'common' clays at, or even below, cost price so as to clear space to store their 'best clays', on which they expected to make larger profits when trade resumed. This drove some small firms, who only produced lower grade clays, out of business, although newcomers might restart these pits when trade picked up again.

Short-term recessions in china-clay demand occurred, however, within the context of a strongly upward trend in sales, associated with a long-term downward movement in selling prices. As we saw in Part One, in the early days of production, china clay fetched as much as £13 a ton, but by the mid-1800s it was selling for around £1.25 or less. As the accompanying table of Values and Prices suggests, the average price fell to a lower level and stayed there during 'the Great Depression', a period of generally falling prices from the later 1870s, although 'best clays' might still sell for two or three times the price of 'common clays', while ball clays might be sold for as little as £0.25 a ton.[5] This continued reduction in price helps to account for the long-term growth in demand for clay, even though competition from other whiteners and fillers sometimes drove china clay out of the market altogether. The importance of keeping prices down was illustrated in china stone as well, when, after one clay merchant put his price up, his sales fell from 2,500 to 1,000 tons a year.[6]

Changing sales destinations

When English potters were the main customers, the bulk of West Country production was shipped to Staffordshire via Runcorn, with lesser quantities going to other British ceramic centres and to nearby continental ports. By the 1870s, however, only a quarter of Cornish china-clay production still went to Runcorn, and only one per cent was sent up the River Severn to Worcester potters . Foreign competition had impacted upon clay sales in two ways. A combination of improved continental canal networks and cheaper shipping freights meant that European clay producers could undercut the West Country claymen in continental markets, as will be shown later. In the same way European ceramicists could sell their products in London and east coast areas of Britain at a lower price than equivalent wares from Staffordshire, Bristol or Worcester. West Country sales of china clay to these traditional English markets slumped by over a third between the 1890s and 1914.[8]

As for ball clay, falling demand from the traditional English potters was offset by some orders from manufacturers of sanitary ware, drain pipes and other bulky earthenware items, discussed in the previous chapter. Among the firms concerned

Rowse's Cooperage at Charlestown.

was Candy & Co. Ltd, founded by Frank Candy of Heathfield, near Newton Abbot, making bricks, drainpipes and salt-glazed wares from the early 1870s. The growth of the ceramics industry in North and South Devon, Dorset and Northern Ireland is discussed in Chapter Eighteen.

Ball-clay shipments overseas

Foreign exports of ball clay accounted for a smaller proportion of total sales than for china clay. They started from a very small base because ball clay exports were hindered by Port of Exeter bureaucracy until 1852. In 1854 foreign sales from South Devon only amounted to 700 tons, but had reached nearly 24,000 by 1904. Watts, Blake, Bearne & Co. shipped clay to many continental customers, including Villeroy & Boch in the Saar, De Porcelayne Fles in Delft, Pickman in Seville and other potteries in France, Germany, Italy, Sweden, St Petersburg, Riga and Egersund.[10] However, export sales of ball clay from South Devon in the early 1900s represented a quarter of the total output of ball clay of around 100,000 tons, compared with three-quarters for china-clay exports.[11]

North Devon clay was shipped from Fremington, near Barnstaple, and from Bideford to Europe and to America, mostly via Liverpool. In 1898 a French barque loaded 1,000 tons of clay at Bideford bound for Delaware, said to have been the biggest cargo ever sent from that port. Coasters also carried clay to Liverpool for transfer to boats on the Trent and Mersey Canal or the Shropshire Canal en route for the Staffordshire potteries.[12] Ball clay sales from North Devon in 1914 amounted to 34,000 tons and from Dorset in 1898 were about 70,000 tons a year, but the proportion of export sales is not known.[13]

The share of china-clay sales going overseas peaked at 81 per cent in 1911. As the table shows, Cornwall's two leading china-clay export territories were also Britain's two major industrial rivals, the United States and Germany. In 1912 Americans took 38 per cent of Cornish china clay exports, while Germany bought more than the figure in the table indicates, since much of the clay imported by Holland and Belgium was destined for Germany.

Destination of china clay exports from Cornwall and Devon in 1912			
To Europe (000 tons)		To rest of world (000 tons)	
Germany	94	USA	253
Netherlands	68	Russia	46
Belgium	59	India	20
France	46	Canada	13
Italy	21	Other	42
Total	**288**	**Total**	**374**

Adapted from Howe, 1914.

In 1875, America only imported 130 tons of West Country china clay. By 1912 the figure had risen to over 253,000 tons a year.[14] Most of the high proportion going to America was for paper making. Already, by 1876, US production of paper equalled Britain's at 350 million pounds by weight per annum, and American imports of paper from Britain had ceased while exports were growing rapidly. British mills were still larger in capacity, with 274 mills in Britain making the same quantity as 812 mills in the USA, and in 1876 Sir Sidney Waterlow, the paper magnate, confidently predicted that Britain would hold its own because its paper was superior in quality.[15] However, labour productivity in American mills was twice that in Britain, and America became predominant. In 1860, UK production of paper and board was 112,000 tons, about level with USA output. By 1912–14, while UK output had risen nearly ten-fold to 1,085,000 tons a year, US production had increased over 40 times, to 4,705,000 tons.[16] USA paper makers took the lion's share of the increase in West Country china-clay sales in the decade before the 1914 War.[17]

An increasingly competitive world

Although china clay sales increased twelve-fold and ball clay sales trebled during the half-century or so up to the 1914 War, the clay merchants were always aware of threats from many different competitors. Rumours abounded of discoveries of cheaper substitutes, new sources of high-quality clay, mounting competition from continental suppliers and declining demand from domestic potters who were losing trade to foreign rivals. As we saw in Part One, porcelain producers in Germany, France and elsewhere had identified and extracted kaolin of their own before it was exploited on a commercial scale in the West Country. However, they jealously guarded their supplies rather than selling them to potential rivals, whereas in free-trade Britain clay from Cornwall and Devon sold on the open market in large quantities. Then, as significant new deposits of kaolin were discovered in Europe, continental producers began to sell clay from Limoges, Normandy, Brittany, the Pyrenees, Meissen, the Rhine Valley and Bohemia, southern Sweden, Spain, northern Portugal, northern Italy and southern Russia.

Hitherto continental clay producers had been handicapped by the high cost of transporting clay overland to the Baltic or Atlantic ports, but the construction of canals in Europe reduced transport costs and made them more competitive. By 1890 the Cornish clay merchant H. Peters, who operated from the north German port of Harburg, was complaining that at least 60 per cent of the china clay sold in West Germany came from Bohemia, whereas 15–20 years earlier it had nearly all originated from Cornwall. West Country clay merchants, he asserted, could only compete in price at or near coastal areas of the continent, and then only for 'best clays', which made up 65 per cent of exports to Germany. 'Medium clays' made up 20 per cent of sales to Germany and 'common clays' accounted for 15 per cent, but soon, he predicted, even 'best clays' would be undercut by top-grade clay from Bohemia.[18]

West Country ball-clay producers, too, were facing strong competition from Germany. Between 1850 and 1914, ball-clay production from Westerwald, north-east of Koblenz on the Rhine, escalated from 10,000 to 480,000 metric tons a year, that is to say about four times the South West's output. However, most of the large Westerwald production went into bricks and drainpipes. From Koblenz some could be taken cheaply by barge to many parts of Europe along the comprehensive canal network, the major customer being France. The best clays were exported in casks worldwide from Rotterdam.[19]

American clays

It was not only in Europe that new deposits of china clay were being exploited. It was produced near Montreal and in dozens of sites in the eastern and southern states of the USA, although in terms of quality these clays tended to contain small amounts of iron and titanium and were mostly used for products where 'best clays' were not needed, such as pipes, ordinary pottery, wallpaper and newsprint. Higher quality clay was said to be available in Alabama, but was too far from the railroad to be competitive.[20] Indeed, transatlantic freight costs from Cornwall to the north-eastern seaboard of America were only half those of rail freight from Georgia, a major producing area, to internal American customers.[21]

While producers in some parts of North America suffered from a seasonal short-age of water (as did areas in India), which held back their progress,[22] there was always a possibility that they would overcome their problems. In 1909 a rumour spread of a new process that could transform five tons of 'rough elements' into one ton of 'finest clay' in just one hour, without the aid of fuel or water,[23] possibly by the use of chemical additives. At about the same time an American company was reported to have secured a concession for 100,000 tons of 'fine quality' china clay a year in British Guiana.[24] China-stone producers, too, were said to have received a 'set back from a foreign produced substitute',[25] probably Scandinavian feldspar.

Chile and Peru were also known to possess china-clay deposits, while small-scale exploitation was occurring in many parts of India, Malaya and various parts of Australia,[26] and in 1895 a headline in the *St Austell Star*, 'The Clay Industry Threatened', reported the discovery of 'thick beds of fine china clay' at Cape Flats in South Africa.[27] As the geologist J.H. Collins remarked in 1911, 'What has been, to a large extent, pretty much a West of England monopoly is being threatened from various quarters... It remains to be seen what will be the result, as import duties against us become more generally imposed and progressively increased.'[28]

Nevertheless, West Country china clay merchants held their own, their 'best clays' retaining their competitive edge, even if lesser clays could be produced more cheaply elsewhere. Their clay continued to arrive at Rouen, from where it was taken to Alsace and Limoges along the canal systems, successfully competing with the local production.[29] As the table below shows, the main foreign producers of clay were also the chief customers for Cornish china clay, shown on page 169. Cornish and Devon claymasters not only produced over 70 per cent of total world output, they completely dominated international trade in the product, since foreign producers of clay were importers rather than exporters, buying twice as much clay from Cornwall as they produced themselves.

Producers of china clay circa 1912 (000 tons per annum)					
		%			%
Cornwall and Devon	861	72	USA	34	3
Bavaria	187	16	Russia	18	2
France	75	6	Italy, Spain, Belgium	13	1

From Howe, 1914.

Marketing and distribution strategies

To control and exploit their domestic and worldwide markets, the larger clay companies established their own sales depots and agents. From South Devon, Watts, Blake, Bearne & Co. appointed agents in Gloucester, Runcorn, London, Tyneside and Brussels. From Cornwall, the West of England Co. opened UK outlets at Runcorn, Garston, Manchester, Leith, Bo'ness, Glasgow and Chatham, while their European depots included Ghent, Aachen, Mannheim, Riesa, Genoa, Dresden and St Petersburg. Parkyn and Peters ran depots in London, Liverpool, Runcorn, Preston and Dundee and overseas at Antwerp, Nantes, Genoa and New York. North & Rose (who will be discussed in Chapter Nineteen) used agents in Holland, Belgium, France, Italy, Finland, Norway and Sweden, selling several varieties of 'Kaolin anglais' and four qualities of stone 'pierres de Cornouailles', as well as other minerals such as plaster of Paris, chalk, ochre and colourings of all kinds. They also exhibited in trade fairs at home and abroad, winning medals at Antwerp in 1885 and at the London Imperial International Exhibition in 1909, and issued brochures in foreign languages.[30]

Some larger, export-oriented clay firms sent their sons and nephews out into the world to market clay. A.L. Rowse, born and bred in the St Austell clay country, regarded this as a weakness, not a strength, a means of finding soft jobs for their wastrel elements,[31] but the commercial success of the clay families suggests that they chose some of their more dynamic members to promote their sales. Small clay producers, who could not support their own sales force, either dealt through the larger companies or used independent clay merchants, like the Nicholls family of

Charlestown and the Fowey China Clay Co., both of whom dealt with German paper-makers.[32]

In the 1870s, around 30 china-clay merchants and producers and china-stone grinders were listed in Cornwall, two-thirds of them in St Austell and the rest in the clay district of St Stephen, St Dennis, Roche and Charlestown. Another ten or so clay merchants were recorded in West Devon.[33] By 1914 the number of Cornish clay merchants had doubled, with rather more than half still based in St Austell, but the rest scattered from Par and Fowey on the south coast to Newquay and Bodmin to the north, Liskeard in the east and Penzance to the west. A dozen merchants operated in the Plymouth region.[34] The larger clay companies also bought and sold from each other and acted as agents for each other, possibly to fulfil large contracts for particular grades of clay of which they were short.

Pit to port transport in Dorset and Devon

During the second half of the nineteenth century, steam locomotives played an increasing part in transporting clay to the sea, which continued to be the main highway to customers in Britain and abroad, while pipelines made their appearance in Mid Cornwall and West Devon. During this period, Dorset ball-clay merchants extended their own tramway networks, the Pikes developing lines to the west at Creech and Povington, while Benjamin Fayle & Co. dug new pits to the east at Newton. To carry their increasing loads, both firms acquired steam locomotives. From 1866 onwards the Pikes bought four, and in 1885 Fayle & Co. purchased one for their old tramway running north to Middlebere and added another in 1905 when they abandoned this line and built a new narrow-gauge track, nearly six miles long, to Newton and along the Goathorn Peninsula to Goathorn Pier in Poole Harbour. They extended their jetty so that steamers could load clay directly for shipment to London and elsewhere.[35]

In North Devon, as in Dorset, the clay merchant W.A.B. Wren, mentioned in Chapter Eleven, had to construct his own narrow-gauge light railway in 1881 because there had not been enough potential for the transport of other commodities. However, he agreed to carry freight for landowners on the route to Torrington from the Marland clayworks. These landowners only charged very low rents and this, together with their contribution to freight revenues, helped to make the railway pay. The original track, with ten wooden viaducts, cost £15,000, nearly half as much again as the estimate. Four locomotives were used and, as trade grew, another three were added in 1908, by which time about 27,000 tons of clay a year were being transported, rising to 34,000 tons in 1913. Also carried were 4,000 tons of bricks, with return loads of 4,500 tons of coal, manure and other commodities for the Marland works and for the landowners. From the 1890s until 1914 several ambitious schemes were proposed to build a railway from the main line of the

London & South Western Railway at Torrington southwards to Okehampton, and a company was formed with a nominal capital of £250,000 for this purpose, but less than 10 per cent of this sum was subscribed and the venture came to nothing.[36]

In the south of the county, although the South Devon Railway from Exeter reached Teignmouth and Newton Abbot in 1846, it had little impact on the carriage of ball clay. In 1866, however, the Moretonhampstead & South Devon Railway was constructed partly on the route of Templer's Haytor tramway and partly alongside the Stover Canal, which was bought for £8,000. Clay-loading facilities were installed at sidings between Bovey Tracey and Newton Abbot. Clay could be sent northwards by rail to Bristol or westwards to Fowey after the Great Western Railway merged with the South Devon Railway and acquired the Cornwall Minerals Railway and eventually the Port of Fowey.

Nevertheless, despite these rail improvements, the bulk of South Devon ball clay continued to be transported by canal, road and water. Watts, Blake, Bearne & Co. leased the Stover Canal from the railway company and continued to use it. Notwithstanding the poor state of the roads, which were worsened by increasing traffic, clay firms relied upon horse-drawn wagons for the short haul to the nearest quay, although Watts, Blake, Bearne & Co. hired steam tractors on a daily basis to deal with the backlog of orders held up by the 1913 ball clay strike, to be discussed in Chapter Seventeen. Clay then went by barge to Teignmouth and by sea to potteries at home and abroad. Some larger loads of 650–750 tons went to Italy or Spain, but most was transported in smaller consignments because carriers preferred to send their cargoes with other, more lucrative, commodities.[37]

Inland china-clay transport

In Cornwall, the momentum of railway development was comparatively slow. Although railways reduced the cost of transport by as much as two-thirds, the initial capital investment was high, often exceeding the promoters' estimates, while revenues were sometimes disappointing. Schemes confidently conceived at the height of a boom tended to come to fruition in the trough of the next slump. Fluctuations in world trade were increasing in intensity and china-clay customers, particularly paper-makers, were especially sensitive to changes in consumer demand. Long-term investment in large-scale railway projects simply to serve china-clay traffic was thus a risky business.

Nevertheless, local groups, aided by London financiers, made a number of unsuccessful attempts to extend earlier rail networks, including two lines in 1863–64 up the Gover and St Austell River valleys from Pentewan to serve districts not covered by railways.[38] A London speculator, W.R. Roebuck, together with a wealthy contractor, Sir Moreton Peto, converted part of Treffry's old horse-drawn tramway in the Luxulyan valley to steam locomotion in 1872. Roebuck's main aim was to make Fowey a great iron-ore exporting port – 'the Middlesbrough of Cornwall' – and to carry out his project he spent a fortune on steam locomotives and rolling stock. Unfortunately he launched his venture in one of the most severe trade depressions of the century and the iron mines closed as swiftly as they had opened. Revenue from the carriage of other commodities, including lead, copper, tin and even clay, was disappointing and the scheme collapsed in 1875, ruining Peto.[39]

The Great Western Railway acquired Peto's line, but a few years later, when the worst of the recession was over, Roebuck was anxious to try again, informing the *West Briton* that the Duchy Peru Mines producing iron and zinc 'will henceforth yield large profits'. The editor of that newspaper declared that 'two or three more Roebuck's and a little more 'One and All' spirit would make a vast difference to our prospects'.[40] But the harsh reality was that clay traffic on its own was not enough to make such a railway line profitable. The revenue from other minerals was dwindling and the clay merchants had to wait until the turn of the century before the GWR invested in the transport and materials handling infrastructure at Fowey, to be discussed later.

Another railway venture of the period that never saw the light of day was a proposed line to Mevagissey. The railway to Pentewan, which had been horse-drawn for many years, by 1870 was plainly out of date, and a steam locomotion system was advocated. In Mevagissey there was talk in 1873 of a line to Pentewan which

Lee Moor locomotive – an 0-4-0 saddle tank built by Peckett of Bristol.

continued to the north of St Austell. Interest appeared to lapse, however, when the Pentewan Railway & Harbour Co. was incorporated in that year to allow steam working, but in 1879 a route from Pentewan to Mevagissey was reported to be under survey. In 1886 a more ambitious scheme was proposed, supported by local interests, using a railway with centre-rail electrification. The idea of a light railway was mooted in the 1890s and in 1910 a huge new deep-water dock was proposed at Pentewan. This was successfully opposed by the GWR, who were seeking powers to build a Trenance Valley branch. And so, although Mevagissey may have seen some schooners loading clay from its inner harbour, the railway never reached that port.

The variety of methods for transporting clay that had developed in Cornwall by 1914 can be illustrated by a publicity brochure of H.D. Pochin & Co. From their works north of Liskeard it flowed nine miles by pipeline to dries near the GWR line at Liskeard, where it was either transferred by rail to the company's private berth at Looe and loaded onto sailing ships or small steamers, or sent by rail to Fowey, Par or Plymouth. Clay from Gothers pit, near St Dennis, was taken on the company's own light railway to a wharf by the main GWR line nearby for onward transmission to Fowey, Newquay or Plymouth. From another Mid Cornwall pit it went by pipeline to dries by the GWR line at Burngullow and then by rail to Fowey, Par, Falmouth, Newquay or Plymouth. Clay from pits in West Cornwall was transported by lorries, traction engines or horse wagons to Penzance. There it was either loaded onto small coasters from the company's own pier, or transferred to larger vessels of up to 1,500 tons by crane from the floating dock.[41]

On Dartmoor, china-clay output had reached levels that were too great for the horse-drawn railway installed by Phillips and described in Chapter Seven, that took casks of clay down to Laira Wharf on the River Plym, renamed 'China Clay Wharf'. Two steam locomotives were purchased by the Martins in 1899, and although they could only pull loads of half their capacity because clay spillage could make the line slippery, this still meant 40 truckloads a day compared with the three or four previously drawn by horses. Tonnage carried increased to between 60,000 and 70,000 a year, which sufficed until just before the 1914 War. In addition, the Sellecks built a pipeline to Marsh Mills, near Plymouth, and Olver & Sons were recorded as filling a dozen small clay ships a year, with an average cargo of 150–160 tons apiece in the pre-war years.[42]

The contrasting fortunes of the Cornish clayports
In the years up to the 1914 War, the relative fortunes of Cornish clay-handling ports changed greatly. Penzance and Hayle continued to deal with small quantities of clay from pits in West Penwith, for instance in 1885 Penzance shipped out 1,570 tons and Hayle 390. Penryn dispatched similarly small amounts from the St Day

area (1,024 tons in 1885). Porthleven did a modest trade in clay and fire bricks from pits in its area. As mentioned in Chapter Seven, Phillip Wheeler & Co. started to export china clay from Porthleven in 1856 and in the next decade the clay handled by that port rose to an average of 700 tons a year.[43]

Another pit to use Porthleven was William Browne's clay works at Tregonning Hill, leased from the Duke of Leeds, which produced 900 tons in 1858. Harvey's of Hayle and Holman Bros of Camborne, two great names in Cornish engineering, later acquired some of these clay works, exporting 2,000 tons a year in the 1890s, rising to a peak of 7,000 tons in 1908. Cheshire was their main destination, although they developed some trade with European ports, only for it to fade away because of difficulties in finding return cargoes to take to Porthleven as local industry declined.[44]

On the north coast, clay continued to be shipped out in small quantities from Newquay (4,152 tons in 1885) and after pits opened on Bodmin Moor an agreement was reached for the 12-mile railway from Wenford Bridge to Wadebridge to transport clay. This line had been constructed in 1834 to carry sea sand containing lime to farmers who used it to improve the quality of the soil, with return loads of iron ore, tin ore, ochre and granite from the De Lank quarry on Bodmin Moor. Shipping clay from Wadebridge offered the advantage, for many destinations, of avoiding the treacherous passage around Lands End used by vessels coming from the south-coast ports. Some smaller ship owners tended to keep their vessels in the Irish Sea, travelling between the Severn Estuary, the Mersey and the Clyde. However, the Camel Estuary presented its own problems. It was so difficult to navigate because of shifting sandbanks that some ship owners charged more to take clay to Runcorn than from south-coast ports and in 1885 the tonnage of clay leaving the Camel Estuary only totalled 1,495.

After 1895, when the Bodmin to Wadebridge line was connected to the GWR system at Bodmin Road, clay could be carried south from Bodmin Moor on the Lostwithiel to Fowey branch line. This branch had been opened in 1869 as a broad-gauge track for the carriage of iron ore but closed from 1890 for financial reasons, only to open again in 1895 with standard gauge. In 1899 the North Cornwall Railway, masterminded by the GWR's rival, the London & South Western Railway, extended its line along the north coast from Wadebridge to Padstow so that clay could be sent by rail from Bodmin Moor directly to the port. In 1903 a new firm, the North Cornwall China Clay Co., developed a large clay works at Stannon Marsh, combining some small pits, and laid pipelines for about eight miles to Wenford Bridge, where they built large clay dries, discussed in Chapter Nineteen.[45]

St Ives also despatched small quantities of clay. In 1858 Towednack (probably Bedlam Green) clayworks was shifting 300 tons a year, with St Ives given as the

Pentewan Harbour c.1914. (Royal Institution of Cornwall)

main shipping port. It seems likely that clay from that source had gone through St Ives since at least 1820, when it was probably loaded onto beached vessels. In a list of 14 ports which appeared under the heading 'Shipments of China Clay and China Stone in 1877', St Ives was shown as having despatched 14 tons of stone, but the port did not appear in the lists for the two previous nor the two subsequent years.[46]

Of the three harbours constructed in St Austell bay, Pentewan fared worst because of silting from the waste materials that flowed down from the clay country. Its share of Cornish clay shipments remained small, at 6–8 per cent, and just before the 1914 War silting held up a Liverpool-bound coaster for two weeks, a mishap from which the port never recovered.[47] Charlestown, on the other hand, retained a 15–20 per cent share of clay exports. In Chapter Seven we described an abortive proposal to construct a branch railway to the port, and in 1864 the owners made another unsuccessful effort. At that time the inner harbour was restricted in size by a ramp for launching sailing ships, which could be constructed much more safely there than on the beach. A Llanelly engineer, M. Rosser, presented estimates for enlarging the harbour at a cost of nearly £4,000 and installing an inclined railway powered by a stationary rotative beam engine for nearly £5,000.[48]

The railway was not built, but the inner harbour was enlarged in the 1870s at a cost of nearly £3,000, and shipbuilding ended.

In the 1890s Stocker's West of England China Clay Co. and John Lovering & Co. were reported to be interested in acquiring the port of Charlestown, as were the Great Western Railway Co., backed by unnamed 'clay merchants'. Had either group succeeded in purchasing the harbour, it might have been more extensively developed but, for unknown reasons, nothing came of the idea. Later, in 1908, Lovering built a pipeline to Charlestown from his pit at Carclaze, to be referred to in Chapter Fourteen.

The deep-water ports of Falmouth and Fowey

Meanwhile, as china-clay output in Cornwall rose rapidly, it became clear that handling costs could be significantly reduced if deep-water all-tides ports were available capable of servicing the newer and larger ships that were being built. The cost of carrying clay in small coastal vessels and then transferring it to large ocean-going ships at Liverpool, for instance, was nearly as great (over £0.20 per ton) as taking it from Liverpool to America. Two front-runners in the competition for recognition as the main all-tides port for the Cornish clay trade emerged: Falmouth and Fowey.

The best anchorage along the Cornish coast, indeed one of the finest in Europe, was at Falmouth and, after the port lost its status as the point of first and last call for the worldwide packet-ship network in the mid-1850s, its merchants sought to develop an alternative. Led by Howard Fox, doyen of one of Falmouth's greatest merchant families and long-time Chairman of Falmouth Docks, they turned to clay exporting. The first load of china clay arrived by rail in January 1864, when ten wagons containing almost 100 tons arrived from the West of England clayworks at Burngullow.[49] Soon Falmouth was shipping out more clay than Pentewan, and the developer W.R. Roebuck included proposals for a line to Falmouth down the Fal Valley from Meledor Mill in a mineral railway network scheme of the 1870s. His project failed spectacularly in what was called 'the Great Depression' that followed, and Falmouth never played more than a relatively minor role as a clay-exporting harbour. The Secretary of Falmouth Chamber of Commerce later claimed that the Great Western Railway Co., after it took over the mineral railway network, deliberately ruined their chances by charging higher freight rates per mile to transport clay there than to Par and Fowey.[50]

Falmouth's loss was Fowey's gain. As we saw in Chapter Seven, Treffry had envisaged developing Fowey in the early-nineteenth century but, after meeting opposition from Rashleigh's executors, built Par Harbour instead. Not until 1869 did a mineral railway open between Lostwithiel and Fowey, and in that year the first clay ship, the 130-ton schooner *Rippling Wave*, left Fowey for Genoa and

Leghorn, bringing back a cargo of black marble to Bristol.[51] Despite the temporary closure of the Fowey to Lostwithiel railway, the port's trade grew, so much so that by 1909 it was criticised as 'cramped, congested and of insufficient depth'.[52] The GWR had gained a monopoly of the transport of clay by rail, and what the clay merchants considered their high-handed manner was the source of continuous friction, exacerbated when the railway took over the main clay port of Fowey and seemed reluctant to improve its facilities. The first China Clay Association, formed in 1906, was in constant correspondence with the GWR about slow forwarding of trucks to Fowey, contamination of clay and high charges for freight handling.[53]

To solve the problem, St Austell Mercantile Association proposed the construction of a deep-water port at Pentewan, at an estimated cost of a quarter of a million pounds, but it did not take the GWR long to persuade Cornwall County Council that these ventures would cost twice the estimated amount and would never pay, and to retaliate with a proposition for a deep-water facility at Fowey.[54] By 1914 the GWR had developed a rail network between Fowey and various parts of the clay district that made it the main channel for china-clay exports, some brought in by rail from Lee Moor.

Fowey clay merchants' shipping scams?
The previous chapter concluded with the story of how china-clay merchants were enshrined in Cornish folklore after they were involved in an illegal activity of adulterating flour. In this chapter we finish with suggestions of a much more disreputable, indeed downright murderous, malpractice. Some unscrupulous nineteenth-century British and foreign ship owners overloaded old and unseaworthy vessels and over-insured their contents. If the ship reached its destination, the owners had reduced the costs of transport per ton, if it sank they were over-compensated for their loss. To stamp out this malpractice, Samuel Plimsoll pressed for a Merchant Shipping Act that required the hulls of vessels to be marked with a 'Plimsoll Line' showing how low in the water they could be loaded. Yet even after the Act was passed in 1876 some owners secretly overloaded their ships by night after they had left port.

This criminal practice was attributed to Cornish clay merchants in the fictional tales of Cornwall's leading Edwardian man of letters, Sir Arthur Quiller-Couch. From his vantage point on the River Fowey, 'Q', as he was called, observed the growing passage of china-clay and stone ships and, as Harbour Commissioner for 21 years and Trinity House Pilotage Sub-Commissioner, he knew the tonnage they carried and indeed quoted it approvingly in one of his articles.[55] He also exploited the tale of a shipping scam in more than one of his stories. In 'Shining Ferry' a rich clay merchant sent a vessel out of Fowey that was 'leaking like a five-bar gate', because china stone (in lump form) had 'fairly knocked her open and the timber all round *[was]* as rotten as cheese'. The ship owner also ordered the mooring light to be extinguished, increasing the risk of collision, to 'save two-pennorth of oil'. Such a man, observed his ship's captain, only employed 'a skipper who'll earn insurance-money and save oil'.[56]

This story, although written in 1905, seemed to be set in a pre-Plimsoll Line era because Q remarked in it that 'nowadays no British ship so scandalously over laden would be put out to sea'.[57] In another of his tales that clearly related to post-Plimsoll times, a retired seafarer on the 'Troytown' (Fowey) Harbour Board tried to stop the 'scandal of robbin' the underwriters and puttin' seamen's lives in danger', only to find, to his shame, that some of his retirement pension came from funds invested in a rotten, overloaded and over-insured clay ship that sank off the American coast.[58] It seems doubtful, though, that clay merchants were willing to consign their fellow Cornishmen to a watery grave. For, according to Ward-Jackson's study of ship registrations, whatever interest the clay merchants might have had in ship ownership had practically disappeared by the end of the nineteenth century. Cargo freights had been low since the 1870s and there was probably more money to be made in investing in the production of china clay and stone than in their transport . Like ball-clay producers, they probably found that wide variations in the size of consignments for different destinations meant that chartering was more convenient than ship owning.[60]

To sum up this account of the marketing and distribution of clay, we note a remarkable combination of the old and the new. The telegraph and the telephone had revolutionised communications, railway networks had penetrated to most corners of the South West, steamships had greatly reduced travel times worldwide and ball-clay merchants were encouraged to export their product. Yet despite these dramatic changes, the horse and wagon, the canal and privately owned light railways remained the principal forms of short-haul transport in Devon and Dorset, and the vast bulk of output was still carried by sailing vessels.

References
[1] B.R. Mitchell, *British Historical Statistics*, 1988, pp.526, 536–67; Stanley Chapman, *Merchant Enterprise in Britain*, Cambridge, repr. 2003.
[2] *NE* 10 October 1913.
[3] *SAS* 1 February 1906.
[4] Derek Giles, 'Some lesser known china clay firms: Martin Brothers', *CCHSN*, 2006, 14, p.3.
[5] *British Clay Worker*, XII, October 1903, p.244.
[6] *SAS* 30 November 1911.
[7] Kenneth Hudson, *The History of English China Clays*, Newton Abbott, n.d. p.24.
[8] Henry F. Collins, 'The China Clay Industry of the West of England', *MM*, 1920, pp.25–32.
[9] E. Bourry, *A Treatise on Ceramic Industries*, English Edition, London, 1926.
[10] L.T.C. Rolt, *The Potters' Field*, Newton Abbot, 1974, p.45.
[11] H.J. Trump, *Westcountry Harbour*, Teignmouth, 1976, pp.101, 106, 139.
[12] M.J. Messenger, *North Devon Clay*, Truro, 1982, p.22.
[13] *The Clay Lands of Dorset*, London, 1960, p.10.
[14] J. Allen Howe, *Handbook to the Collection of China-clay and China-stone*, HMSO, 1914, p.235.
[15] *Cornish Times* 2 December 1876.
[16] Magee, 1997, p.204.

[17] Henry F. Collins, 'The China Clay Industry of the West of England', *MM*, 1920, pp.25–32.

[18] *SAS* 12 September 1890.

[19] Klaus-Dieter Mayen, *Tongräber im Westerwald*, Montabaur, 1985, p.9.

[20] *BCW*, XL, 1903, p.28.

[21] *Associated China Clays Report*, 1920.

[22] A.D. Selleck, *The China Clays of Devon and Cornwall*, unpublished manuscript, Plymouth Central Library, 1948.

[23] *SAS* 28 January 1909.

[24] Howe, 1914, p.134.

[25] *SAS* 29 September 1910.

[26] Howe, 1914, pp.41–135.

[27] *SAS* 25 October 1895.

[28] J.H. Collins, 'The China Clay industry of Cornwall', *MJ*, 1911, p.455.

[29] *Kelly's Directory of Cornwall*, 1914; Collins, 1920. In 1894, 38 china-clay ships and five ball-clay ships unloaded at Rouen. *BCW*, IV, August 1895, p.121.

[30] North and Rose, *La Production du Kaolin Anglais*, St Austell, 1910.

[31] A.L. Rowse, *A Cornish Childhood*, London, repr. 1993.

[32] *SAS* 9 August 1900.

[33] *Morris & Co's Commercial Directory and Gazetteer of Devonshire*, 1870. William White, *History, Gazetter and Directory of the County of Devon*, Sheffield, 1878–79.

[34] *Kelly's Devonshire Directory*, 1914.

[35] Rodney Legg, *Purbeck's Heath*, Sherborne, 1987, pp.81–92; R.W. Kidner, *The Railways of Purbeck*, Oakwood, 2000, p.51.

[36] Messenger, 1982, pp.25–32; Rod Garner, *The Torrington and Marland Light Railway*, Southampton, 2006, pp.27–48.

[37] Trump, 1976, pp.101, 106, 139; Rolt, 1974, pp.140–46.

[38] M.J.T. Lewis, *The Pentewan Railway*, Truro, 1981, pp.32, 40.

[39] Alan Bennett, *The Great Western Railway in Mid Cornwall*, Cheltenham, 1992; Maurice Dart, *Cornish China Clay Trains*, Shepperton, 2000, *East Cornwall Mineral Railways*, Midhurst, 2004; R.J. Woodfin, *The Cornwall Railway*, Truro, 1972; John Vaughan, *The Newquay Branch and its Branches*, Yeovil, 1991.

[40] *WB* 9 May, 14, 21 July 1881.

[41] H.D. Pochin & Co., *Pochin's China Clay*, Manchester, c.1915.

[42] R.M.S. Hall, *The Lee Moor Tramway*, Blandford, 1963; B. Gibson, *The Lee Moor Tramway*, Plymouth, 1993; Roy E. Taylor, *The Lee Moor Tramway*, Truro, 1999. In 1921 a pipeline was built to take clay slurry to a drying plant by the GWR line. William Lethbridge, *One Man's Moor*, Tiverton, 2006, p.197.

[43] R.M. Barton, *A History of the Cornish China Clay Industry*, Truro, 1966, p.133; Stuart N. Pascoe, *The Early History of Porthleven*, Redruth, 1989, p.24.

[44] Richard Pearse, *The Ports and Harbours of Cornwall*, St Austell, 1963, pp.55–56, 137, 144; D.M. Trethowan, 'Porthleven Harbour', in H.E.S. Fisher, *Ports and Shipping in the South West*, Exeter, 1971, pp.78–79.

[45] We are indebted to Maurice Dart for some of this information. Other sources are C.F.D. Whetmath, *The Bodmin and Wadebridge Railway*, Wokingham, 1994, p.16; Barton, 1966, pp.108–09, 133.

[46] C.H. Ward-Jackson, *Ships and Shipbuilding of West Country Seaports*, Truro, 1986. *Hunt's Mineral Statistics* for 1858 and 1877 both make some reference to St Ives being used for shipment of china clay in relatively small amounts.

[47] *QJS*, 1877, pp.500–12; *RCG* 13 August 1886; Pearse, 1963.

[48] Richard and Bridget Larn, *Charlestown*, Charlestown, 1994, pp.108–11.

[49] *WB* 13 August 1864.

[50] *WB* 12 April 1894.

[51] Michael Bouquet, *West Country Sail*, Newton Abbot, 1971, p.29.

[52] *SAS* 23 September 1909.

[53] *China Clay Association Minute Book*, China Clay History Society Archives.

[54] *SAS* 9, 23 September, 25 November 1909, 6 January, 29 December 1910.

[55] Sir Arthur Quiller-Couch, 'Introduction', to *Borrow's Guide to Cornwall, Liverpool*, n.d. (c.1913).

[56] Quiller-Couch, *Shining Ferry*, London, 1905, pp.24, 288.

[57] Quiller-Couch, 1905, p.281.

[58] Quiller-Couch, *Hocken and Hunken*, London, 1912, repr. 1921, pp.238, 242, 261.

[59] Ward-Jackson, 1986, pp.63, 98.

[60] Trump, 1976, p.56.

Chapter Fourteen
How Innovative were the Claymen?

This chapter examines the technological progress made by the clay merchants up to 1914. The findings of the previous chapter, that despite revolutionary improvements in land and sea transport, some clay families remained wedded to the horse and wagon and the sailing ship, might suggest technical inertia on their part. Certainly this was a view shared by most observers, as least as far as the china-clay industry was concerned. For instance, in 1867 the influential *Mining Journal* asserted that anyone with £1,000 to spare could run a china clayworks: 'the modus operandi is simple enough for a boy of 14 to understand'.[1] This was not the only influential and apparently well-informed observer to belittle the achievements of the clay merchants, for as the geologist J.H. Collins remarked in 1878, 'visitors to the china clay districts are sometimes hasty in condemning the rudeness and inefficiency of the methods employed'. He then added, 'this hasty judgement is usually modified rapidly as a better acquaintance with these methods is acquired'.[2]

Unfortunately, he was too optimistic, for the history of the china-clay industry is littered with accusations of technological backwardness. Joseph M. Coon, china-clay surveyor, judged that their working methods 'changed somewhat slowly' during the Victorian and Edwardian periods.[3] Kenneth Hudson, historian of English China Clays, concluded that, blessed with 'an inexhaustible supply of strong men in search of jobs', they were 'always more inclined to reduce expenditure than to think of modernisation or growth'.[4] The Cornish historian A.K. Hamilton Jenkin, while full of praise for the skills of underground miners, set little store by clayworkers, merely regarding them as 'wonderful men at shifting sand'.[5] Even Philip Varcoe, descendant of one of the oldest china-clay families, commented that all the clay merchants of the nineteenth century were doing was 'making mud pies'.[6] Not until the path-breaking study by Rita Barton did the achievements of the nineteenth-century china-clay leaders receive appropriate recognition.[7]

Ball clay mining in Devon and Dorset

A number of writers have recorded the technical development of ball-clay working in some detail.[8] By and large, while the major improvements in china-clay processing occurred in refining and drying, the main advance in ball-clay production took place in methods of extraction. As we saw in Chapter One, ball clay was originally cut out of small pits or trenches after removing a layer of topsoil a few feet in depth, and as deeper seams were dug a 'square pit' technique evolved, with clay hauled up in buckets by a horse-powered crane.

As the ball-clay trade developed, different grades of material were recognised which could be divided into two main groups. The first, called 'whiteware clays', yielded a white or cream-coloured body and imparted plasticity and bonding power before firing. They were essential constituents of most types of pottery, wall and fireplace tiles, sanitary ware and other household items. The second and more common type, known as 'stoneware clays', contained appreciable amounts of quartz and were used in the manufacture of salt-glazed and other kinds of common stoneware. Clays belonging to both groups occurred one above the other and could be extracted from the same works, but to get at the whiteware clays, which sold at a higher price, underground workings were introduced towards the end of the nineteenth century.[9] The discovery of a particularly fine white-firing clay at depth near Kingsteignton in South Devon may have led to the first use of underground mining methods.

In South Devon the method used was to sink a rectangular timbered shaft directly to intersect the seam. From this shaft a series of side tunnels, or 'drives', were dug in a fan shape for about 100 feet in different directions. The shaft and the 'drives' were enclosed in larch timbers, this wood being used because it would gradually bend under pressure, indicating that repairs had to be made, whereas some other types of timber tended to crack suddenly. The pressure on underground timber supports, because of the plastic and mobile character of the ground to be worked, meant that in some cases it was only possible to keep excavations open for a short period of time. Underground workings therefore had to be limited to comparatively small areas, with an average production of about 6,000 tons, and individual mines only operated for two years or so. When the mine was worked out the wooden supports were withdrawn for possible further use elsewhere and the tunnels would gradually collapse. Clay was taken to the bottom of the main shaft by wheelbarrows.[10]

In the Marland works of North Devon, underground mining also developed at the end of the nineteenth century, using sloping shafts, sunk at an angle of 75° from the horizontal to intersect inclined layers of ball clay. These sloping shafts levelled out some 50 feet from the surface to reach a total length of up to 600 feet, and horizontal tunnels ran off the main shaft on either side to a distance of 60–90 feet. Wagons were raised and lowered by hoisting engines at the surface.[11]

China-clay working using pumping shafts
A widely used development of adit working from the mid-nineteenth century was to have a pumping shaft on one side of a clay pit and connect this to an adit lead-

ing to a vertical washing shaft, or 'rise', in a low point of the pit. A diagram in Chapter Ten illustrates this system. The washing shaft contained a vertical wooden launder as in adit working. The washed clay and sand would run towards the washing shaft and most of the sand would be settled in sand pits. The sand was generally removed using an inclined railway which might share the same source of power as the pumps but was usually separate. The pumping shaft contained plunger pumps which were operated by a reciprocating rod connected to a power source such as a beam engine or water wheel. The pumps delivered clay and fine sand to the 'mica' refining channels on the surface. The use of pumping shafts was particularly convenient when the area of ground available for working was limited. The tunnelling work, carried out by 'shaft' men and recognised as the most arduous task in a clayworks, was carried out by men with mining experience.

Radical advances in refining and drying china clay

While it is clear that ball-clay merchants were making significant changes in their methods of extraction, to a superficial observer the activities of many china-clay pits may not have seemed to alter very much, except that they were getting larger and deeper and that, as we shall see, female and child workers were being dispensed with. Yet if the industry had really remained archaic, characterised by technological inertia, how did the clay merchants achieve such a striking increase in production? The implication of Hudson's and Hamilton Jenkin's remarks is that employers took on proportionally greater numbers of the men who were so good at shifting sand. In fact output had risen 12 times between the mid-1800s and 1914, while the labour force only increased around two and a half times. In other words, production per worker must have more than quadrupled. This might have been achieved if new sources of clay had been exploited that were significantly easier to win, but in fact the opposite was the case. Pits had to be made deeper to find the better clays, which meant that more effort was needed to bring them to the surface and dispose of the eight to nine tons of waste material that now came with every ton of saleable clay. Indeed, in some cases the debris ratio was even higher as 'leaner' or less clay-rich ground was worked.

Given these increasing difficulties, if the same production methods had been used throughout the period, surely output per worker would have gone down, instead of rising four-fold in the second half of the nineteenth century? Clearly, greatly improved working methods had evolved, and to understand how they came about we need to examine the clay merchants' problems more closely. The key factor in the superiority of West Country china clays was their whiteness, and producers had to ensure that the 'best' clays were not extracted indiscriminately with lesser 'medium' or 'common' clays. At the same time they needed to work clays of inferior quality, which had limited uses, to allow pits to be opened or developed.

There was always a temptation to 'pick the eyes out' of a pit, as in metal mining, by working only the best areas, to meet the demands for higher quality from paper makers, and to justify the increased costs of removing sand and rock and of tipping them elsewhere as pits grew in extent. A delicate balance was thus needed between two potentially conflicting goals: achieving larger-scale production and improving standards of quality.

The development of mica drags
In the second half of the nineteenth century the overall increase in sales led to considerable increases in mica separation capacity, especially for pits where the mica-

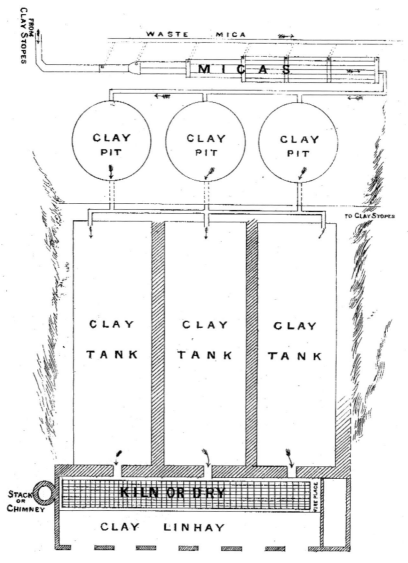

Typical plan of micas, thickening pits and tanks, drying kiln and linhay, from A Treatise on China Clay, *by David Cock, 1880.*

ceous content was high. Longer 'runs' of mica drags were built by duplicating or extending existing capacity. Catch pits for sand removal would also have been extended to meet demand. These were often built close to mica drags but occasionally were sited closer to clay pits. Attempts were made to find improved controls for setting the mica drag 'traps', which replaced the bars of timber used in early micas.[12,13] Micas could be as long as 300ft, with a series of parallel channels, each about two feet in width. Pipework for draining the residues once or twice a shift allowed residues to be discharged into local 'white' rivers, or sometimes dried separately as a low-grade 'mica' clay. China clay was always subject to contamination, and gauze screens were installed at the end of the channels to remove vegetable fibres and wind-blown leaves. These were cleaned by rotating brushes or by hand.

Systems also evolved to thicken clay before the drying process. These replaced the multitude of shallow sun pans that covered often valuable areas of ground around clayworks. Deeper tanks were built with drystone or lime cement walls that took up less space. These were more expensive, but used rock debris from the clayworks, and it was found that a two-step process produced a thickened clay more quickly than a single stage. These processes were carried out in the walled structures, the first often circular in shape and referred to as 'pits', the second usually rectangular and known as 'tanks'. They were generally adjacent to the new coal-fired drying floors that were being built.

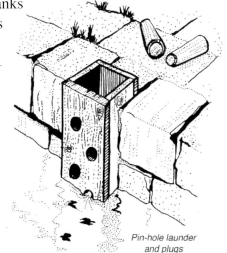

Pin-hole launder and plugs

Coal-fired drying kilns
Perhaps the greatest improvement in productivity during the second half of the nineteenth century arose from the replacement of the traditional drying of clay in the sun and wind by the use of coal-fired drying kilns. These kilns began to replace open-air drying systems in the latter half of the nineteenth century, but small kilns had long been used in the Staffordshire potteries to dry batches of clay slip. These kilns were often square in shape, used under-floor heating and were paved with firebricks or tiles.[14] They burnt local coal, which was readily available, cheap in price and not subject to the duty levied on coal that was shipped by sea and amounted to £0.22 a ton, which increased its price by a third.[15] This heavy duty may have been a factor limiting the use of coal for drying clay until the tax was abolished in 1831.

As early as 1811, however, a lease has been identified for the possible use of kilns in Cornwall, and another reference to drying by 'natural or artificial' heat dates from 1821.[16] The first kilns definitely recorded in the china-clay industry date from 1845. They had a heated floor of about 80 square feet. It is not clear whether they were associated with thickening tanks or whether they dried clay blocks from sun pans.

These were superseded by larger rectangular kilns with thickening tanks close to the drying pan. There were, in fact, two types of pan kiln. The earliest types became known as 'run in' kilns and a later type was called a 'tramming' kiln. Practically all accounts of early clayworking overlook the differences between these two types. Run-in kilns were designed to use the gravity flow of clay from thickening tanks to the drying pan. They were usually built on sloping sites where thickening pits were adjacent to deep, short-sided tanks placed immediately alongside the length of the pans, which ranged in width from 9 to 18 feet and in length from 40 to 200 feet. They were paved with tiles 18 inches by 12 inches of varying thicknesses.

Hatchway for run-in kiln at Wheal Martyn.

The tanks were connected to the pan by two or three narrow access hatches, each about two feet square, equipped with a guillotine-type door to control the flow of clay from the bottom of the tank. Clay was thickened in these tanks to the point where it contained just enough water to flow satisfactorily, rather like custard, when the door was raised.

A particular safety hazard was associated with these kilns. Thickened china clay is thixotropic so it will flow more easily when stirred. On occasions, when clay ceased to run in, a man would enter the tank to 'pull in' or 'hack down' the clay. This could be helpful but dangerous if the clay started to flow out of control, when a man in a tank could be swept into and even through the hatchway to the kiln, sometimes with fatal consequences.[17]

Tramming kilns, a development of run-in kilns, were used from around the 1870s. They were generally on a larger scale and served by larger tanks, most of which were sited alongside the pan. A major difference was that, although a little thickened clay might be run onto the pan, when the tank was broached most of the clay was shovelled or scooped into wagons and trammed on rails to the kiln pan. These rails were laid temporarily on the floor of the tank as it was being emptied, and when a tank was empty the rails were lifted for re-use.

Another important difference in this system was that a wider hatch was used in the tanks, with a series of horizontal boards to seal the hatchway. When ready for drying, the boards were removed one at a time from the top of the hatchway, a safer process. The wagon needed a gap of about four feet to enter the pan area. Here it crossed to the pan, still on rails, and thence to a travelling bridge which spanned the width of the kiln, rather like a workshop travelling crane but at floor level. The bridge allowed clay to be transferred to different points in the kiln. The 'tramming' kiln was more widely used than the 'run-in' kiln.

The tramming system required clay to be thickened for longer in relatively shallow tanks, resulting in less coal usage, up to 15 tons of clay being dried by one ton of coal, compared with nine or ten tons by run-in kilns. On the other hand, extra labour costs were incurred in handling, as well as capital costs for wagons, rails and bridges.[18]

Apart from these improvements, the breaking and washing down of clay had also been advanced in some pits by the use of water hoses, under a pressure of 50–150 lbs per square inch (obtained by gravity in deeper pits, by pumps in shallow ones).[19] Howe[20] reports that although 'hydraulicing' as used in Californian gold mining from the 1850s seems a suitable method where there is a sufficient head of water, it had not been regularly used except at Stannon Marsh.

Technological change and the disappearance of the female worker

How rapid were the changes in drying and refining processes we have just described? Some possible evidence comes from the swift disappearance of women workers. Lynne Mayers has shown that the number of females in Mid Cornwall recorded in the Census as china-clay workers had dropped by 1881 to a mere fraction, fewer than 10 per cent, of the total 20 years earlier. By 1901 they had virtually disappeared. The main activity for which they were employed was the handling, stacking and cleaning of clay blocks dried in the open air. Does the absence of these workers mean that the switch from open-air to coal-fired drying was largely accomplished by 1881 and complete by 1901? One problem is the parallel existence of old and new methods for many decades – water-wheels and steam engines, for instance, and horse-drawn and steam-powered wagons – which will be discussed later. In the same way, some small pits continued to dry clay in the traditional way: Marshel Arthur describes in an autobiography how he worked in one in the early 1900s run by the generally progressive Medland Stocker.[21]

How do we reconcile the continuation of manual drying methods with the disappearance of the women who carried out this operation? One possibility is that women were employed but not recorded in the census. One study has concluded that in the mid-1800s in England as a whole the number of female farm hands was actually double that enumerated in the Census. In Cornwall Mayers has found that some enumerators simply wrote in the occupation of the male head of household and described the rest as clayworkers' wives, sons or daughters. Moreover, Mayers has observed employment of women and children in claywork varied with the state of the labour market. In times of labour shortage, clay merchants would first take on adolescent males, then older women, then boys and girls. Possibly the flooding of the labour market with redundant workers after the copper crash of the 1860s contributed to the rapid disappearance of the female worker. Child employment might also have been reduced for the same reason and by legislation, especially after the 1870 Education Act.[22] In the pits where the old methods lingered on, females may have been replaced by boys. Marshel Arthur was allowed to start work before reaching the official school-leaving age in order to support his widowed mother.

'Practical tinkerers' and technical education

The conclusion to be drawn from this description of change seems inescapable. Working methods, at least in the larger pits, had not merely 'changed somewhat slowly', as Coon remarked. Clayworkers were not merely 'shifting sand', as Jenkin described. Nor were clay merchants simply 'making mud pies', as Varcoe suggested. On the contrary, a transformation in extraction, refining and materials handling had taken place. Who were the authors of this technological progress? Here we must contrast the well-developed scientific institutions of West Cornwall with the absence of any formal system of technical instruction in the china-clay country. In the early 1800s a burst of scientific and cultural creativity led to meetings in Bodmin, Truro, Penzance and Falmouth and the formation of what became the Royal Cornwall Agricultural Association, the Royal Institution of Cornwall, the Royal Geological Society of Cornwall and the Royal Cornwall Polytechnic Society, together with mining academies at Camborne, Redruth, Penzance and Truro.

The china-clay district was largely untouched by this intellectual fervour, and even when the clay industry was booming in the 1890s its leaders turned down an offer from the great Cornish benefactor John Passmore Edwards to fund a library, technical and art school, in contrast to most other Cornish towns, which accepted it. Ratepayers of the St Austell Rural District Council, where the bulk of the rateable value from china clay was concentrated, saw no reason to pay for an institution in the town of St Austell which, they assumed, their clayworkers would not trudge miles to attend.

The china-clay area thus lacked specialised scientific and technological bodies to research, codify and spread knowledge of improved working practices and geological and chemical discoveries. What, then, provided its propellant force for change? Three main factors were important: a continuous series of marginal adjustments made on the spot, the introduction of basic technical instruction funded by central government, and the adaptation of processes and machinery from other industries, both local and international.

As in many other industries of the eighteenth and nineteenth centuries, a great deal of technical progress owed its origins to 'practical tinkerers' or 'scientific mechanics', as economic historians have called them. These men carried out a series of small adjustments that, over a period of time, had a cumulative effect that paralleled the results of more famous strokes of individual genius.[23] Drawing upon his own family history in the china-clay industry, W.J. Tonkin has described the way in which technical knowledge was acquired, accumulated and passed on. Although no formal training system existed, mechanics and key tradesmen were 'issued' with boys as cheap labour for specific jobs. They would latch on to a lad who was especially 'keenly', as they called it, and request him for future jobs until he became an unofficial apprentice, eventually taking the lead as his mentor grew older. One example was a lad who left school in the later 1860s when he was only nine years old and worked as a tin streamer, but later designed and made waterwheels and surveyed large tank kilns and rail sidings, having made his own surveying instruments.[24]

However, by the early 1890s the British Government, alarmed that the country was falling behind its better-educated rivals, allocated modest sums to local authorities for technical training. Cornwall County Council set up a Technical Instruction Committee and, although the two china-clay Councillors were not included,[25] much to their annoyance, the St Austell District enrolled nearly 700 students, putting it third among the 20 or so districts of Cornwall, after more densely populated Camborne and Redruth, whereas Truro only came thirteenth, with fewer than 200.[26]

A number of clay-country men helped to spread technical knowledge. The geologist J.H. Collins published a book and articles on the clay industry, and William Langdon Martin, who ran Lee Moor clayworks, was a keen amateur geologist. Frederick Augustus Coon, prominent councillor and writer and speaker on economic and political affairs, acted as secretary and lecturer on clay matters for the local Technical Education Committee. Joseph M. Coon tested samples in his own laboratory and David Cock wrote a treatise on clay working. F.R. Ray, headmaster of a local 'academy', county councillor and editor of the *St Austell Star*, although not a clay man, was an enthusiast for further education and a member of the County Technical Instruction Committee.

Day and evening classes in English, arithmetic, bookkeeping, surveying and science helped produce efficient supervisors of the many specialised operations that were evolving in the clay industry, and it was these men who, with their combination of natural intelligence and technical education, made many of the improvements that increased labour productivity. Ingenuity was also displayed by Dorset and Devon claymen in developing special tools and devices for digging, mining and handling ball clay and, interestingly, almost identical equipment and methods were devised by ball-clay workers in Westerwald, Germany, some at about the same time. According to a German historian of that industry, technical innovations that 'changed the face of the rapidly industrialising Westerwald ball-clay works were overwhelmingly the fruit of technically gifted practical men who, independently of one another, worked on labour-saving equipment and methods.[27]

Adapting worldwide technology

Apart from their ingenuity in devising and modifying their own practices, china claymen had always adopted knowledge, techniques and machinery from workers in other industries. Cornish tin streamers and miners had long experience of handling water, sand and slimes in similar conditions to those of china clayworks. In the very earliest days it was the client, the master potter, who was the expert on the quality of clay that was needed. Ceramic producers had developed knowledge of stone-grinding mills and the construction and firing of kilns for drying ground flint Later, technical knowhow was imported from further afield as Victorian and Edwardian Cornwall witnessed a 'Great Emigration' to all parts of the globe, and some returning migrants brought back techniques of great value. Cornishmen who had joined the 'Gold Rush' brought back knowledge of the 'California system' of hosing down material on a vast scale, and pipeline technology also benefited from innovators in the oil fields of Pennsylvania.

Pipelines were not new to the clay industry, as discussed in Part One, since the journey from most pits to the clay dries was downhill and, in the early clayworks, conduits, or 'launders', as they were called, took clay for short distances down hollowed out logs, or channels made of planks of wood. More sophisticated versions were suggested in 1858 by William Pease, once Treffry's right-hand man, to the clay landlord Fortescue and they were also advocated about that time by Stocker. However, they may have been influenced by news of improved techniques from the United States, where, by 1866, pipelines were linked to oil refineries from most of the existing oil wells and proved much more efficient and cheaper than the wagoners or 'teamsters' who formerly carried it in barrels.[28]

Pipelines were developed in Cornwall by Pochin, Lovering, Selleck and others. Lovering's took four years to complete, ran from his Carclaze works, two miles inland and 600 feet higher, down to his clay dries that were cut from slate and shale at Charlestown and were among the largest in the clay district.[29] These new dries, also built in 1908, allowed dried clay to be taken in wagons along what was, effectively, an underground railway about a quarter of a mile long and then tipped directly into sea-going vessels via chutes. The whole system thus greatly reduced the labour costs of materials handling. On Dartmoor, Selleck built a pipeline from his works at Wotter down to Marsh Mills near Plymouth, as described in Chapter Twelve. Excavating equipment, possibly used for the construction of the Manchester Ship Canal, found its way to Cornwall[30] and became known locally as Stocker's 'steam navvies' and 'steam grabs'.[31]

China-stone producers may also have been encouraged to try the new rock drills and high explosives that were being introduced into Cornish metal mines.

In the early 1900s some works became major users of electricity, installing their own gas engines as prime movers to drive dynamos which powered centrifugal gravel pumps in shallow pits and replaced steam pumping. One example was the Carne Stents Co. near St Austell, while on Bodmin Moor the St Neots China Clay Co. and Northwood Clayworks introduced improved water-wheels to generate hydroelectric power.[32] At Ponts Mill in the Luxulyan Valley, water turbines powered china-stone grinding mills. Simple wooden filter presses, which had been patented in 1856 for use in the brewing industry, were soon adopted by potters and suggested by J.H. Collins for use in the clay industry. Improved cast-iron presses were installed at Hendra Old Dry at Nanpean in around 1911.[33] Clay producers also ran boot-making shops, cooperages, stables to house their horses and farms to provide feedstuffs for them, owned their own railway wagons and operated light railways. They also opened brick and tile works to use some waste materials, as discussed in Chapter Twelve.

While the china-clay district imported technology from America and Germany, who were also its leading rivals in china-clay production, it owed its greatest technical debt to Cornish metal-mining engineers, who had raised the efficiency of steam-pumping engines to unparalleled heights.[34] After copper output collapsed, Cornish beam engines were available at reduced prices, but gifted engineers had made their way to the clay country even before mining declined, including William West at St Blazey and James Thomas at Charlestown. Apart from exporting equipment worldwide, these men produced beam-engines, china-stone grinding machinery, water wheels and other items for local use.[35]

Parallel technologies

In assessing technological advances in china-clay working, however, we must always recognise that, in common with many other sectors, progress was far from uniformly spread throughout the industry. Large, geologically well-endowed clayworks using state-of-the-art techniques and equipment and involving investment of £100,000 or more were surrounded by small pits with a capital of a few thousand pounds and following traditional methods. Even in 1922, after some important amalgamations had taken place between the leading firms, employment was still equally divided between the five biggest companies controlling 2,276 workers, while the remaining 64 firms employed 2,626 men.[36]

Coexistence of old and new persisted over long periods, and while steam power was installed in a combined clay and tin works as early as 1837, it took half a century for all pits to convert to steam. The last of these steam pumping engines did not cease work until 1959, when it was over 100 years old.[37] This overlap of technology was not necessarily due to lethargy. The high cost of carting coal to some clay pits delayed the introduction of steam power. John Lovering, for example, who kept a tight control on expenses, calculated that it would increase his running costs by a quarter.[38] Coal-fired kilns also took a long time to come into common use for similar reasons.

Because of a tendency to equate the use of steam power with industrial progress, water-wheels have also been associated by some observers with technical backwardness, despite the vast improvements made in water-wheel technology by Rennie, Smeaton and the Frenchman Poncelet. Water power long remained an efficient source of energy in many parts of Britain where coal was not locally available. The last water-powered mill for grinding flour to be erected in Cornwall, Trelowarren New Mill on the Lizard peninsula, did not come into operation until as late as 1923,[39] while the last clay-country water-wheel was not taken out of use until around 1960, about the same time as the last steam engine.

Sometimes the reluctance of workers to change due to a fear of redundancy held up the introduction of new processes: Stocker found himself with a strike on his works when he first brought in pressure hoses to wash down clay.[40] Resistance to change was sometimes justified, however. Lovering's patented control systems for traps on mica-drags were not always used. Stocker's electrically-driven 'bridge' that traversed clay dries was ruled unsafe by the Quarry Inspectorate. His producer gas system for drying kilns proved to be impractical and was soon abandoned.[41] As late as the 1940s the industry was still being criticised for 'primitive handling of materials' in china-stone extraction,[42] but wheelbarrows remained an effective way of loading railway trucks until small mechanical shovels were introduced, and 'dubbers', or hand picks, were sometimes the best tools for breaking down clay-bearing ground.

Contradictory views of the china-clay country

An observer who received regular news of technical progress in the china-clay industry but also attended exhibitions of the works of the Newlyn and Lamorna artists, or read novels written by authors with a clay-country background, might well have been puzzled by the conflicting impressions he gained. Newspapers such as the Truro-based *West Briton* and *Royal Cornwall Gazette* and the Plymouth *Western Morning News*, as well as local press, including the *St Austell*

Star and *Newquay Express*, reported briefly on the introduction of new machinery and methods for extracting, refining and drying clay. The china-clay merchants' expertise was also recognised from time to time by other clay producers. In Part One we saw how, in the 1850s, Lord Morley and the Plymouth proprietors of the Lee Moor clayworks brought in Cornishmen to advise on the efficiency of their methods. Half a century later Henry Holwill, manager of the North Devon Clay Co., was sent to Cornwall to see if anything could be learnt from the china-clay industry.[43]

Edwardian artists' portrayals included Laura Knight's watercolour 'Men Working in a Clay Pit' of 1912 and an illustration in a children's book, *In the Potteries*, by William J. Claxton of 1913, where a black and white photograph was rose-tinted to present a romantic view of a clayworks at sunset (see Chapter Ten).

These interpretations were valid in that, as we have seen, modern and traditional methods existed cheek by jowl, but they gave a misleading impression of technological backwardness. Indeed, the inert nature of the material they extracted seemed to be associated with the character of the claymen themselves, according to some writers. Sir Arthur Quiller-Couch, Cornwall's leading man of letters in the early-twentieth century, was one such author. A character in one of his short stories explained that clay work 'deadens the touch' for the finer things in life, unlike prospecting for copper or tin, where 'fire's in the veins and you follow it like a lover... with clay, what is it but ashes, burnt out and dead?'. Only Eden Phillpotts, in one of his Dartmoor novels, portrayed characters who appreciated china clay's superior qualities. One of them, from Sticklepath in Devon, explained that: 'There's nought grand or pushful or hard about me. I'm just common mud, I am, and 'tis no good axing common mud to be china clay.'[44] Curiously, Thomas Hardy, although he must have witnessed ball-clay working in Dorset, never referred to it in his writings.

Sound commercial reasons for secrecy

However to some extent the clay merchants only had themselves to blame for such perceptions of dullness, because of their reticence about their technical achievements. A.L. Rowse, born and bred in the St Austell area, was fond of remarking that, just as the Chinese had kept secret the use of china clay and china stone for centuries, so the Cornish clay merchants had imposed a wall of silence around their activities. 'The claymasters kept to themselves,' he maintained, 'married among themselves and did not let out any of their secrets.'[45] Earlier, the *Mining Journal* had commented, 'they made their fortune so quietly'[46] and in Chapters Eighteen and Nineteen we shall discuss how they held their cards close to their chests during the clay strike of 1913, and also when outsiders were attempting to take over the industry.

What Rowse and other commentators failed to recognise, however, was that the clay merchants had sound commercial reasons for reticence in disclosing details about their dealings with their customers, their landlords and their fellow clay producers. As we saw in Chapter Twelve, public mistrust lingered on about manufacturers who used china clay in such products as cotton textiles, paper and paint. Some of them even bought clay from third parties to conceal their use of it. Oilcloth makers also had their own 'secret processes' and did not want their rivals to know about the materials that they used.[47]

The clay merchants were reluctant to reveal too much about their operations because it might affect the level of dues demanded by their landlords when their leases came up for renewal. Geological complexity affected the market value of a pit, and producers might encounter clays of a better quality, or a lower proportion of waste materials, thus enhancing profitability. Again the installation of new methods might involve major capital expenditure, and clay merchants preferred to keep details of improved performance to themselves. The price of clay also depended upon its whiteness, which could be enhanced by the addition of blueing agents such as ultramarine from 1840 or aniline dyes from the 1860s. These additions were made discreetly within the very smallest windowless buildings to be found in the clayworks, apart from the tiny tip-men's huts on the top of the sand heaps. However, certain tints might react unfavourably with chemicals used in cotton cloth or paper manufacture and such knowledge was also kept secret.[48]

To sum up, the ball claymen had radically transformed their methods of extraction, whereas china clay merchants had transformed refinement, drying and materials handling into a complex process that was, by the standards of the day, technologically advanced. Yet their achievements went almost unnoticed, even by some apparently well-informed historians, perhaps because no scientific or mechanical geniuses changed the face of the industry, and progress was evolutionary rather than revolutionary. Much of it resulted from a long series of marginal improvements by anonymous, on-the-job practitioners, combined with adaptations of processes and machinery developed in other places and for other purposes. Clay merchants and their clients, though, were sometimes careful to conceal innovations. They often, however, had sound motives for keeping their successes to themselves, even at the cost of appearing technologically backward. Finally, we may note one radical change that some observers did remark upon, although they did not associate it with the technical advances that took place. Women and children virtually disappeared from the workforce.[49]

References

[1] *MJ* 13 April 1867.

[2] J.H. Collins, *The Hensbarrow Granite District*, Truro, 1878, repr. St Austell, 1992, p.26.

[3] Joseph M. Coon, 'The China Clay Industry', *RRCPS*, 1927, pp.56–58. Coon, a clay industry surveyor and shareholder, was also manager of a local electricity works and a telephone exchange.

[4] Kenneth Hudson, *The History of English China Clays*, Newton Abbot, n.d., pp.11–12, 29, 42.

[5] A.K. Hamilton Jenkin, *The Story of Cornwall*, London, 1934, p.114.

[6] Philip Varcoe, *China clay, the Early Years*, St Austell, 1978, p.7.

[7] R.M. Barton, *A History of the Cornish China Clay Industry*, Truro, 1966.

[8] For example, Cyril Brackenbury, 'Clay Mining in South Devon', *TIMM*, XL, 1931, pp.238–77; *Watts, Blake & Bearne, Devon Ball Clays and China Clays*, Newton Abbot, 1954; *The Clay Mines of Dorset*, London, 1960; Terence Davis, *Wareham: Gateway to Purbeck*, Wincanton, n.d.; L.T.C. Rolt, *The Potters' Field: A History of the South Devon Ball Clay Industry*, Newton Abbot, 1974; M.J. Messenger, *North Devon Clay*, Truro, 1982; Rodney Legg, *Purbeck's Heath*, Sherborne, 1987; Ball Clay Heritage Society, *The Ball Clays of Devon and Dorset*, Newton Abbot, 2003.

[9] Board of Trade, *Report of the Enquiry on the Ball-Clay Industry*, HMSO, 1946, p.4.

[10] Brackenburt, 1931, pp.266–67.

[11] Messenger, 1982, p.22.

[12] J. Allen Howe, *A Handbook to the Collection of Kaolin, China Clay and China Stone*, HMSO, 1914, p.30.

[13] Patent 3102 of 1882.

[14] Lorna Weatherill, *The Pottery Trade*, New York, 1971, p.20.

[15] Bridget Howard, 'The Duty on Coal', *JTS*, 26, 1999, pp.30–35. The coal tax was first imposed in 1698. The drawback, or repayment of tax charged, for coal used in smelting tin and copper ore was introduced in 1710.

[16] R.M. Barton, *A History of the Cornish China Clay Industry*, Truro, 1966, p.121.

[17] On the 1880 series of Ordnance Survey maps, run-in kilns can sometimes be recognised. Pan floors run adjacent to one or two short tanks that tend to be smaller than tramming kiln tanks.

[18] Tramming kilns can be identified on larger scale OS maps by their usually greater size and by the length of tanks. Most kiln relics to be seen today are of tramming kilns because older run-in kiln sites have been used for other purposes.

[19] R.N. Worth, *WDM*, 18 September 1903, p.8; Percy Bean, *The Chemistry and Practice of Sizing*, Manchester, 1910, pp.134–50; *SAS* 11 March 1897, article on Burngullow Clay Works.

[20] Howe, p.30.

[21] Marshel Arthur, *Autobiography of a China Clayworker*, Federation of Old Cornwall Societies, 1995, p.13. Edward Stocker had resisted the introduction of early models of coal-fired kilns because of the uneven way in which they dried clay.

[22] Lynne Mayers, *Balmaidens*, Penzance, 2004. We are indebted to Lynne Mayers for much valuable additional information on this topic.

[23] John Rowe, *Cornwall in the Age of the Industrial Revolution*, 1953, repr. St Austell, 1993; Pat Hudson, *Industrial Revolution*, London, 1998.

[24] Private communication from W.J. Tonkin.

[25] Alderman Varcoe and Councillor William King Baker from Penzance.

[26] Ronald Perry, 'Monument to a Victorian Giant', *ABK*, 1998, 91, pp.13–15.

[27] Translated by the authors from Klaus-Dieter Mayen, *Tongräber im Westerwald*, Montabaur, 1985, p.6.

[28] Daniel Yergin, *The Prize*, London, 1993.

[29] *SAS* 22 October 1908. Lovering used stoneware pipes manufactured in Devon.

[30] Photographs in the authors' possession of the West of England Co.'s operations in the 1890s suggest it was using Dunbar-Ruston 10h.p. steam excavators and horse-drawn rail wagons designed for the canal contractors.

[31] *RCG* 20 September 1907.

[32] *BCW*, 1904, 1913.

[33] John Tonkin archives; see also J.M. Coon, 'Development of Mechanical Appliances on China Clay works', *RRCPS*, III, 2, 1916, p.110.

[34] Von Tunzelman, 1994, pp.284–87.

[35] Anon., *Sketch of the Life of William West, CE of Tredenham*, Redruth, 1873; Bernard Broad, 'One Hundred and Twenty Years Engineering at Charlestown', *English China Clays Review*, 1958, pp.2–4. In the 1880s the Charlestown Engineering Works, started by Thomas, was acquired by a syndicate headed by Thomas Martin, clay merchant at Lee Moor.

[36] Norman Pounds, *China Clay*, Wheal Martyn Archives, n.d., p.103.

[37] *WMN* 25 February 1959.

[38] Similar reasons delayed the introduction of steam-powered drills and cranes in the Cornish granite industry for decades: Peter Stanier, *South West Granite*, St Austell, 1999, pp.67–71, 85–95.

[39] Anthony Hitchens Unwin, 'Some Early Watermill Techniques', *JTS*, 29, 2002, pp.113, 119.

[40] *WB* 20 September 1877.

[41] Howe, 1914, p.53; John Tonkin archives.

[42] English Clays Lovering and Pochin, *Report of the China Stone Committee*, 1943.

[43] Messemger, 1982, pp.21–22.

[44] Sir Arthur Quiller-Couch, 'Step o'One Side', in *Q's Mystery Stories*, London, 1937, pp.323–41; Eden Phillpotts, *The Beacon*, London. 1915, p.157.

[45] A.L. Rowse, 'The Cornish china clay industry', *History Today*, 1967, pp.483–86.

[46] *MJ* 13 April 1871.

[47] Charles Thurlow, 'The early use of china clay by the cotton and paper industry', Paper presented to *China Clay History Society*, 2004.

[48] Charles Thurlow, 'The use of china clay for aluminous compounds', Paper presented to *Trevithick Society*, 1990.

[49] Mayers, 2004, pp.93–98. The employment of children was illegal, but clay captains turned a blind eye if a family was in desperate need of the income; see Arthur, 1995.

Chapter Fifteen
The Price of Success: Accidents, Congestion and Pollution

In any extractive process, as output rises its social cost upon local communities becomes more obvious. As far as the impact of the ball-clay industry on life in the remote villages of Victorian Dorset and North Devon was concerned, social costs seemed minimal. It brought welcome opportunities for relatively well-paid employment, pits were located in isolated and often unproductive land, the bulk of the clay was transported by private tramways, only a few miles long, across uninhabited heaths and unloaded from private quays to barges and ships without disturbing the neighbourhood. In South Devon, as will be discussed later, some deterioration of roads occurred.

In Mid Cornwall and on Dartmoor, on the other hand, the development of the china-clay industry on a larger scale caused more of a stir. Since the best clay was often found at the bottom of the pits, for every ton of clay extracted, more and more waste material had to be raised and piled high, on non-kaolinised ground if possible, sometimes as much as 12 tons of debris for every ton of clay.[1] Although picturesque from a distance, the dust that blew in dry weather from the china-clay tips, burrows or 'sky-tips', may have troubled those who worked or lived nearby, but they accepted it as the price of their jobs.

Skytips near Bugle. Clusters of pits and tips like this led to the title of The Cornish Alps.

The increasing volume of waste water containing clay and mica, which flowed down to the sea, formed white rivers. Although attractive to passing railway passengers, they were accepted by many residents as 'white rivers', like the 'red rivers' of West Cornwall. As trade developed over the nineteenth century, Par and Pentewan harbours were prone to silting, and clay and mica allegedly killed off fish. As more clay and stone were carried, country roads in Mid Cornwall and Dartmoor were churned up and became impassable in wet weather, while the streets of St Austell were blocked by heavy clay wagons taking clay to the ports. Finally, as pits became deeper, the potential risk of accidents to men working in or near them tended to increase.

Working conditions and industrial accidents
Whenever china claymen demanded higher pay, the clay merchants never failed to remind them of their good fortune. They contrasted their healthy, open-air life with conditions in the grim factories of the period. For instance, skin cancer (epithelioma) caused by the handling of lubricating oils among young cotton spinners was one of the first occupational diseases of the Industrial Revolution to be identified. By late Victorian times workers in the shale oil and coal tar industries also suffered from this disease.[2] Nearer at hand, the clay employers could point to the disastrous fall of rock in the Delabole slate quarry in North Cornwall that resulted in 15 fatalities and six serious injuries. Slate quarrying seemed a hazardous occupation, for between 1875 and 1893, at least 21 accidental deaths occurred at Delabole.[3]

Lung diseases were another common health hazard. Again, employers could contrast the working conditions of some clay workers with the cramped, unhealthy existence of the underground metal miner, his life-span shortened by collapsing tunnels or silicosis caused by the fine quartz dust created by the pneumatic drills, soon to be dubbed 'widow makers', that were coming into general use. This comparison did not apply to ball clay miners. They worked by candlelight, sometimes ankle deep in water, with the danger of a sudden inrush of water or the ignition of 'fire damp' methane gas. Buckets containing about 300 kilograms of mined clay or waste were hoisted by wire ropes via overhead pulleys powered by steam engines and, from the early 1900s, steam-generated DC electricity. Despite these hazards, however, occurrence of serious accidents was surprisingly rare, compared with those in metal or coal mining.[4]

On the western side of the St Austell granite mass, china-stone extraction, using dry quarrying methods and slow hand-drilling, could result in pulmonary problems, although the incidence of illness was not high until pneumatic drills were introduced in the twentieth century. A similar hazard appeared when sawing machinery was installed in confined spaces in the Delabole slate quarry.[5] In china-clay working, the highest earning workers were the 'burden men' or 'sand men', who were paid by the cubic fathom to remove soil from the top of the pit and waste materials from the bottom and who, as we shall see, were at risk from falling debris. Their work was also physically exhausting, and few could carry it on for the whole of their working lives. Indeed, special lower rates of hourly pay were agreed for older men and, under the paternalistic regime in many pits, as long as an employee of long standing could manage to walk to work he received this lower rate.

Others whose working conditions were far from ideal, and who were paid at a higher rate, were the 'shaftsmen' who dug and maintained the shafts, levels and rises through which clay slurry passed, and the 'washers' and 'breakers' who broke up the clay that was washed down. 'Kiln men' were another category who were paid higher wages because they had to work outside in tanks and inside in hot conditions, removing the dried clay while the kiln pan was still drying clay. In addition, men working at run-in kilns faced the danger of being swept into the kiln by the flow of clay slurry, as mentioned in the Chapter Fourteen. However, they generally worked a shorter day, perhaps five hours. In contrast, at Redlake in the heart of Dartmoor, men endured 'Klondyke-like conditions' in winter compared with the clay areas.[6]

Health and safety regulations

Social reformers of the Victorian era brought in a series of Factory Acts and Mining Acts to regulate industrial health and safety, and not without success, for the UK death rate from industrial accidents and disease fell from four per thousand workers per annum in the early 1850s to 1.3 per thousand in the later 1890s.[7] It was not until this time that clay and stone workers were covered by the Quarrying Acts, which came into operation in 1895, but enforcement was a massive operation. Although regulations only applied to workings over 20 feet deep, more than 100,000 quarries were registered throughout Britain. Government inspectors visited clay pits to ensure that regulations were observed, but since they gave advance notice of their arrival, employers could tidy up operations temporarily. Moreover, when the Acts interfered with established local working practices they were sometimes amended. For example, a deputation of china clay merchants was able to persuade Asquith, the Home Secretary in 1894, to allow youths to work early and late shifts, hitherto forbidden by the Acts, although not the night shift, from 8p.m. to 4a.m.[8]

During the first ten years after the Act, 1,150 fatalities and another 12,000 serious quarrying accidents were reported for the country as a whole, roughly one death per 1,000 workers per annum.[9] In Cornwall, according to Peter Stanier, the safety record of granite quarries was good because the main operators were safety conscious. In the decade from 1882 to 1892 only two fatalities were reported among a granite workforce of 800 men,[10] only a fifth of the average death rate of the quarrying industry as a whole.

China clay accidents
The incidence of deaths through industrial accidents in the china-clay and stone industry in Devon and Cornwall was similarly low. Only 15 fatal accidents were reported in Cornwall in the *St Austell Star*, which gave extensive coverage to such matters, between 1890 and 1913. With an accident rate averaging one per year in a workforce rising from 4,000 towards 5,000, this only amounted to a quarter of the national figure for quarries.[11] At Lee Moor in Devon the first fatal accident in 50 years was recorded in 1911, when Walter Baskerville, aged 19, was suffocated by a fall of clay.[12] This was a remarkable record, given that the output of clay was then approaching 70,000 tons a year. To put this into perspective, the Cornish metal-mining disaster at West Wheal Owles in the 1890s resulted in 20 deaths, and 15 Cornish miners were killed in the year 1906 alone.[13]

Death from falling debris
As might be expected, the most dangerous place to work in china-clay pits was on or near the almost vertical sides, either removing the 'overburden' or topsoil, or breaking up the clay. Two-thirds of the reported fatalities occurred because of falling debris, and Sir Arthur Quiller-Couch referred to this hazard in one of his short stories. In this a clayworker pointed out a dangerous cornice of overhanging overburden to a clay merchant, who ignored his warning. Two days later 'nine or ten tons of that very cornice came down and crushed a couple of workers'.[14]

This possibly referred to 'the terrible deaths of William Henry Geach and Benjamin Andrews, a deaf and dumb man', reported at the Trethowel clayworks leased by Samuel Dyer, James Perry, F.J. Gaved and others. The two men were digging out clay 10–12 feet from the top of the pit when overburden fell on them, sweeping them down another 35–40 feet. Dyer and his son, Captain Thomas Dyer, were criticised for employing a deaf and dumb man in such a place and admitted that falls had already occurred at this spot and that the men were reluctant to work there.

The Coroner concluded that Samuel Dyer, a self-made man and Methodist lay preacher, who claimed that he had never had a fatal experience in 50 years of works management, would never have let men dig in that spot if he thought it was really dangerous, since he had allowed his son Thomas to work alongside the men. Captain Edward Truscott of Trenance clayworks, leased by the Varcoes, backed up Dyer, testifying that he himself would not have hesitated to work there. However, the jury decided that insufficient care had been taken to prevent falls of overburden, and Samuel and Thomas Dyer were each fined £5 plus costs, the maximum penalty being £20.[15]

Less than a year later, Thomas Dyer was involved in another fatality when

Frederick Charles Snell, who was working below him, and who had been injured in the previous accident, fell down the stope after being hit by a falling stone and died the following day.[16] Samuel Dyer himself was hurt in an accident some years later.[17] Deaths in Lovering pits included that of John Bray, who was buried alive when sand fell on him while washing down clay. W.T. and John Lovering attended the inquest but no blame was attached to them.[18] In other Lovering pits, John Truscott was buried in clay while working at night after heavy rains,[19] and Jacob Hancock died from 'poisoning' weeks after being hit by a piece of clay.[20]

In a Varcoe pit, Bernard Strongman, aged 19, was crushed to death by falling clay. Captain Tonkin, representing the clay merchant, was supported by Captain Lobb and other witnesses who said the location was not dangerous, but a representative of the Workers' Union and the youth's father, Joseph Strongman, maintained the side of the pit was 'too straight', that is, not sloped enough. A verdict of accidental death was recorded.[21] At Dr R.F. Stephens' clayworks, Richard Dawe was 'tramming' overburden on the top of a pit when he tipped his wagon over the edge and fell 17 feet.[22] Parkyn and Peters were sued by the widow of William Lean for £250 damages after he crashed to his death when the rope broke that was lowering him down the pit in a kibble, or bucket.[23] At the Cornish Kaolin Co. pit, Fred Hearn was seriously injured when a stone fell on his head from a wheelbarrow being moved above him.[24] In a rare accident in a china-stone quarry, Arthur Richards suffered an injury to his skull when he fell from a ledge at a West of England pit.[25] A similarly rare ball-clay case occurred when a ten-hundredweight lump of clay fell on Charles Gibbins, a clay labourer, causing fatal injuries.[26]

Workers' compensation for accidents

Until the passing of the Workmen's Compensation Act of 1897, employees had to prove that their employers were responsible for an accident. This sometimes proved difficult, and even after the Act came into force it could take a year to reach a settlement. Claymen who survived accidents were sometimes awarded compensation if they could no longer work, or were employed on less arduous tasks at reduced wages. At St Austell County Court William Varcoe, a clay labourer, was awarded 12s. (60p) a week, or about half average earnings, as compensation from Wheal Anna Clay Mining Co. after falling eight feet into a linhay, about 1,000 yards from the clay pit. The company appealed to the High Court of the House of Lords, who overturned the award because the spot where the accident occurred was not covered by the Quarries Act.[27] In another case, William Bartlett of the Great Halviggan clay company claimed for damages after the loss of an eye in an accident. Formerly earning £1.25 a week on a contract to remove overburden, he was then only making £1.05 weekly as a labourer, including overtime. Captain Richard Payne, of Carne Stents works, testified that he only employed able-bodied men for contract work, while Captain James Grigg, for 25 years captain at Great Halviggan, said he would not take on one-eyed men because of the strictness of the law about accidents.[28] Nevertheless, Bartlett failed in his claim.

Accidents involving machinery and fencing

Most machinery was unguarded, and accidents might be caused by carelessness, fatigue or over-familiarity with the operation. At Singlerose Clayworks the engine man was in the habit of taking a short-cut under an angle-bob between strokes of the engine and, misjudging his time, broke both legs. A lad was killed while repairing a machine belt at one of Stocker's pits,[29] and H. Ruse was injured by a flywheel when trying to adjust a pumping engine at Goonamarris clayworks while the machinery was still working.[30] Ernest John Peake fell off a scaffold and broke his arm[31] and Charles Lake, a carpenter's labourer, died from a haemorrhage while constructing a drying kiln at Par.[32] Another engine man at a pit near Greensplatt was luckier: he used to step through the spokes of the large outdoor flywheel when it was in motion to entertain onlookers and was never injured.[33]

Despite a Quarry (Fencing) Act enforced by local authorities, tourists wandered about unfenced pits in the clay district to see what was happening, and local people took short cuts. The body was found of a farm labourer who had fallen into a pit while on his way home through a clayworks.[34] Samuel Morley Trudgian, aged 14, of Caudledown clayworks, fell ten feet to his death while on an errand to another works leased by John Lovering. At the inquest the inspector expressed a hope that clay producers would pay more attention to paths.[35]

Although ball-clay work seemed relatively free of serious accidents, operating clay barges from Newton Abbot to Teignmouth owned by Watts, Blake, Bearne & Co. appeared to be marginally more dangerous. James Tibbs strained himself trying to turn round a barge and died later of heart failure caused by the incident,[36] and John Veale drowned after falling headlong into the waters of Teignmouth harbour when the pole he used to turn the barge snapped.[37]

Occasional accidents also occurred among labourers loading china clay and stone onto ships. George Tippett was crushed to death by casks in the hold of a clay ship owned by Bennetts of Truro,[38] and a similar incident may have inspired a rather cruel comic poem by Quiller-Couch, written to raise funds for the cottage hospital at Fowey, about a labourer who fell into a ship and fractured a bone, the first verse of which went:

'*From dock to hold he bounced,*
And there in anguish lay,
Unpleasantly surrounded
By tons of china-clay.'[39]

China clay and river pollution

While industrial accidents aroused public interest and were widely reported in the local press, river pollution was at the time mainly a bone of contention between clay merchants or mine owners and other industrial and commercial users of the waterways. A River Pollution Prevention Act was administered by local Sanitary Authorities and, as Le Neve Foster explained, 'the miner often incurs the wrath of the fisherman, who stirs up the Sanitary Authority or River Conservancy Board to action'.[40] Silting at Pentewan was discussed in earlier chapters, and the sand dunes south of Par and Carlyon Bay offer other examples of the enormous quantity of material that flowed down from clay works and created new beaches, later to be regarded as valuable amenities. As more material flowed downriver from clay-works, often preceded by mining waste of earlier times, other entrepreneurs working downstream of china-clay production began to complain about the pollution of their water.

As early as 1854, Creed and St Just in Roseland Turnpike Trustees had to make a new cut in the river at Tregony because of silting from St Austell and St Stephen clayworks. During the later 1860s disputes arose between the landowner Carlyon and various clay merchants, such as the Martins and Loverings and J. Olver, about pollution of the St Austell River, and by 1894 James W. Brown of Bristol estimated that every day 900 tons of 'clay' were carried downstream,[41] although it is possible that much of this was actually fine sand and mica rather than saleable clay. Sand from Carclaze pit was washed through a deep adit to Par Moor, where some

West Bridge at St Austell, with clay residues making a 'white river'.

was used for concrete block making. The remainder went down an old mine shaft and through an adit to help form the beach at Carlyon Bay. Concrete blocks were also made at Pentewan from sand washed down the St Austell river to the beach.

However, in 1908 St Austell Rural District Council investigated a complaint from a Pentewan firm about sand running down the river from Carne Stents China Clay Co. Shortly afterwards a clay storage tank at these works broke, releasing 700–800 tons of materials downstream,[42] and disputes continued until 1911.[43] In the meantime, a lake of water that the Lansalson clayworks kept for summer use as a water source burst its banks, and hundreds of tons of stone and sand swept along the road to the Wheal Martyn clayworks, where a cyclist had to pedal furiously to keep ahead of the torrent.[44]

According to Matthew Dunn of Mevagissey, the high content of clay in the sea drove pilchards away, and he blamed this for the demise of seine fishing vessels, which numbered 44 in 1818, but had disappeared completely by the 1870s. Many horse troughs were provided on the routes followed by wagoners, some financed by charitable organisations such as the Metropolitan Drinking Fountain and Cattle Trough Association, but Dunn also claimed that horses were so used to drinking 'white water' that they refused clear water and wagoners had to 'muddle' it with a handful of clay before they would drink.[45]

Oyster breeders on the Fal River below Tregothnan, where the Ruan River from the clay district joined it, claimed that the clay was ruining their breeding beds, and one producer said he had lost 3,000,000 oysters in this way. When reporting this claim, the editor of the *St Austell Star*, aware that public concern about the adulteration of flour with china clay still persisted, assured his readers that 'Fortunately people need not fear china clay. It may even be eaten, in limited quantities, without anxiety'.[46] The Cornish guide-writer and novelist J. Henry Harris went further, suggesting that china clay was good for the claymen: 'the amount which they absorb with their pasties must be fatal to microbes'.[47]

Protests were not confined to Cornwall, and a successful claim for compensation was brought against Captain C.R.H. Selleck, the clay producer on Dartmoor, by the owner of a water mill below his works. The opening of another clayworks on Duchy of Cornwall land on Dartmoor in 1906 was delayed by objections from the owner of a paper mill downstream at Ivybridge, who also asserted that, apart from polluting his water supply, the clayworks would also poach his labour force, although this seemed unlikely, since most of them were women.[48]

Transport congestion

Of more concern to the general public than water pollution was the enormous rise in china-clay traffic, both on the country roads and in the thoroughfares of St Austell. Pipelines, mentioned in Chapter Fourteen, were one answer to this problem, and clay merchants also recognised their value in enhancing the image of 'purity' of their product. Parkyn & Peters publicised the dispatch of their clay by pipeline, untouched by human hand, directly to their own drying kilns located at railway sidings, where it was loaded onto their own railway wagons.[49]

However, despite the improvements made in local networks of railways and pipelines, the output of china clay was growing at such a pace that large quantities were still being carried along lanes that were not capable of dealing with such a heavy traffic. From the 1890s the newly formed County Council was responsible for main roads and the new District Councils for secondary roads. We have seen how the local Council on Dartmoor made Selleck repair roads that his clay wagons had ruined. When steam-powered traction engines were introduced in 1881,[50] pulling heavy trucks along the roads around St Austell, they 'cut up the roads in a terrible way, making them impassable'. St Austell Parish Councillors complained about the 'great inconvenience, annoyance and danger' caused to the public and an increase in the cost of maintaining roads of £600 a year, and by-laws were introduced to increase the width of wagon tires[51] so that they did not sink so deeply into the roads.

These problems were not confined to Cornwall, and in 1901 Devon County Council proposed to compel carts exceeding 21 hundredweight capacity to use tires not less than four inches wide. Since ball-clay wagons carried at least $21\frac{1}{2}$ hundredweight on three-inch tires, this was vigorously and successfully opposed by Watts, Blake, Bearne & Co. in South Devon.[52] In North Devon and Dorset the main clay merchants built their own railways, so road congestion did not seem a problem.

Apart from the damage that the wagons did to the thoroughfares, they were also a menace to other road users. One Cornish observer described how, in St Austell, people dodged the clay wagons 'in the narrow crooked labyrinths called streets. Everyone gives way to the clay teams – butcher-boys and motor drivers screw themselves into nothingness, or back down side-streets when the clayman is in view'.[53] An exception was Charlestown Road, a four-laned horse-and cart-way, but the sharp corner of Bodmin Road into Truro Road, known as Lovering's corner, was a danger spot, and a wagon crashed into one of the rare motor cars seen on the streets in those days, damaging its horn, lamp and radiator.[54]

Wagons, each pulled by two or three horses, proceeded in pairs through Fore Street, the main thoroughfare, as far as the White Hart Hotel, where they doubled up to four or six horses before climbing the steep incline of East Hill. 'The drivers were always distinctly audible,' recalled Philip Varcoe, whose office was nearby, 'and they would make a run at the hill, with cracking whips and shouts to the horses, "Come 'ere, Dell", or "Tiger" or "Lion".' Loaded wagons ploughing through congested, narrow and steep streets were a menace. An elderly relative of

Team of horses leaving a dry at Nanpean with china clay for shipment.

Philip Varcoe told of how she was 'deprived of my skirt and petticoats by the wheel of a passing clay wagon'.[55]

In 1908 stringent regulations were applied to reduce the chaos that reigned, with eight to ten wagons drawn up awaiting their turn to ascend in the commercial heart of St Austell, where four banks were situated.[56] 'The poor driver of a clay wagon is driven from pillar to post,' complained one clayman, 'and looked on in the town almost as an undesirable.' There was at that time no alternative road, however, and in 1909 St Austell Council brought pressure to bear on the Great Western Railway to build branch lines to take the traffic off the roads, which eventually the GWR did.[57] Even so, in 1910, according to the clay merchant and attorney Higman, there was a growing tendency for boys to be employed to drive the wagons. 'St Austell clay wagons travelled mostly on the wrong side of the road,' he said. After one accident caused by a lad of 18, the judge declared the use of such youths to be a public danger, and called for some check to be put on it.[58]

Environmentalists and romanticisers

Although, like other extractive activities, the clay industry radically changed the landscape, environmental campaigners in Mid Cornwall and Dartmoor were at first more concerned about tourist growth than china-clay expansion. A rash of buildings began to cover stretches of the coast and moorland with hotels, villas and terraced boarding-houses. The Cornish literati, led by Quiller-Couch, the maritime historian Arthur Norway and the best-selling novelist Silas Hocking, warned that soon every field with a sea view would be littered with 'monster hotels' and 'jerry-built' speculative housing.[59] Quiller-Couch likewise expressed, in sentimental terms, his indignation at the physical and social effect of clay working on Bodmin Moor: 'The stream was polluted,' he wrote, 'the fish died, the fairies were evicted from their rings beneath the oak, morals underwent change.'[60]

Dartmoor, like Cornwall, had been scarred by metalliferous mining for centuries, and the mid-1800s were a period of relative prosperity, with about 50 mines dotted over the moors. By 1900 only half a dozen remained, but photographs of the period show extensive workings, with water wheels, engine houses, stamping machinery and other equipment on Duchy of Cornwall land.[61] Perhaps because of its ancient traditions, metal mining escaped the attacks of early environmentalists, but clay producers were not so spared.

The prospect of river pollution caused by a new clayworks at Redlake on Dartmoor, alarmed hoteliers and parish councillors, who feared it would damage the area's image as a place for the solitary traveller wishing to escape the congestion and pollution of the big towns. Owners of country houses and private hotels were promoting the neighbourhood as a haven for well-to-do enthusiasts of river fishing as well as nature lovers, and their opposition to the clayworks was supported by the Dartmoor Preservation Society, who complained that it would ruin the flora and fauna and peaceful ambience of the area. Delays caused by these objections seriously stretched the finances of the adventurers.[62]

On the other hand, as discussed in the previous chapter, in the years before the 1914 War a few writers and artists, including Anne Treneer and Laura Knight, were already viewing the man-made clay landscape in a romantic light. Treneer, who was educated at a boarding-school in St Austell and who practised as a teacher in the clay district, recalled in later years its 'glittering pyramids... seeming to sail between land and air' above 'deep pools of turquoise and white streams'. Knight painted two clay pit watercolours; one, in 1912, showing sand removal and washing in the lower part of a clay pit, and another, in 1914, showing the removal of overburden from the top of a clay pit.[63] The guide writer J. Henry Harris enthused about the 'simply delicious' walk with 'charms everywhere', to Roche, 'one of the portals to the land of the white men – a wonderful land'.[64] However, even in 1932, when guide-writers and artists were vying with each other to romanticise the clay landscape, the novelist Joseph Hocking, brother of Silas, only saw 'dreary moors', punctuated by 'huge heaps of debris from the clay works'.[65]

The ball-clay district of Dorset also possessed some desolate moors, but few people were disturbed by clay working on this poor and thinly inhabited area apart perhaps from furze gatherers. One loss, however, was the destruction of a medieval mill north of Corfe Castle.[66] This solitary heath was part of Thomas Hardy country, but the author did not seem to be troubled by the invasion of clay merchants, nor did he try to romanticise their workings. It was left to later writers to admire abandoned pits, transformed into 'remote and lovely' lakes with 'wide sheets of water smothered with the colour of water lilies in July and August'.[67]

One of these lakes is worthy of mention because of its connection with the romanticism of Cornish clay pits by Laura Knight. Around 1846 a pit some 50 feet deep and three acres in extent had been dug by Watts, Hatherley & Burn of Devon and later acquired by Walter Pike of Dorset. When abandoned it filled with water, and the high concentration of minerals and the absence of oxygen kept the 'Blue Pool', as it was called, clear of animal and vegetable life and it was transformed into a popular beauty spot by a later owner, with a tea house, souvenir shop, ball-clay museum and woodland walks. Before this, however, in the early 1900s, Augustus John painted it. He had earlier stayed in Lamorna and had possibly been influenced by the groups' romantic paintings of clay pits and disused granite quarries.[68]

To conclude, in the period up to the 1914 War, although some consequences of the rapid growth of the clay industry were beginning to affect local communities, they did not weigh as heavily upon the public consciousness as they were to do in the twentieth century. Health and safety aspects were not treated as a serious problem: china-clay work, and some ball-clay work, was healthier than many other mining and industrial activities, and the incidence of fatalities and serious accidents was comparatively low. On the other hand, traffic congestion was causing mounting problems in Mid Cornwall, Dartmoor and South Devon, and the impact of clay wagons on rural roads was an increasing inconvenience to farmers.

Polluted rivers were also a matter of concern to fishermen and to downstream users of water, while silting ruined Pentewan as a harbour. On the other hand, there were some gainers as well as losers from silting. In Devon, operators of 'sand barges', as they were called, made a living by loading free of charge waste material that had been washed down the River Plym from the Lee Moor clayworks and taking it to Sutton Harbour at Plymouth, where they sold it as builders' sand. Already, by the 1890s, according to Robert Burnard, partner in a nearby chemical factory, from 50,000 to a 100,000 tons had been recovered in this way.[69]

The visual impact and the instability of the china-clay tips were later to lead clay producers to 'green' the white 'Cornish alps'. But in Edwardian times neither environmental issues nor romantic notions troubled many clay-country communities. They were only too aware that, with metal mining and granite quarrying in decline, they depended upon the industry for their very existence. If 'burras' or 'sky-tips', as the Cornish called them, rose higher and higher above them, their attitude was 'As long as the burras keep growing, boy, we'll eat!'.[70]

References
[1] J.H. Collins, 'The china-clay industry of Cornwall', *MM*, 1911, p.455.
[2] Terry Wyke, 'Mule Spinners' Cancer', in Alan Fowler and Terry Wyke (eds), *The Barefoot Aristocrats*, Littleborough,

1987, p.184.

[3] Catherine Lorigan, *Delabole*, Reading, 2007, p.83.

[4] Ball Clay Heritage Society, *The Ball Clays of Devon and Dorset*, St Austell, 2003, pp.24–27.

[5] Lorigan, 2007, pp.89–90.

[6] According to the Rector of Ivybridge. *WMN* 24 October 1911.

[7] Sir Clement Le Neve Foster, *A Treatise on Ore and Stone Mining*, London, 1910, p.735.

[8] *SAS* 27 April 1894.

[9] Health and Safety Executive, *A Hundred Years of Law*, 1995, p.5. An accident was classified as fatal if death occurred within a year and a day as a result of it.

[10] Peter Stanier, *South West Granite*, St Austell, 1999, p.173.

[11] Le Neve Foster, London, 1910, p.735.

[12] *SAS* 21 September 1911.

[13] Her Majesty's Commissioners, *Health and Safety in Mines*, HMSO, 1864, pp.1–11; *Accidents in Mines*, HMSO, 1886, p.6; Cyril Noall, *Cornish Mine Disasters*, Redruth, 1989; Peter Joseph, 'Mining Accidents in the St Just District', *JTS*, 1999. Between 1874 and 1883 the Cornish mining death rate had been a third below the British colliery average, by 1910 it was double the national rate. D.B. Barton, *A History of Tin Mining and Smelting in Cornwall*, Exeter, 1989, pp.183, 244.

[14] Sir Arthur Quiller-Couch, 'Step o'One Side', in *Q's Mystery Stories*, London, 1937, pp.325–41.

[15] *SAS* 21 November 1900, 10 January 1901; *BCW*, IX, December 1900, p.354.

[16] *SAS* 24 October 1901.

[17] *SAS* 14 February 1907.

[18] *SAS* 28 May, 4 Jun 1903, *BCW*, XII, June 1903, p.106.

[19] *SAS* 5, 12 September 1907.

[20] *SAS* 29 September 1910.

[21] *SAS* 4 July 1912.

[22] *SAS* 6 July 1906.

[23] *SAS* 12 December 1890.

[24] *NE* 23 September 1913.

[25] *SAS* 23 July 1908.

[26] *BCW*, September 1913.

[27] *SAS* 21 May 1908; *BCW*, 1907, p. 241.

[28] *SAS* 23 May 1912.

[29] *SAS* 20 September 1889.

[30] *BCW*, 1907, p.240.

[31] *NE* 23 September 1913.

[32] *SAS* 24 July 1913.

[33] We are indebted to John Tonkin for these examples.

[34] *SAS* 13 April 1911.

[35] *SAS* 18 August 1905.

[36] *BCW*, December 1909, p.244.

[37] *MDA*, 20 March 1912.

[38] SAS 18 July 1907.

[39] Quoted in A.L. Rowse, *Quiller-Couch*, London, 1988, p.65.

[40] Le Neve Foster, 1910, p.700.

[41] *BCW*, III, December 1894, p.201.

[42] *SAS* 29 October, 10 December 1908.

[43] CRO, DDG/1657 of 1854–45, DDCF/3951 of 1865, DDCN/ 2657/8 of 1867 and 1869, 2678 of 1911.

[44] *BCW*, March 1910, p.318.

[45] *MJ*, 9 April 1881, p.425.

[46] *SAS* 14 January 1897.

[47] J. Henry Harris, *Cornish Saints and Sinners*, London, Second edition, 1910, p.249.

[48] E.A. Wade, *The Redlake Tramway and China Clay Works*, Truro, 1982, pp.20–22.

[49] *Kelly's Directory of Cornwall*, London, 1914. The GWR constructed sidings for several clay companies, including Trewheela at Melangoose Mill, West Goonbarrow at Bugle, Treskilling at Luxulyan and Wheal Remfry near Meledor.

[50] *RCG* 9 September 1881.

[51] *RCG* 12 October 1881.

[52] L.T.C. Rolt, *The Potters' Field*, Newton Abbot, 1974, pp.140–41.

[53] Harris, 1906, p.249.

[54] *BCW*, November 1909, p.225.

[55] Philip Varcoe, *China Clay: the Early Years*, St Austell, 1978, p.38.

[56] *CG* 10 July 1908.

[57] *SAS* 11 February, 25 November 1909, 11 August 1910.

[58] *BCW*, August 1910, p.123.

[59] Silas Hocking, Address to the Midland Cornish Association, *RCG* 24 February 1898. For further discussion of this topic, see Ronald Perry, 'Silvanus Trevail and the development of modern tourism in Cornwall', *JRIC*, 1999, pp.33–43.

[60] Sir Arthur Quiller-Couch, *Cornish Stories*, London, 1944, p.111.

[61] Tom Greeves, *Tin Mines and Miners of Dartmoor*, Newton Abbot, 1986.

[62] Wade, 1982, pp.20–22.

[63] Anne Treneer, *Cornish Years*, London, 1949, pp.15, 100; Dame Laura Knight paintings, 'The Clay Pit', 1912, and 'The China Clay Pit', 1914, both exhibited at Penlee House, Penzance; William J. Claxton, *In the Potteries*, London, 1913, p.13. (See Chapter Twenty.)

[64] Harris, 1906, p.242.

[65] Joseph Hocking, *Caleb's Conquest*, London, 1932, p.5.

[66] Rodney Legg, *Purbeck's Heath*, Sherborne, 1987, p.94.

[67] Legg, 1987, p.82.

[68] Michael Holroyd, *Augustus John*, London, 1974, II, p.38. His painting 'The Blue Pool' hangs in the Aberdeen Art Gallery.

[69] Crispin Gill, *Plymouth River*, Tiverton, 1997, p.130.

[70] Private communication from John Tonkin.

Chapter Sixteen
Industrial Relations 1860–1912

In Chapter Ten we described how popular unrest in the West Country up to the mid-1800s largely took the form of rioting about a lack of food, directed against merchants and farmers who were suspected of hoarding grain or diverting it to other regions where they could sell it at a higher price. In the second half of the century, discontent in the industrial areas of Britain increasingly involved disputes between employers and employees about rates of pay, hours of work and the right to form trade unions. Strikes by workers and lockouts by employers did not occur evenly throughout the period, but concentrated in clusters of activity in the mid-1860s, the early 1870s, around 1889–91 and from 1910 to 1913. The earlier disputes involved skilled men, but later lesser skilled workers such as railwaymen and dockers became more militant.

This chapter explores the extent to which these trends affected relations between employers and employees in the clay districts of the South West. An absence of any primary or secondary sources of evidence of strikes or lockouts in the Dorset clay areas suggests that any disputes that might have taken place there were localised and settled on the spot. The Dorset clay workforce was fragmented and isolated, and few other groups of workers could have provided a nucleus for industrial action. Wage levels for Dorset labourers were low. In the Swanage area near a ball clay district over 90 stone quarries were worked in 1877, but by small groups of owner-occupiers.[1]

In Devon and Cornwall the situation was different. Relatively large concentrations of clay workers existed around Newton Abbot and in the St Austell area, and the claymen sometimes lived cheek-by-jowl with militants in other industries. The main unskilled labour organisations that moved into these groups were the Gasworkers' Union and the General Workers' Union. This chapter will discuss industrial relations from the 1860s until the early 1900s, and the major clay strikes of 1913 will be the subject of the following chapter.[2]

Unrest in the 1860s
The 1860s were desperate times in some related industries such as brickworks, where blacklegs were beaten up, bricks were destroyed and a policeman was shot dead in Ashton-under-Lyme. Bombs were thrown into brickmasters' homes, machinery was broken and engine houses were blown up in the Manchester and Sheffield areas and at Bridgewater in Somerset in 1864.[3] The decade was also a

turbulent period for the building trades of Devon, including Plymouth and Newton Abbot in the south and Torrington in the north. The main unrest centred on Exeter, where carpenters and joiners, plasterers and plumbers formed unions but found that their efforts to improve pay and reduce hours of work were undermined by employers bringing in non-union labourers who accepted lower wages. When the local tradesmen reacted, sometimes violently, non-union men took them to court on charges of assault and intimidation, and magistrates dealt severely with the unionists, supported by establishment figures such as the Dean of Exeter.[4]

Unrest also grew in the metal-mining industry, a Parliamentary Report of 1864 having raised awareness of the miners' poor pay, ill health and bad working conditions. A dispute occurred in Devon Great Consols Mine at Tavistock in 1866, while in Cornwall lay preachers and teachers organised meetings of miners. These led to the creation of the Miners' Mutual Benefit Association, run by Arthur Bray and Henry Cliff, both Nonconformist lay preachers, financial donations being handled by two local West Cornwall Anglican vicars, the Revd R. Prettyman Berkeley and the Revd Wadislaw Lach-Szryma.[5] In 1866 the mining employers presented a united front in dismissing workers who joined the Association because, although ostensibly organised as an accident and pension scheme, its rules included provision for strike pay at £0.60 a week plus £0.05 per child. The workers retaliated by calling a strike, troops were brought in and the employers declared a lock-out. The dispute soon ended, however, because the Association had not been in existence long enough to build up a reserve, and the only support it had came from quarrymen in the Stone Cutters' and Excavators' Society.

These events coincided with the 'copper crash' of the 1860s, a financial panic that transformed a West Country copper mining recession into a catastrophe, shown by a graph in Chapter Nine. Employers held the whip hand in the labour market and dealt with workers on a mine-by-mine basis, determined to nip any attempt at collective action in the bud. When, therefore, in 1870 the Franco-Prussian War cut china-clay producers off from important continental customers, and clay merchants laid off workers, the claymen had no option but to accept their fate.[6]

The 1875 Mid Cornwall clay strike: victory for the workers

Soon, though, postwar recovery in Europe, coupled with an American railway boom, created an international upturn that swung the balance of power towards most workers, setting off a wave of strikes throughout Britain. Disputes spread through Devon among navvies building railways at Okehampton, coal 'lumpers' at Torquay and Sidmouth, shipwrights at Brixham and Salcombe and foundry workers at Exeter.[7] Strike fever did not, however, seem to affect the ball-clay workers.

When it reached Cornwall the tin miners, whose industry had touched its historical peak of production, downed tools from St Just in the far west to the Devon border. Mining adventurers, anxious to keep output up, quickly agreed to wage rises averaging 25 per cent, and where miners led, others followed, including engineers, builders, tradesmen and tailors. In Mid Cornwall, demand for labour was enhanced by railway building, including an attempt by outside speculators to make Fowey a great iron-exporting harbour, discussed in Chapter Thirteen.[8]

In 1872, with the state of labour demand and supply temporarily favouring workers, 200 men in Thomas Stocker's West of England clayworks struck for more pay and were swiftly rewarded with increases of 25–50 per cent. Men's wages were increased from 12s. (60p) for a basic week to between 15s. and 18s. (75p to 90p). Youths received 11s. (55p), women 5s. (25p) and children 3–4s. (15–20p). Young and fit 'burtheners' and 'sand men', who removed the topsoil and waste materials, earned large sums, more than some skilled tradesmen or underground miners.[9] However, the good times did not last long, and were followed by the deepest and longest depression of the Victorian age. The tin price slid down and mining employers cut surface workers' wages by 12 per cent, although they did not touch underground wages because they were short of skilled men.

By the summer of 1875 china-clay prices had dropped by 20–25 per cent and the clay merchants decided to take a leaf out of the mining adventurers' book and announced a cut of 10 per cent in the basic wage of 2s.6d. (12$\frac{1}{2}$p) a day, effective from 1 July. On Friday, 2 July the men at the Martyn's works downed tools and marched from pit to pit gathering support. That evening they held a mass meeting in St Austell and the following morning they put out the fires in drying kilns, took away the horses from the clay wagons and tried to stop the shipment of clay from Charlestown harbour. Colonel Walter Raleigh Gilbert, the Chief Constable, ex-Royal Artillery officer and member of a well-known local family from Menheniot, called out 50 men to quell any possible riot.[10]

However, the leader of the strikers, the clay labourer Joseph Matthews, was no firebrand and persuaded the men to listen to the advice of local dignitaries, including the County Magistrate, Robert Gould Lakes, respected by clay merchants and men alike,[11] who exhorted them to keep the peace, while Bale, one of the clay merchants, reported amid cheers that he and some other employers would not be cutting wages. At a meeting at Bugle the strikers, on receiving written assurances from some clay merchants that their wages were not be reduced, passed a resolution that all those who received their full wages should return to work. This shrewd move sowed dissension among employers, since those who had not cut wages would be able to pick up orders from those who had, for, as was sometimes the case in the china-clay industry, while some firms were running at a loss, certain of the larger firms were in full swing, with their men working overtime and averaging 21s. (£1.05) a week. Some pits in the Roche area were also paying 3s. (15p) a week higher than the basic wage. This anomaly is further discussed in Chapter Seventeen.

A deputation of strikers met a committee of clay merchants chaired by Thomas Martin and including Edward Martin, John Lovering, Higman and Truscott. Martin explained that large stocks of clay were piling up at Runcorn because potters and cotton manufacturers were not buying, and that no more clay was needed for six months. He offered a choice: work a nine-hour instead of an eight-hour day for the same money or be laid off. However, the strikers took what they saw as hard-luck stories with a pinch of salt, for while they could not see the mountains of unsold clay said to exist in Cheshire, they could certainly witness 'best clays' still being shipped out of the Cornish clay ports. These two occurrences were not mutually incompatible, since inferior grades of clay could only be sold at a price of 60–75p a ton on board ship, representing a loss of 10–15p a ton, whereas finer qualities might still fetch £1.50p a ton.

Even if the employers' statements were true, many men believed that the threat of foreign competition would force the employers to avoid a strike by giving in to their demands, and so they decided to call their bluff and refused to return to work. For a few days the pits were idle, but without any serious disturbance, intimidation or damage to property. Then the clay merchants capitulated, restoring wages to their previous level and promising to allow men to make up their loss of pay by working overtime.[12] Impetuous clay employers like Thomas Stocker (father of Medland Stocker), it would appear, had overstepped the mark, and misjudged the attitude of their more cautious fellows who were not ready to join in an all-out confrontation.

The 1876 strike: the employers turn the tables

Encouraged by this bloodless triumph, Matthews campaigned for the formation of a workers' combination, like those being set up in many other parts of Britain, to combat the power of the employers' federation. It was called the St Austell China Clay, China Stone and Surface Labourers' Friendly Union. The clay merchants had no objection to the 'Friendly' part of the union, since it made sickness and accident payments to members. This reduced the burden on the poor rates and, as the employers were major contributors to the rates, it saved them money. On the other hand, the Union Rule Book also made provision for strike pay and an official 'Black List' of non-union claymen. To men like Thomas Stocker the Friendly Union seemed a very unfriendly 'Strike Club'.

This change in workers' tactics from a simple protest against a pay cut to an attempt to form a union proved to be their undoing. On the one hand, leading clay merchants were still anxious to avoid trouble by steering clear of wage reductions, and when Thomas Stocker announced his intention to reduce pay, and the union threatened to strike, he had to back down because of lack of support from fellow employers. On the other hand, Stocker had their approval in stifling the growth of unionism. Matthews, the union leader, was sacked, and Stocker and some other employers began to weed out any men known to be union members. The union then gave notice of a strike if the men were not reinstated, but employers continued to dismiss union men and publicly stated that a strike against any individual clay merchant would be followed by a lockout of the entire industry.[13]

The stage was set for a trial of strength, and at the end of November 1876 the *London Times* recorded that: 'The whole of the men in the china clay district of St Stephen and St Dennis are now out on strike, about 2,000 men altogether... the struggle is really on the part of the men for the existence of their Union.'[14] The *London Times* had reported the 1875 strike in terms that were sympathetic to the workers' case, but the battleground had now shifted. From a strike for more pay it was now a fight for union recognition, and middle-class opinion, in Cornwall as in Britain, was divided upon the merits of the case. The Liskeard-based *Cornish Times* claimed the strikers' behaviour 'equalled the trade union tyranny in some of the worst parts of England... dictating to the masters that they shall have no man in their works whose name does not appear on the roll of the union'.[15] The *London Times* was also much more critical of strikers who had 'broken the promises given to Mr John Tremayne MP and Mr Lakes, County Magistrate, that they would not commit any illegal act, and... forcibly driven the non-unionists from the various clayworks in which they were employed'.[16]

The workers' case

In contrast, the *Royal Cornwall Gazette* favoured the workers' case. 'The men seem to have commenced with the best possible intentions,' it commented, 'and the masters have only themselves to thank for the present condition of affairs.' If, the *Gazette* argued, the employers had not cut the 'already small pittance' they paid the workers, they would not have formed unions, and if there had been no union members to sack there would have been no strike.[17] The more radical *West Briton* also condemned 'the heartless manner in which the claymasters treated the men on strike', with an attitude of 'we will starve you out and break up the union'.[18] R.G. Lakes and John Tremayne informed a mass meeting of strikers that they had every right to form a union,[19] and the Revd Thomas J. Bennett, vicar of Treverbyn, spoke up on behalf of the strikers on many occasions. 'If claymasters are allowed to form an association,' he demanded, 'why not the men?'[20] In contrast to the orderly and peaceful nature of the 1875 strike, however, relations between masters and men soon turned violent. Exaggerated rumours, claims and counter-claims about violence and intimidation were the order of the day and, under the headline 'A Reign of Terror', the *Cornish Times* reported that 130 police had been drafted in from all parts of Cornwall to quell 'a lawless mob' armed with sticks.[21]

The *Western Morning News* described frequent acts of violence and intimidation by strikers against men who continued to work, in which one worker's collarbone was fractured and another was badly treated.[22] Even the *West Briton*, more sympathetic to the workers' case, reported in critical terms how Moses Kent, an 'engineman' of St Dennis Consols pit, who refused to join the union, was 'seized, beaten with sticks in spite of the efforts of the captain to rescue him, mounted on a stick and dragged for some hundreds of yards' before fainting. When he came to he was told he would be killed unless he joined the union, which he did. At Retew pit one worker was threatened with hanging and another with death by drowning.[23] Thomas Stocker complained that a stone had been thrown through the window of a chapel in which he was giving the evening sermon, and Higman declared that his works had been seriously damaged.[24]

Public support for the strikers waned as news of the turbulence spread, but by no means all of the intimidation and threats came from the workers' side. According to statements in the *Royal Cornwall Gazette* and the *West Briton*, one employer shouted, 'Bring the soldiers down from Plymouth and shoot the devils down', and clay captains called at the homes of union members, warning them they would never get another job unless they left the union.[25] Indeed, one correspondent blamed the employers' attitudes upon the clay captains, 'tyrannous despots who behaved like little kings and inflamed the minds of the masters against their men'.[26] This was hardly fair to the average clay captain, however, whose loyalties were divided. He came from the same families and communities as the men whom he supervised, worshipped at the same chapel, sometimes as a lay preacher, and joined in the same sporting and social activities, yet he was expected to be, and was, on the side of the employers.

Those members of the Cornish establishment who tried to retain a balance between the maintenance of law and order and the avoidance of aggravation in a tense social situation had to make difficult decisions. A force of 150 County Constabulary was assembled at St Austell. A military contingent under Major Lloyd was said to be on the move from Plymouth barracks, but Tremayne and Lakes and the Chief Constable, perhaps fearing a repeat of earlier 'Camborne Riots', where the intervention of 200 policemen only made things worse, sent back the reinforcements. In order to avoid provocation they delayed the arrest of two men involved in assaults on workers, Rundle and a well-known wrestler Hawkey, until three o'clock in the morning when they were in their homes. Gilbert personally conducted the strike leader, Matthews, to the police station, Matthews himself urging his men not to intervene. Matthews was sentenced to ten weeks' hard labour and the other two leaders were sentenced to six weeks' imprisonment and a fine of £50 respectively.

A call for firm action

The temperate and cautious nature of the authorities' reaction to worker unrest infuriated some clay merchants, who demanded swifter and tougher action. Writing from Plymouth, William Langdon Martin, who ran the Martin Bros clay-works at Lee Moor, accused the magistrates of 'fatuous imbecility' in allowing 'an unruly mob to run riot through the district for two days without check or interruption'. Later, when Lakes wrote a letter, which became public, asserting that he had 'not heard of any bad conduct' and that 'not a five shillings' worth of property had been damaged', Martin wrote to the Chairman of the Magistrates Board demanding Lakes' instant dismissal. Martin, as we saw in Chapter Eleven, was said to be a good employer at Lee Moor, and no Dartmoor men joined in the strike, but union officers asserted that he wanted to stir up trouble in the St Austell district so that the strike continued and his Lee Moor clayworks could pick up orders.[27]

This seems unlikely in the light of a letter he wrote to Lord Morley, owner of the mineral rights at Lee Moor. Condemning alleged violence against non-union men at St Austell, Martin claimed that 'the Masters have all combined... and will not take back any but those who do not belong to the Union. The magistrates behaved very badly and... it is most fortunate no lives were lost. We are quiet at Lee Moor, the men not having joined the Union'.[28] Perhaps they were less militant because the Martin brothers had set up a General Labourers' Benefit Society managed jointly by masters and men.[29]

The *Western Morning News* suggested the formation of a similar body for St Austell as a way out of the impasse, and Thomas Stocker offered £150, spread over five years, to such an organisation, provided it omitted strike clauses. Union officials derided his suggestion as 'Stocker's grandmotherly government', an obvious attempt to draw the teeth from the union, and spread the rumour that he had already bet his fellow employers £150 that he could end the strike before Christmas, thus immediately recouping his money.[30] However, after the dispute was over Stocker formed the Claymen's Friendly Society (without any strike clauses) with himself, the rector of St Stephens, Arthur Coode JP and the clay merchant Simon Truscott among the Trustees.[31]

The strikers run out of funds

The strike did not end at Christmas, and as it moved into its fifth week the plight of the strikers was growing more and more desperate. To add a little Christmas cheer the union paid three weeks' strike allowance in advance, although strike pay was only half of the absolute minimum earnings, and the Penny Bank, administered by the magistrate R.G. Lakes, also repaid small sums to its depositors.[32] Local donations from shopkeepers and other well-wishers were pitifully small at £20 – hardly surprising, since some clayworkers existed from hand to mouth at the best of times and had run up large debts with local traders, some of which were never paid back.[33]

In the early weeks of the strike, Matthews had made great play of a promise of £3,000 from trade unions 'up north', which would have kept the strike going for several weeks.[34] It was also rumoured that the Stone Cutters' and Excavators' Society, which helped Cornish miners in a strike of the 1860s, had promised £650 a week.[35] Yet it was six weeks before Burnard, a local union official, went off in search of aid only to return empty handed, after which J. Caleb Bullen, the Union Secretary, set off to see if he could do better. Halliday, President of the Amalgamated Association of Miners, came down from Manchester to promise funding,[36] although it was his union that had been hampered a few years earlier by Cornish miners brought in to break a coal miners' strike.

He seemed reluctant, however, to send money to strikers accused of riotous behaviour, and it was to reassure him of their good behaviour that Lakes, the magistrate, wrote the letter of support that so enraged Martin. In the end the Mineworkers sent £800, a generous gesture, although it barely covered a week's strike allowances. For a while 'no surrender' remained the watchword of the thousand or more men still on strike when they triumphantly brought back Edwin Hawkey from jail.[37] However, when Bullen returned from a trip to Scotland, Manchester, Bristol and Plymouth with declarations of solidarity from other unions but no cash, the remaining strikers went back to work on Monday, 22 January 1877.

Why did the strike fail?

What had the strikers achieved? They and their families had suffered great hardship for two months – and had run up debts that would take years to repay – without gaining any increase in wages. Indeed, some had lost their jobs, since employers refused to sack labourers they had taken on during the strike to make way for the strikers, and the *West Briton* said: 'Not a particle of blame attached to employers who cannot be expected to take on more men than they need'.[38]

Why had the strikers won in 1875 and lost in 1876? Bullen claimed the strike had failed because he had not approached other unions early enough for their help, but in the midst of a great depression other unions had their own problems, and Halliday's Amalgamated Association of Miners was soon embroiled in a struggle which led to its collapse. Bullen must also have been aware of the reputation of Cornishmen in some parts of northern England as strike-breakers, but mistrust between Cornish claymen and industrial workers in England was mutual. Decades later, stories still circulated about 'truck loads of golden sovereigns from the north' that were either promised but never came or which lined the pockets of local leaders.[39] Not only had Halliday 'come down to promise bread and gave us none', he had charged £20 for his visit, while Burnard and Bullen's fruitless trips had taken another £30 out of the slender strike funds.[40]

This lack of solidarity with unions outside Cornwall was paralleled by an absence of support from clayworkers in other parts of the West Country. The several hundred claymen on Bodmin Moor and Dartmoor did not come out in sympathy, nor did those in the small clayworks in West Cornwall. Indeed, the Penzance-based *Cornish Telegraph* only published the briefest of reports on it, albeit presenting it from the St Austell workers' point of view, mentioning their willingness to reach an agreement and their rejoicing when imprisoned strikers were released.[41]

A more serious reason for the collapse of the 1876 strike was that union officers had chosen the worst possible time to stop work. When they downed tools in the summer of 1875 the employers were not sure whether the general recession in trade would continue, while some workers could tend their smallholdings to produce extra food or find temporary employment on farms. By the winter of 1876, unsold clay was piling up in linhays and the workers had no second strings to their bows. If Stocker and others had been trying to manipulate events to force a strike at this moment, their tactics certainly paid off. Finally, whereas in 1875 some employers were willing to pay higher wages, in 1876 they were united in opposing a combination that might permanently tip the balance of negotiating power towards the men.

Localised action

During the generally depressed conditions of the later 1870s and the 1880s, significant and long-running disputes were rare in Britain. In South Devon, short strikes were scattered and spasmodic, involving Devon Great Consols miners and Brixham trawlermen in 1878, Newton Abbot tanners a year later and Torquay coal lumpers in 1885.[42] In Cornwall, claymens' memories of their humiliations cast a long shadow over their readiness to strike in a period of labour history dominated by the run down of metal mining. By 1880 the output of copper ore had fallen to only a quarter of its level of 15 years earlier, and while the value of tin sales increased, tin production did not rise much and so few new jobs were created for men thrown out of work from copper mines. China-clay workers were in a weak bargaining position and, according to the *Mining Journal*, were 'only men of the common labouring class' who were already earning a reasonable wage. They received 2s.6d. (12½p) for a 7½ hour day, while those on piece work could average 3s.0d.–3s.6d. (15–17½p) a day, and with rents of £3–£4 a year and a wife and children also employed, the *Mining Journal* argued that they could 'rank in the social scale of skilled artisans in many other parts of England'.[43] However, as noted in Chapter Fourteen, opportunities for females to work in clay pits had fallen greatly by this time.

Despite the collapse of copper mining, for a time the miners' tradition of popular unrest continued, and as late as 1885 they won a famous victory by electing the radical Conybeare for the Parliamentary Division of Camborne, Redruth and Hayle in opposition to the official Liberal candidate in what was supposed to be a safe Liberal seat.[44] This, though, seemed to mark the last gasp of mining militancy in West Cornwall. In Devon, ball-clay employers also cut wages during the continuing trade depression of the later 1870s without violent resistance. C.D. Blake, partner in the important Watts, Blake, Bearne & Co., when informing a potter in 1879 that they were reducing wages, remarked that half the Cornish china-clay producers were losing money.[45]

Militancy spreads from the east

During the 1880s a Devon farmer had complained that 'agitators' were infiltrating his hitherto peaceful area and stirring up discontent. Trade-union organisation was spreading to rural areas, and to less skilled workers, encouraged by the successful action of 'match girls' and dockers in London in 1889. A year later Thomas Martin, who was running Lee Moor, informed his landowner, Lord Morley, that 'the agitators' had visited Lee Moor and other Devonshire clayworks, but that no industrial action had taken place.[46] Martin may have been referring to the Dockers' Union, who tried unsuccessfully to impose a closed shop at Devonport docks in 1890, but it is more likely that he was concerned with the activities of the Gas and General Workers' Union, who were recruiting in the region. In 1891 their local leader, Jack Gardiner, led a demand for union recognition by textile mill workers at Buckfastleigh. The employer, John Berry, threatened to sack all those who joined the union and brought in 70–80 non-union workers to replace them. After a lockout lasting 24 weeks from December 1891 to May 1892, which cost the union £1,690 in strike pay (50p a week for men, 25p for women), Gardiner had to call off the strike and the employer only took back the workers as vacancies occurred among the replacements. Another strike that he organised was equally unsuccessful when workers at the Bovey Pottery asked for union recognition but again the employer recruited replacements from Staffordshire.

These failures illustrated how unions could only succeed if they were tightly organised and chose the right time to strike. The Association of Carpenters and Joiners, for instance, struck in Exeter on 31 May 1890 and continued until the end of July, a peak period for building employers, and took the precaution of posting pickets at railway stations to intercept non-union men. Some union men were fined £2 plus £1.05 costs (about 3 weeks' wages) for intimidation, but they gained half the pay increase they had demanded. Other strikes occurred among Exeter bricklayers in 1891, and Barnstaple carpenters and joiners and Newton Abbot railwaymen in 1893. The results of these strikes are not clear, but they were probably at least partially successful because their unions were well organised.[47]

Wages disputed at Newton Abbot

Perhaps it is not surprising that ball-clay workers were affected by the surrounding militancy. In May 1891 the ball claymen of Watts, Blake, Bearne & Co. presented a petition for wage rises and a few days later a large number gave a week's notice to stop work. The employers, while not accepting that the men were underpaid, made what they called 'satisfactory' proposals on pay. These included 2s.8d. a day for those on fixed rates, that is to say 2d. (less than 1p) more than the basic china-clay wage at the time of the 1875 strike. By this time underground

MEMORANDUM

From Watts, Blake, Bearne & Co.,

NEWTON ABBOT.

To their Workmen at Kingsteignton.

Heading of memorandum to ball-clay workers at Kingsteignton.

ball-clay mining was taking over from open pit working and a complex structure of piece rates operated for sinking shafts, getting and wheeling clay underground and removing timber from disused 'white clay pits'. The employers also took the opportunity, in a memorandum, to publicly advise the men's leader, Samuel Whitear, who, according to them, had been 'complaining without just cause', to look elsewhere for work with 'masters more agreeable to him'.[48]

Strike action in Cornwall

Meanwhile, in Mid Cornwall, the mining influence gradually disappeared. By 1851, in the clay parishes of St Stephen, Roche and St Dennis, around a quarter of male workers were employed in local mines and a similar proportion in china clay and china stone works. Thirty years later, clay and stone accounted for nearly half the male workforce and mining only seven per cent. While stoppages of work occurred from time to time, they were short, sharp and settled on the spot. What strikes there were took place in ancillary trades. In 1891 dockers at Pentewan came out because of alleged unfairness in the promotion of a supervisor.[49] Two years later, coopers, who made casks for shipping clay, successfully downed tools to demand higher piece-rates per cask produced, although their activities while on strike caused the editor of the *St Austell Star* to suggest they should form a union to prevent 'wild cat action, excessive drinking and abusive behaviour'.[50] When Phillips Bros of St Austell tried to cut back this rise in pay in 1901, the 40 men in their Trewoon cooperage came out on strike and were swiftly joined by coopers from Mount Charles and Charlestown, even although their piece-rates had not been reduced. Their meetings showed some enthusiasm for joining a union, which died away, however, when the dispute was resolved in their favour.[51]

Then, in 1907, signs of a more concerted action appeared. A rise in tin prices had brought about a short-lived recovery in demand for labour and small numbers of claymen at the West of England Co. works at Lower Halviggan struck for an increase from £0.90 to £1 a week, which they claimed other clay merchants, such as the Martins, were paying. They walked three miles to the company's main office, where the new managing director, T. Medland Stocker, promised to look into their claim if they returned to work.[52] Shortly afterwards another group of Stocker's workers at Trethosa, who already received £1 a week, stopped work and confronted the clay captain, David Bassett, with a demand for £1.05. Medland Stocker immediately went to meet the men and agreed to their claim provided they gave up the custom of being paid when work was halted by bad weather.[53] The wage demands of claymen aroused much debate at several regional conferences of Co-operative Societies, where Cornish delegates pressed for trade-union organisation in the St Austell area.[54]

A new wave of strikes country wide
While wages had risen fairly steadily through the nineteenth century, in the early 1900s they stayed constant while prices rose. Seizing upon the discontent caused by this, some radical labour leaders were calling for mass action by non-unionised labourers to overturn the entire capitalist system. Local newspapers reported these

THE ST. AUSTELL STAR.

THE Best, Largest and most widely-circulated Newspaper printed and published in the extensive Mining and China Clay district of which St. Austell is the centre.

The ST. AUSTELL STAR enjoys an unrivalled position in Mid Cornwall, having a Circulation many times larger than that of any other Newspaper published in the district ; it is therefore the very best Advertising Medium in the locality.

PRICE ONE PENNY. EIGHT PAGES.
PUBLISHED AT ST. AUSTELL EVERY FRIDAY.

All communications should be addressed to the PROPRIETOR,

"ST. AUSTELL STAR" OFFICES, ST. AUSTELL, CORNWALL.

Advertisement for the St Austell Star *from Kelly's Directory, 1897. The editor was F.R. Ray.*

events in dramatic and militant terms. Headlines in the *St Austell Star* told of 'The Great Shipbuilding Lockout' of 1910, followed by 'The Great Industrial War' between cotton textile workers and employers, 'The Great Railway War' and 'The Great Miners' Strike' of 1911 and 1912. No details were spared of the bloody riots and of violent deaths through shootings and explosions in South Wales, involving the strike-breaking police force 'The Tonypandy Men', who were to play a decisive role in the 1913 clay strike.[55]

Far from being a remote occurrence, moreover, the clay families felt the direct impact of these industrial actions. The lockout of 120,000 cotton textile workers threatened demand for the china clay used in their processes, while the strike of railwaymen, dockers and seamen brought other industries that used china clay and stone to a halt. At a more domestic level, Cornish coal merchants, who had not been able to replenish their stocks after the seamens' and dockers' strikes, began to run out of fuel when 850,000 coal miners downed tools. Coal prices rose by as much as 60 per cent, but the clay merchants who had wisely built up their reserves came to the rescue of clayworkers' families by supplying them with coal at the old price. The clay employers also had enough fuel to help the Great Western Railway with coal to run trains to the clayports. In this way Higman, Lovering and others were able to build up goodwill while encouraging mistrust of union activities in the minds of their employees.

Signs of unrest were also appearing in Cornish granite and slate quarrying. In the mid-1800s this had been an activity of value roughly equal to that of china-clay and stone extraction. The granite quarrymen's pay was higher than that of the claymen, at £1.50–£1.80 a week,[56] and the first major strike, which did not occur until 1898, was called to show solidarity with Plymouth limestone quarrymen who were on strike, rather than because of any grievance with Cornish employers. Then, in the early 1900s, labour relations deteriorated. Paternalistic quarry owners the Freemans were no longer dominant, the industry shifted from small semi-independent groups to large mechanised operations and local producers were rapidly losing ground to Norwegian competitors. Employment in Cornwall fell from 1,600 to 600 and when the men struck for a minimum wage instead of piece-rates they were in no position to call the tune.[57]

At the Delabole slate quarry, which employed about 400 workers, a few stoppages of work occurred, but only in 1882 and 1910 did they involve a significant number of workers. In 1882 some 'tippers', who removed slate and rubble, went on strike because of a disagreement about piece-rates which brought the whole quarry to a standstill, although the rest of the labour force was willing to continue working. The employers brought charges of intimidation against the tippers but these were dismissed due to lack of evidence. The strikers were then dismissed and replaced by new men and work resumed after about a week, with no increase in piece rates. In 1910 the employers declared a lockout over a dispute about Saturday afternoon working, but this was resolved after only one day. The employers were generally reluctant to make any concessions, and wages were low, skilled tradesmen only receiving 90p a week and others earning from 65p a week. Even this was reduced in 1886 during a recession.[58]

The unions move their big guns into Devon and Cornwall

Meanwhile, after railwaymen and dockers had brought industrial strife into the West Country, two rival unions began to recruit among the ball-clay men of Devon and the china-clay men of Cornwall. One was the Gasworkers, and General Labourers, Union, founded by Will Thorne, the son and grandson of brickmakers, who had been elected MP,[59] and one of its leaders at this time, Jack Jones, became active in the South West. The other was the Workers' Union, advised by Tom Mann, architect of the historic London dockers' strike of 1889 and the triumphal Liverpool dockers' strike of 1911.[60] The Workers' Union Secretary was Charles Duncan, MP for Barrow in Furness. At first both unions seemed to make equal progress among Cornish clayworkers, but after the Workers' Union appointed Charles Robert Vincent as their organiser they enjoyed greater success and the Gasworkers withdrew to concentrate on Devon. Both unions focused on places where clayworks were thick on the ground, the Gasworkers in the Kingsteignton ball-clay area and the Workers' Union in the St Austell district.

Vincent had set up as a bookseller and stationer in Truro in around 1894, at a time that the earliest radical organisation to penetrate the West Country, the Socialist Democratic Federation, established a branch there. The Plymouth branch of the Independent Labour Party was formed a year later. Vincent probably aided Jack Jones when the latter stood, unsuccessfully, as Parliamentary candidate in Cornwall in 1906, supported by the Social Democratic Federation. Jones, however, failed to repeat Conybeare's earlier triumph at Camborne. He was not helped by a rumour, circulated by his Liberal opponents, that he had 'taken Tory gold' to split the Liberal vote and let the Conservatives in, and he and his followers were lucky to escape a turbulent reception at a meeting when this story spread.[61]

It was after this that Vincent must have switched his allegiance from Jack Jones' Gasworkers' Union to the Workers' Union. At first he met with suspicion from claymen who questioned why, as a union officer, he was paid two or three times their pay. His reply was that he had given up a trade that made him twice as much again, although this is open to doubt, as he was said to have been saved from eviction from his bookshop for not paying his rent only through the clemency of the landowner, Lord Clifden.[62] Vincent received powerful support from Tom Mann and Charles Duncan, as well as from Charles Beard, local government councillor and union agent in the Midlands, Matt Giles, South West organiser and member of the Union's National Executive Council, A.E. Ellery, Bristol agent of the union, and Joe Harris, an Irishman and the union's Cornwall organiser. Harris had experience in trade-union work in Dublin, Belfast and the North of England. Miss Varley, a union organiser in the Black Country, came down later to rally the claymen's wives.

National strike tactics

Meanwhile, a great wave of largely successful strikes in Britain had seen union membership pass the two million mark in 1900 and it was now approaching the four million level, nearly 30 per cent of the working population. To capitalise on their greater strength, workers were expected to present a united front and back up each others' strikes. Tom Mann was to show the importance of collective action in 1913, when Liverpool dockers refused to handle farm produce from Lancashire and contributed to the farmhands' strike funds. Hitherto, the farmhands had been too weak to stand up to local landowners, but through inter-union solidarity they were able to gain wage increases.[63] What was now called the Amalgamated Society of Gasworkers, Brickmakers and General Labourers had also organised a successful strike of firebrick workers, gaining pay increases, including a 10s. (50p) minimum wage for women.[64]

Both the Gasworkers' Union in Devon and the Workers' Union in Cornwall pursued the classic strategy of combining action by workers in production and distribution, signing up dockers and cooperating with railway unions. When Cornish railwaymen embarked upon a six-week strike in 1912, the dockers of Par and Fowey joined them, and the claymen were also expected to block deliveries. However, they refused to cooperate, although Mann warned them that they would suffer the consequences if they struck and asked for help.

Vincent and Matt Giles were also following another well-known tactic: using threats of strikes to gain small increases in pay which won favour with the workers. They would announce at mass meetings that they had agents in every pit, ready to respond to an immediate call to arms, but then at the last minute postpone the strike. Sometimes they claimed that the employers were about to offer a marginal increase, on other occasions they said they were waiting for approval from their union executive. Meanwhile, they were building up good relations with some clay merchants: Vincent said that he and Tom Mann dined with one of them, possibly Medland Stocker, who allowed union officials into his pits to address his workers. Vincent also appeared to be modifying his pay demands. At first the rallying cry was a 25 per cent increase, later he announced this would be negotiated in several stages rather than by one all-out strike. However, what followed were the two biggest strikes in the early clay industry history, to be described in Chapter Seventeen.

References

[1] We are indebted to Peter Stanier for help in examining sources of information about industry in Dorset. The sources consulted include the Dorset Record Office, the Dorset County Library and the references cited in end note 1 in Chapter Eight.

[2] Apart from references cited in the text, industrial relations in Cornwall are treated in A.K. Hamilton Jenkin, *The Cornish Miner*, 1927, repr. Newton Abbot, 1972; D.B. Barton, *Essays in Cornish Mining History*, Truro, 1986; John Rowe, *Cornwall in the Age of the Industrial Revolution*, 1953, repr. St Austell, 1993; Bernard Deacon, 'Attempts at unionism by Cornish Metal Miners in 1886', in Philip Payton (ed.) *Cornish Studies Ten*, Exeter, 1982, pp.27–36; 'Heroic Industrialism? The Cornish Miners and the Five Week Month 1872–4', in Philip Payton (ed.) *Cornish Studies Fourteen*, Exeter, 1986, pp.48–51; Philip Payton, *The Making of Modern Cornwall*, Redruth, 1992, pp.141–47. Devon and Dorset references are given in the text.

[3] P.S. Brown and Dorothy N. Brown, 'Industrial Disputes in Victorian Brickyards', *BBS*, 99, 2006, pp.6–9.

[4] J.B. Harley and E.A. Stuart, 'Strikes and Intimidation in the Building Trades of Late Victorian Exeter', *DCNQ*, XXXV, 1986, pp.58–63.

[5] Bernard Deacon, 'An Idea Moves West', in Gordon McCallum (ed.), *A Study of Adult Education in Cornwall*, St Ives, 1991, pp.23–52; Philip Payton, 1992, pp.141–47; *Cornwall, a History*, Fowey, 2004, pp.245–47. For a discussion of Lach-Szryma's other economic activities (he chaired a committee of Newlyn fishermen to raise funds for their harbour), see Ronald Perry, 'The changing Face of Celtic Tourism', in Philip Payton (ed.) *Cornish Studies Seven*, Exeter, 1999, pp.94–106.

[6] *RCG* 13 August 1870.

[7] J.H. Porter, 'The Incidence of Industrial Conflict in Devon, 1860–1900', *TDA*, 1984, 116, pp.63–75.

[8] Maurice Dart, *Cornish China Clay Transport*, Shepperton, 2000; Alan Bennett, *The Great Western Railway in Mid Cornwall*, Cheltenham, 1992.

[9] *RCG* 4 May 1872.

[10] Gilbert served as Chief Constable from 1857 until he retired in 1890 at the age of 83.

[11] Clay merchants with china stone interests made a presentation to him in gratitude for his arbitration in allocating sales quotas of china stone.

[12] *RCG* 10 July to 7 August 1875.

[13] *WB* 21 December 1876; *RCG* 9 December 1876.

[14] *The Times* 24 November 1876. We are indebted to John Tonkin for this reference.

[15] *CT* 25 November 1876.

[16] *The Times* 25 November 1876.

[17] *RCG* 1 December 1876.

[18] *WB* 14 December 1876.

[19] *WB* 23 November 1876.

[20] *WB* 14 December 1876.

[21] *CT* 25 November.

[22] *WMN* 24 November 1876.

[23] *WB* 23 November 1876.

[24] In fact a valve had been turned on to drain a clay tank. *RCG* 30 December 1876.

[25] *WB* 23 November, *RCG* 25 November 1876.

[26] *RCG* 30 December 1876.

[27] *WB* 8 19 January 1877.

[28] *PDRO*, Accession 69, letter of William Martin of 5 December 1876. We are indebted to John Tonkin for this reference.

[29] *WMN* quoted in *RCG* 16 December 1876.

[30] *RCG* 29 December 1876; John Reed, 'Centenary of the china clay workers' strike', *CG* 16 December 1976.

[31] *RCG* 26 July 1878.

[32] *RCG* 23 December 1876.

[33] 'The China Clay Trade', *MJ*, 10 July 1880, pp.790–91, letter from R. Symons of Truro.

[34] *WB* 30 November 1876; *The Times*, 24 November 1876.

[35] J.R. Ravensdale, 'The China Clay Labourers' Union', *HS*, 1986, pp.51–62; *RCG* 9 December 1876.

[36] *WB* 4 January 1877.

[37] *RCG* 12 January 1877.

[38] *WB* 1 February 1877.

[39] *SAS* 21 September, 7 December 1911; J.R. Ravensdale, 'The 1913 China Clay Strike and the Workers' Union', *Exeter Papers in Economic History*, 6, 1972, pp.53–73.

[40] *WB* 1 February 1877.

[41] *CTel* 26 December 1876, 9 January, 6 February 1877.

[42] Porter, 1984.

[43] *MJ* 9 April 1881, p.425.

[44] Bernard Deacon, 'Conybeare Forever', in Terry Knight (ed.), Old Redruth, Redruth, 1992, pp.37–43.

[45] William Lethbridge, *One Man's Moor*, Tiverton, 2006, p.214.

[46] *PDRO*, 1423 of April 1890. We are indebted to John Tonkin for this reference.

[47] Porter, 1984.

[48] Watts, Blake, Bearne & Co., *Memorandum to their Workmen at Kingsteignton*, Newton Abbot, 21 May 1891.

[49] *SAS* 27 March 1891.

[50] *SAS* 7 October 1893.

[51] *SAS* 10 January 1901.

[52] *SAS* 8 August 1907.

[53] *SAS* 22 August 1907.

[54] Catherine Lorigan, *Delabole*, Reading, 2007, p.196.

[55] *SAS* 19, 26 September 1907, 17 August 1911, 14, 21, 28 March, 4 June 1912.

[56] *WB* 17 August 1869.

[57] Peter Stanier, *South West Granite*, St Austell, 2000, pp.169–70.

[58] Lorigan, 2007, pp.71–83.

[59] Will Thorne, *My Life's Battles*, London, 1925, repr. 1989.

[60] Eric Taplin, 'Unionism among seamen and dockers', *NWLHS*, 1987.

[61] Private letters of 27 August, 29 September 1983 from Alfred Jenkin MA in the RIC Library.

[62] Alfred Jenkin Collection, Courtney Library, RIC, Truro; *Cornish Life* July 1981.

[63] A. Mutch, 'Lancashire's Revolt of the Field', *NWLHS*, 1982, pp.56–67.

[64] *Report of Chief Labour Correspondent on Strikes and Lockouts in the UK in 1913*, HMSO c5809.

Chapter Seventeen
The 1913 Ball-Clay and China-Clay Strikes

T he year 1913 set a record for strikes in many industries in Britain, including building workers who won a long-drawn-out dispute at Barnstaple in North Devon. The Gasworkers' Union, led by Jack Jones, had also been equally active among the ball-clay workers of South Devon, and early in July, claymen at Whiteway & Co., in the centre of the ball-clay industry at Kingsteignton, struck over union recognition. They marched to Newton Abbot, where Jack Jones addressed them. He described himself as an 'angel of mercy', come to the strikers' rescue but willing to 'bury the hatchet' with employers and negotiate a reasonable bargain.

However, the clay merchants did not respond, and the strike soon spread to other works, including Goddard & Co., Hexter & Humpherson, where William Hexter published a list of the wages he paid, showing average earnings of £1.11s.4d. (£1.57) a week. 'Gangers', in charge of small groups, were paid another shilling (5p) a week. Learning of this, some of the clayworker's wives, astonished to find that their menfolk earned so much, came to his office to demand proof of his statement. It is not surprising that they were astounded by these earnings, which were already higher than the maximum demands of the china-clay strikers. On the whole the atmosphere remained peaceful, except at some Watts, Blake, Bearne & Co. works, where ' crab cranes', used for hoisting ball clay, were thrown down the mine shafts and the firm offered £20 reward for information on the offenders.[1]

The news that clayworkers in Cornwall were demanding a minimum wage of 25s. (£1.25) reached the area, followed by a newspaper headline 'Five thousand men out in Mid Cornwall'. The Plymouth-based *Western Morning News* gave extensive coverage to this item and even more attention to a violent dispute in South Africa involving West Country miners, but little to what it called 'The Mid Devon Strike'. No supporting action was reported from ball-clay men in North Devon or Dorset, and two (unidentified) South Devon companies, who paid higher wages, were reported as working normally. However, Jack Jones claimed financial support from union men in other industries in Torquay and Plymouth. The vicar of Kingsteignton urged moderation on both sides, but an anonymous letter, believed to be from the Watts, Blake, Bearne & Co. chairman, C.D. Blake, announced the suspension of appearances by that company's brass band because 'socialism has been preached from the platform' during performances.[2]

The ball-clay strike continued

As the strike entered its second month, cracks appeared in the solidarity of the clay merchants. Jones negotiated union recognition by the Devon & Courtenay Co., whose men returned to work, and he claimed that only two (unidentified) firms were blocking a general settlement. Over 70 men, he asserted, had left for work elsewhere, including some to North Devon ball-clay works where there was no strike. All his union's men in the area were on strike, he boasted: lightermen on the clay barges, 'lumpers', who handled the cargoes, ships' 'firemen' (or stokers) and other seamen. When strikers were accused of cutting mooring ropes on barges to stop employers from using them, he retorted that his men 'had not been quite so wild as those in Cornwall'.

This suggested that the news of the violent actions of china-clay strikers had reached Devon and shocked public opinion. The employers then offered an extra 2d. (1p) a day but no union recognition. Jones turned this down and cheered the men with the news that Barnstaple building workers had won a rise of 7d. (3p) a day after a 15 week stoppage. Will Thorne, the union general secretary, promised the strikers that their fund would benefit from the union's 120,000 members, and Anderton & Rowlands, the travelling fairground proprietors, said they would give some of the profits from their local funfair to the fund.[3]

As August drew to a close, Watts, Blake, Bearne & Co., the strongest opponents, at last showed a willingness to negotiate, and Jack Jones, who was not in the area, announced that he trusted his men to deal with the employers on their own. The Watts, Blake, Bearne & Co. workers met one of the directors, W. Watts, and other strikers negotiated with members of the Whiteway Wilkinson, Hexter & Humpherson and Goddard families. A fortnight later, after complicated discussions on rates for day labourers, clay miners, clay cutters, outworkers, bargemen, sawyers and other occupations, a ballot was held among union members, who only made up about a quarter of the total of over 1,000 men on strike. Well over 80 per cent of those entitled to vote chose to return to work, and the ten-week dispute came to an end. Not all the employers had officially agreed to recognise the union's right to negotiate, but nevertheless they had all tacitly accepted the union's role, and the 750 or so non-union men were asked by the union officials to join it within a week.[4]

What effect the successful outcome of the Devon ball-clay strike had upon the Cornish china-clay strike, which began and ended two weeks after the start and finish of the Devon dispute, is far from clear. The lengthy reports of Cornish union activity in some Cornish papers did not seem to mention the ball-clay strike. Nor did the Workers' Union officials, who were organising the Mid Cornwall strike and who passed through the heartland of the ball-clay strike on their way to and from Cornwall, appear to take public notice of it. The contrast between the two disputes, and the differing roles played by strikers, clay merchants and union officials, will be discussed at the end of this chapter.

The 'white country' strike

On Monday morning, 21 July 1913, the 50 or so men who worked in Carne Stents clay pit, a couple of miles west of St Austell, met their clay captain. They expected him to confirm that their employers would agree to pay them on a fortnightly basis instead of the unpopular 'five week month' system, under which they only received their pay packets 12 times a year. This meant that they sometimes had to wait five weeks for their wages. When the captain told them that the employers had decided not to go ahead with fortnightly pay, the men were so aggrieved that 30 of them downed tools at once. Led by C.R. Vincent, the union organiser, they set off for neighbouring pits to call other workers to join in.

In this apparently fortuitous way began one of the most widely documented events in West Country clay history. In its day it was extensively covered in the local, regional and even national press and recounted in great detail in a 62-stanza ballad said to be written by one of the strikers and sold to raise funds.[5] The Chief Industrial Commissioner of the Board of Trade, Sir George (later Lord) Askwith, was sent from Whitehall to write a report on the strike at the instigation of the St Austell MP, the Hon. T.C.R. Agar-Robartes. In succeeding decades it was commented on, described and analysed by such historians and writers as J.M. Coon, R.M. Barton, Kenneth Hudson, A.L. Rowse, John Penderill-Church, Marshel Arthur, R.S. Best, whose father was one of the Carne Stents strikers, the Tory MP David Mudd and the Workers' Education Association lecturer J.R. Ravensdale. In the early 1970s it was the basis of a full-length BBC Television Film, *Stocker's Copper*, which is still sometimes shown in the clay district.[6]

As the Carne Stents men made their way to the north-west of St Austell they persuaded workers from most, but not all, of the pits they visited to come out in sympathy. However, when they ventured to the north-east of the town they met with a less enthusiastic response, especially from what the 'White Country' poem called the 'lily-livered and jelly-limbed Bugle men'. Throughout the dispute, men from that area were the least enthusiastic to strike, the first to return to work at the earliest opportunity and the most bitterly criticised by the strikers. One reason for this lack of support was that the strikers had taken as their rallying cry, 'Making the Twenty Five', that is achieving a minimum weekly wage of 25s. (£1.25) instead of the then current minimum of less than a pound a week that had not risen for a decade or more. One of the reasons for the reluctance to strike by the Bugle men may lie in the variations in demand for china clay at this time.

The east/west earnings differential

The reluctance to strike by the Bugle men may have been caused by a difference in earnings at the time of the dispute between workers in the potting clay pits from the western part of the clay district and those on the eastern side, near Bugle, who were producing clay largely for the cotton and paper industries. Sales of potting clay increased gradually, but sales to the cotton and paper industries grew at a much faster rate from the mid-nineteenth century, leading to increased opportunities for extended working to meet demands. This could have meant that workers from the Bugle area may have been earning wages well in excess of the strikers' demands. The levels of secrecy surrounding the use of china clay in cotton and paper may have helped to obscure the east/west earnings differentials and encouraged loyalty by workers to those merchants who sold clay to the cotton and paper industries,

Actual pay levels

However, what most of the accounts of the strike fail to point out is that a substantial number, particularly in the more productive pits, already earned at least as much as, if not more than, the strikers' target. The *Cornish Guardian*, based in Bodmin, which was sympathetic to the workers' cause, carried out what it called 'a dispassionate and impartial survey'.[7] It found that about half the workforce, including the 'washers' and 'breakers' of clay, earned between 21s. and 23s. (£1.05 and £1.15) a week for an 8-hour day. If they were unable to work the full day because of bad weather (on average about 50 days a year) they still received this basic wage, but if demand for clay was brisk and the weather good they might work 'a day and a half', that is 12 hours a day, and make £1.575– £1.725 a week. 'Engine men', who looked after the pumping engines, earned on average £1.15 a week, but they had to come to work on Sundays to keep the engines operating.

The rest of the labour force worked at contract rates. 'Burden men', who removed topsoil, and 'sand men', who shifted sand from the bottom of the pit, earned between £1.275 and £1.575 a week for a 7-hour day. 'Loaders', who put clay into casks or bags, averaged £1.50 a week, based on a 7½ hour day. 'Drymen' at the kilns made £1.50–£2 a week for a day that sometimes lasted only 5 hours. 'Shift bosses', who were working foremen and operators of mica drags, received £1.25–£1.35 a week. Trained carpenters, masons and blacksmiths were paid at tradesmen's wages of £1.25 a week or more.

The strike continues

Many workers were thus being asked to lose pay by coming out on strike. Some were reluctant to do so, for although the trade union leaders also demanded extra pay for the higher earners, the differentials they proposed were quite small compared with those currently operating. Nevertheless, after a couple of weeks of vigorous campaigning by Workers' Union officials, involving mass meetings and some 'peaceful persuasion', which those who opposed the strike called intimidation, nearly the whole of the 4,800-strong workforce[8] downed tools, although some reports emphasised that a sizeable proportion of them lacked enthusiasm. The dispute coincided with an unusually long spell of dry and sunny weather, and something of a traditional feast day atmosphere reigned, although the lack of rain upset the strikers' plans to put clayworks out of action by stopping the pumping engines and causing the pits to flood.

At this point the strikers might have achieved their aims if they had confined them to 'reaching the twenty-five', for some leading clay merchants, including William John North, John F. Rose and Medland Stocker, offered pay rises, and indeed conceded the claim soon after the strike was over. However, the men, led by Vincent, had linked their pay demands to an insistence upon union recognition, and this proved to be a sticking point for many employers. As we have seen, they were noted for keeping their cards very close to their chests in their dealings with the outside world, and collectively they maintained a wall of silence, only reporting their decisions to the press through their own trade association. According to a government report on strikes 70 firms were involved, but we do not know what internal discussions took place between them.[9]

Perhaps a majority shared a fear common to many industrialists that the growth of trade unionism heralded the overthrow of the entire capitalist system, as indeed some union leaders openly proclaimed. Whatever their private feelings were, however, their public strategy was one of masterly inactivity. The collective response to all questions about the dispute was that any suffering it caused was entirely the responsibility of the strikers. The employers had not declared a lockout and all men willing to work were free to return, when any grievances could be discussed with individual clay merchants, but of course without the presence of union officials.

The isolation of the St Austell claymen

At this stage a crucial point needs to be made. The voluminous coverage of the dispute has been remarkably parochial, and none of it, except to some extent Ravensdale's account, has noted that the St Austell claymen received almost no financial or practical support from outside, either from their fellow workers in other china-clay areas, or from trade unionists in other industries of the South West or Britain as a whole. Nor did they receive backing from a number of prominent Liberals and Methodists who, they might have expected, would be on their side.

Workers in the clay pits of West Cornwall, Bodmin Moor and Lee Moor did not come out in sympathy, although the Devon men collected a few pounds for the St Austell strike fund. As in 1876, the strike leaders held out great hopes of financial help from other unions, but once again little was forthcoming. Motions of fraternal solidarity were passed at the Trades Union Conference, and railwaymen at Penzance made a modest contribution to the strike fund but, much more critically, neither railwaymen nor dockers at the clayports blocked shipments of china clay. As mentioned in Chapter Sixteen, Tom Mann had warned the claymen who failed to back the dockers and railwaymen that they could not expect help when their turn came.

The result was that the clay merchants themselves, with assistance from some clay captains, loaded several consignments of 'best clay' destined for American paper-makers onto wagons and shipped it out from the clay ports without hindrance. The author of the 'White Country' poem mocked one of the clay handlers for wearing kid gloves, but he was a jeweller who wished to avoid damaging his hands. The strikers had the last laugh, however, when the clay shipped out was blocked by Bristol dockers. Unlike the Cornish dockers, the Bristol men showed solidarity by refusing to transfer it to ocean-going vessels bound for America. Matt Giles, South West Organiser of the Workers' Union, said that his union was on the brink of amalgamating with the powerful Dockers' Union and that no dockers would handle clay in any UK port.

Curiously, even though the strike of Devon ball claymen was proceeding at the same time, there appeared to be no liaison between the two groups. Amalgamation of lesser-skilled unions to strengthen their bargaining power was in the air at that time – three unions merged to form the National Union of Railwaymen. Towards the end of September, Matt Giles claimed that the Workers' Union was about to merge with the much larger Gasworkers' Union, but this did not bring any financial benefit to the china claymen. As for the general public, support was lukewarm. A few local leaders, including the clay landlord Sir William Sarjeant, the Liberal leader Sir Francis Layland Barratt and the Liberal MP 'Tommy' Robartes, contributed to a strike fund set up by another prominent Liberal, Walter John Nicholls, who had succeeded Ray as editor of the *St Austell Star* and who was also Chairman of the Urban District Council.

Financial assistance

The cooperative movement provided almost the only political and financial support for the china claymen. In recent years cooperative societies had been collaborating more closely with trade unions,[10] and Plymouth Cooperative Society was one of the largest in England, with over 35,000 members in 1900. Together with the over 10,000 strong Devonport dock workers, this provided the main focus for action in the South West. Indeed, in 1909 this society proposed, though without success, that the national Cooperative Wholesale Society should introduce a 'Trade Union' label on the goods it sold that were made by firms that recognised trade unions.[11]

During the china-clay strike union leaders spoke at labour rallies in Plymouth, enlisting financial support, and in September a regional conference of Cooperative Societies was held at Delabole, where a proposal from the St Columb Society to relieve distress among clay families was supported by all 30 delegates attending.[12] However, the grand total of all these efforts only came to a few hundred pounds, a drop in the ocean when nearly 5,000 men were out of work. Nor could the strikers expect much in the way of aid from local traders. Some clay families existed on the margin at the best of times, living on credit from shopkeepers towards the end of the month and, as more and more claymen ran up bills, the shopkeepers found themselves short of cash and some faced bankruptcy.

Another significant facet of the strike that has been overlooked by previous writers is the financial status of the Workers' Union itself. Possibly because of its impressive title and the rhetoric of its leaders, it was assumed to be a powerful organisation like the Railwaymen's Association, with 188,000, or the Gasworkers, who organised the successful ball-clay strike in Devon, with 120,000. In fact it was very small, with only about 20,000 members when it began campaigning in Cornwall. What is more, it was of fairly recent origin and so had not had time to build up a large strike fund. Although it promised that all who joined would receive strike pay of 50p–63p a week, according to how long they had been members, it was questionable whether it could continue to pay out thousands of pounds a week for a lengthy period of time, a situation that must have been known to the clay merchants,[13] encouraging them to sit tight and wait for the strikers to go back to work.

Politics and religion

In a Liberal–Methodist stronghold like St Austell, sympathy for the strikers' cause might have been expected. Certainly the Liberal editors of the *St Austell Star*, the *Newquay Gazette* and the Bodmin-based *Cornish Guardian* supported the right of the claymen to form a union. If the employers were allowed to have their association, these newspapers argued, the workers were entitled to theirs. On the other hand, the Liberals opposed forcing every worker to join a union. Among local politicians, 'Tommy' Robartes, the local Liberal MP, although son of a landowner and clay landlord, was popular with the claymen, who felt he was on their side. Others who claimed to be the strikers' friend, however, like the clay landlord Sir William Sarjeant and the clay company director F.A. Coon, were treated with suspicion when they urged the strikers to return to work and discuss their grievances with their individual employers.

Coon was known for his radical views, advocating abolition of the House of Lords and nationalisation of key industries, but he was chased by the strikers through the streets of St Austell and had to take refuge in the White Hart Hotel after he made a speech advising the men to end the strike. The local Liberal Party

Secretary and Chairman, respectively H. Syd Hancock and Henry Hodge, noted for their forceful speeches on other issues, were criticised by Liberal-minded editors for keeping a low profile. Hancock, as a clayworks manager and agent for the clay landlord Sir Charles Graves-Sawle, was thought to side with the employers.

The Revd Booth Coventry.

Methodist leaders were equally divided. At one end of the spectrum the outspoken self-made clay merchant Samuel Dyer, never one to mince his words, caused an uproar when he used the opportunity of a speech at the opening of a Methodist Chapel to make a bitter attack on the union leaders. Honest, hard-working and loyal clayworkers and their families, he declared, were being impoverished to satisfy the ambition of one man, Vincent. In the opposite camp was the Revd Booth Coventry, superintendent of 29 local Wesleyan churches. In the film *Stocker's Copper* he was portrayed as a Christ-like figure, preaching sermons on the rights of man, yet he was always in the thick of the fray and closely involved in the tactical direction of the strike, and was one of those who made impassioned pleas at a labour meeting in Plymouth, demanding a 'living wage' for the china claymen. His reported speeches were those of a firebrand, skilled in the arts of stirring up envy and distrust of the clay merchants, and right-wing newspapers attacked him as a dangerous socialist, the *Western Morning News* condemning his 'fanatical outpourings'.[14]

The majority of Methodist ministers, though, took a middle-of-the-road approach. They sympathised with the plight of the poorer clay families and appealed, unsuccessfully, to the clay merchants to compromise and bring the dispute to an end, but this was as far as they would go. A Methodist Synod refused to endorse the strikers' claims, simply calling for mutual restraint,[15] and Bible Christians allowed the use of their Sunday school to house the Glamorgan police.[16] The radical editor of the *West Briton* criticised the way that some prominent Methodists 'laid low' during the dispute, accusing them of being 'too much in with the big bugs', while the *Newquay Express* complained of their 'rigid silence'.[17] As for the Anglican Church, the newly-elected Bishop of Truro, Wilfred Burrows, while asking his congregation for sympathy towards the stevedores who shipped the clay and the small shopkeepers who were facing bankruptcy, seemed to show little concern for the clay strikers themselves.[18]

A battle between interlopers

Although they put a brave face on it, the strike leaders must have been dismayed by the lack of support or the ambivalence of fellow miners, trade unionists, Liberals and Methodists. As the strike went into its fifth and sixth weeks, men drifted back to work in increasing numbers, especially in the Bugle area. Indeed, left to itself, the dispute might well have petered out if it had not been for an event which stirred up feelings to fever pitch again and made headlines in the national press: the arrival of the Tonypandy strike-breakers.

Law and order had been maintained up to that point by some 200 policemen drafted in from all parts of Cornwall by the County Council, headed by the Chief Constable, Major H.B. Protheroe-Smith, son of Sir Philip Protheroe-Smith of Tremorvah.[19] However, in their desperation to win the dispute before the workers lost heart, the strike leaders had begun to attack engine houses at night to try to put them out of action and also to threaten violence against the 'cowardly scabs', as they called them, who were going back to work. To protect the growing numbers of returning workers from the strikers' pickets, the council called upon outside reinforcements, including 30 from neighbouring Devonport, 60 from Bristol and, in a move that alarmed the strikers, 100 from South Wales, among them the 'Tonypandy men', famous for breaking up violent coal-mining and railway disputes. They arrived impressively equipped with bicycles, truncheons, new shields and powerful electric torches to spotlight the night raiders.

Police with sticks captured from strikers. (From the Daily Graphic, *3 September 1913.*

From then on the strike became virtually a battle between outsiders, with union leaders from London, Birmingham and the North pitted against non-Cornish policemen. In *Stocker's Copper* this confrontation was simplified as a single decisive battle, but in reality a series of skirmishes took place between gangs of pickets armed with staves and policemen equipped with truncheons, in which all the union leaders were manhandled and injured: Councillor Beard and Miss Varley at Bugle, Matt Giles at Roche, Harris at St Stephen and Vincent at St Dennis.

Injured striker on a stretcher being taken for treatment.

The violence was not all on one side. At Charlestown, strikers threw stones at policemen protecting men loading clay onto a ship, but by the time the 'Glamorgan men' arrived the crowd had dispersed. The atmosphere in the claylands after these incidents was described by the wealthy landowner and former mining magnate J.F. Williams, who supported the strikers' case. 'The district is quiet – terrorised into quiet. But the whole countryside is full of rumours... that a man and a little girl have been killed by the Welsh strikebreakers... good-natured civil enough men in their way [who] look on the strikers as natural enemies and jump at any occasion of hammering them with batons.'[20] The rumours of killings were untrue and unjustified.

These events were widely reported and at first only served to stiffen the strikers' resolve. When, in a poll of workers, 2,258 voted to continue the strike, Vincent was carried in triumph through the streets of St Austell. Opponents of the strike, however, could claim that more than half the men had not voted in favour and, with returning workers well protected from pickets, more than 1,000 were back at work in the middle of September. Giles and Vincent tried to raise the strikers' spirits with talk of Bugle men rejoining the strike, an unlikely occurrence, and some of the most vociferous of the remaining supporters of the strike were women. They hurled abuse and threw earth and buckets of water at men going back to work, and at Trewoon, the 'lady of the tin pan' shouted 'You dirty scabs' at them and beat upon her pan as they passed by. But she was fighting a losing battle. By the end of the following week 1,500 were working and, as September came to an end, Vincent and Giles, in consultation with Booth Coventry, decided to call the strike off. It ended officially on 4 October, having lasted almost 12 weeks.

The contrasting fortunes of the ball clay and china clay strikers

Meanwhile, as we saw earlier, the Devon ball claymen had also ended their strike, but as winners, not losers, gaining both a pay rise and union recognition. Why did they succeed when the St Austell china clay workers failed? In Chapter Sixteen we argued that a number of interrelated factors determined the result of the 1875 and 1876 china-clay disputes: the time of the year that the strike occurred; the state of the market for clay; the relative importance attached by the employers to short-term demands for higher pay and the longer-term implications of union recognition; and the comparative solidarity of the workers and the employers. To these we could add in 1913 the relative strength and financial resources of the unions involved.

In 1913 both the ball-clay and the china-clay strikes took place in the summer, and at a time when demand for clay was brisk, two factors which favoured the strikers. On the other hand, while the ball claymen received strong support from local dockworkers and seamen, the china claymen enjoyed no such help. Moreover, the Gasworkers' Union that led the ball-clay strike had 120,000 members and only had to provide strike funds for fewer than 1,000 men in total. Indeed, a mere quarter of these men actually joined the union while the strike was on, but the Gasworkers' Union gave them all strike pay. The General Workers' Union, on the other hand, who organised the china-clay strike, only had 20,000 members but had to finance 5,000 strikers.[21] Curiously enough, while only a quarter of the ball claymen joined the union, they won the strike, whereas even though half the china claymen joined the union, they lost. As for the employers, while the ball clay merchants were divided and broke ranks to pursue independent negotiations, the china clay merchants, whatever their internal differences, presented a united front, refused all attempts at negotiation and brought in police reinforcements to underline their determination to win.

What were the clay strikes about?
The year 1913 was unequalled in the previous history of UK strikes. There had been great disputes before, as seen in Chapter Sixteen, but never had there been so many of them – 1,497 in all. Nor had they been spread over such a wide range of commercial and industrial activities. Most of them were small in scale, only about one per cent involving as many workers as the china-clay strike, and few lasted as long as the ball-clay or china-clay disputes. Most of them were concerned with increases in pay, and in nearly 80 per cent of the disputes the strikers were partially or wholly successful in getting what they wanted.[22] Within this context, the failure of the china-clay workers is noteworthy. Prosperous industries were usually associated with victory for the strikers, and the clay trade was booming, so why did the workers lose?

At this point it is useful to examine some conflicting interpretations of the origins and purpose of the strike that have been presented by contemporary observers and the writers and historians mentioned at the beginning of this chapter. Kenneth Hudson, R.S. Best, A.L. Rowse and the producer of the television film saw it as a spontaneous, unpremeditated uprising by men who were discontented with their pay. David Mudd, in complete contrast, regarded it as a small part of a vast socialist conspiracy to overthrow the entire capitalist system. F.A. Coon, R.M. Barton and J.R. Ravensdale, as well as the Board of Trade investigator Askwith, the editors of the *Western Morning News* and the *Royal Cornwall Gazette* and such clay merchants as Martin and Samuel Dyer all took positions somewhere between these extremes. They believed that most men had no grievances strong enough to bring them out on strike, but had been stirred into action by the intervention of outside 'agitators'. However, to men like Dyer these officials were only interested in furthering their own personal ambitions, whereas to Ravensdale they were enabling unskilled labourers to stand up to the might of the employers.

Supporting Mudd's conspiracy theory were the facts that the ball-clay and china-clay strikes ran almost in parallel, that they followed industrial action by Devon and Cornwall railwaymen and dockers and Devon building workers, and that union officials tried (successfully in Devon) to organise inter-industry liaison to achieve their goals and demanded the same minimum wage. It is interesting to note that further along the South Coast, the trade-union militant David Naysmith came to Portsmouth dockyards in 1913 and increased the membership of the Amalgamated Society of Engineers by 300 to a total of over 1,000. In the same year Tom Mann, in a speech to Portsmouth dockers, advocated taking 'entire control' of the naval dockyards.[23]

Some leaders of the ball-clay and china-clay strikes were certainly involved with revolutionary socialists. In 1913 Tom Mann was arrested for incitement to mutiny after urging soldiers to disobey orders about confronting strikers. Later in 1913, after the clay strikes were over, both Jack Jones and Tom Mann were active supporters of a transport workers' strike in Dublin, led by James Connelly, the republican trade unionist who was executed for treason after the Easter uprising in Dublin in 1916.[24]

Despite this association with revolutionary politics, no contemporary observers in Cornwall openly advanced a conspiracy theory, although the clay merchants recognised a concerted movement on the part of the workers to achieve union recognition. Nor was there any sign of effective cooperation between South Devon ball claymen and St Austell china claymen, or of real support from the other ball-clay areas of North Devon or Dorset or the china-clay areas of West Cornwall, Bodmin Moor or Dartmoor. No help came for the St Austell men from other trade unions, locally or nationally. If the St Austell men were really part of an international movement, they seemed to have ruined their chances by their independent attitude during the railwaymen's and dockers' strikes, whereas the ball clayworkers benefited from local inter-union solidarity.

What happened to the main protagonists?

The ball-clay dispute was not marred by much violence and the two sides seemed to renew relationships without undue bitterness or ill feeling, while the union leaders, Jack Jones and Will Thorne, went on to fresh triumphs elsewhere. As for the china-clay strike, local government enquiries in Cornwall and Bristol cleared the policemen who were drafted in from outside Cornwall of any blame. The officer in charge of the Glamorgan police received a letter, signed by over 40 householders with whom the policemen had lodged, stating that the men had 'acted as gentlemen and their conduct had in every way been beyond reproach'.

The landowner J.F. Williams, sympathetic to the strikers, questioned the police action, and the Home Office characteristically denied any responsibility for the actions of the Chief Constable of Cornwall.[25] Cornwall County Council, on the other hand, 'to show appreciation for their hard and difficult work', gave every policeman in the force three days' leave.[26] The county's ratepayers had to meet the cost of £5,800 for the employment of Glamorgan, Bristol and Devonport police, which required a county rate of a penny farthing ($^1/_2$p) in the pound,[27] and councillors in some areas, including Falmouth, unaffected by the dispute, complained at having to share in its cost.

Booth Coventry left for Yorkshire and then Scotland, Joe Harris stayed to become a popular labour leader in Mid Cornwall, while Vincent seemed to disappear from the scene altogether. Although Workers' Union membership plummeted temporarily, Matt Giles and Joe Harris claimed a famous victory for their union, for not long after the strike was over they agreed with Stocker a three-year, no-strike contract that gave labourers a basic wage of more than the original £1.25 a week they had asked for.

The clayworkers, it was generally assumed, had suffered greatly, forfeiting between £60,000 and £70,000 of lost wages which they never regained (David Mudd later suggested £100,000). However, these calculations fail to take account of the patchy nature of support for the strike. According to Penderill-Church, the strike committee let the men at Stocker's Dorothy and Dubbers pits work in peace throughout the entire dispute. At another of his works, Kernick, they at first refused to strike, then came out under threats of violence, but drifted quietly back at the earliest opportunity. At his Trethosa pit, on the other hand, they downed tools under pressure from the strikers and stayed out. Lovering's Carclaze workers only struck after threats from a large body of men armed with sticks and stones, but soon resumed work without any further pressure from the strike committee. In contrast, some pits were solid supporters of the strike, including the Martin Bros' Virginia works, the St Austell China Clay Co.'s works and the Meledor and Rockhill pits of North and Rose, where the men turned down an offer of a pay rise.[28]

China-clay output for the year 1913 was only four per cent below the record level of 1912, and since much the same number of man-hours were needed to obtain this amount, the clayworkers' income must have been about the same over the calendar year as in the previous year. As far as the employers were concerned, performance was similarly patchy, with some clay merchants suffering a complete loss of production while others were unscathed. Yet clay shipments via Fowey were only two per cent below the 1912 figure, suggesting that they had made up much of the shortfall by running down stocks.

What did the union gain from the strike?

The historian Alfred Jenkin claimed that the Workers' Union officials had so impressed the entire Cornish labour force with their leadership that membership of the union across the county had rocketed to 15,000 by 1918. However, government pressure for compulsory collective bargaining at an industry-wide level trans-formed worker–employer relations during this period, and this would appear to be the main reason for the increase. Interestingly, a small group of slate quarrymen at Delabole invited a representative of the Workers' Union to address them in August 1914.[29] This might suggest that the quarrymen were impressed by the union's organisation of the clay strike, but no Delabole branch was formed until the end of the 1914 War. Stuart Dalley has suggested that many of the china claymen were so disenchanted with the way that the Cornish establishment had treated them that they refused to volunteer for the Armed Forces when war broke out.[30] Certainly the Revd Booth Coventry, formerly the strikers' hero, was booed when he asked them to support the war effort.

By 1919, however, he was once more acclaimed when he made a brief reap-pearance before going overseas. He even shared a platform at a Union May Day Rally with Colonel W.T. Lovering, the first time that a clay merchant had attend-ed such a meeting, and was full of praise for the clay merchant, urging his audi-ence to 'stand together' with their employers in 'one great spirit of cooperation' at a time when the 'whole world was trembling on the brink of bankruptcy'.[31] Lovering congratulated Booth Coventry on his 'eloquent address', expressing sat-isfaction that relations between employers and employees were so good, and urged the workers to send their best men to consult with employers on the Industrial Council that regulated pay and working conditions. The tumultuous events of 1913 seemed to be a thing of the past, but this was far from the case, for the oral histo-rians Garry Tregidga and Lucy Ellis have shown how, nearly a century later, any mention of the strike still caused heated quarrels among descendants of those who took part in it.[32]

References
[1] *MDA* 5, 12 July 1913.
[2] *MDA* 19, 26 July, 2, 7 August; *WMN* 31 July, 1, 2, 7 August 1913.
[3] *MDA* 9, 16, 23 August; *WMN* 11, 12, 16, 22, 27 August 1913.
[4] *MDA* 30 August, 6, 13 September; *WMN* 28, 29 August, 1, 2 September 1913.
[5] Apart from the sources cited, this account of the china-clay strike is based upon reports in the *WB*, *RCG*, *WMN*, *WDM*, *NE*, *SAS* and *CG*. The ballad entitled 'The Cornish Clay Strike: The White Country Dispute' was published in two parts,

the first, by E.J.R Bawden, *History of the Clay Strike in Song*, St Austell, August 1913 and the second, by 'Unskilled Labourer', *A Souvenir of the China Clay Strike*, St Austell, 1913.

[6] J.M. Coon, 'The China Clay Industry', *RRCPS*, 1927, p.664; R.M. Barton, *A History of the Cornish China Clay Industry*, Truro, 1966, p.94, 130 et seq; Kenneth Hudson, *The History of English China Clays*, n.d., Newton Abbot, p.40; A.L. Rowse, *A Cornish Childhood*, London, 1942, repr. 1993, p.124 and *St Austell: Church, Town, Parish*, St Austell, 1960, p.82; John Penderill-Church, *The Clayports of Devon and Cornwall*, Wheal Martyn Archives, n.d., pp.11–26; Marshel Arthur, *The Autobiography of a China Clay Worker*, Federation of Old Cornwall Societies, 1995, p.31; R.S. Best, 'Clayworkers Sacrifice in Vain', *WMN* 18 July 1983; David Mudd, *Cornwall in Uproar*, Bodmin, 1983, p.4; J.R. Ravensdale, 'The 1913 China Clay Strike and the Workers' Union', *Exeter Papers in Economic History*, 6, 1972, pp.53–73.

[7] *CG* 26 September 1913. Surveys by the *RCG* and the *SAS* presented much the same figures.

[8] The figure of 4,800 is that contained in the *Report on Strikes and Lockouts*, HMSO, 1914, p.114.

[9] *Report on Strikes*, 1914.

[10] Sidney Pollard, 'The Foundation of the Cooperative Party', in A. Briggs and J. Saville (eds) *Essays in Labour History*, II, London, 1975, p.201.

[11] Mary Hilson, 'Consumers in Politics. The Cooperative Movement in Plymouth, 1890-1920', *LHR*, 67, 2002, pp.7–28.

[12] Catherine Lorigan, *Delabole*, Reading, 2007, p.197.

[13] In 1911 the *SAS* reported a membership of 20,000, in 1913 the strike leaders claimed 27,000, presumably including 2,000 or more clayworkers. *SAS* 23 November 1911, 27 February 1913; *CG* 12 September 1913.

[14] *WMN* 25 August 1913.

[15] Among those ministers who tried to persuade the employers to negotiate were the Wesleyans Armstrong Bennett and T. Walter Cook, the United Methodists T.S. Lea and Henry Gilbert Low, the Baptist H.C. Bailey, the Primitive Methodist W.A. Bryant, the Quaker J.H. Fardon and the Anglicans J.E. Carey and Frederick Thomas.

[16] In Trevarthian Road, near the railway station, not far from A.L. Rowse's home, who witnessed the occupation. See Rowse, 1942, repr. 1993, p..124.

[17] *WB* 11, 18, 25 September, *NE* 22, 29 August 1913.

[18] H. Miles Brown, *A Century for Cornwall*, Truro, 1976, p.71.

[19] Like the Chief Constable in the 1875–76 clay strike, Protheroe-Smith was a military man who had retired from the 21st Lancers in 1896 at the age of 37 and served Cornwall until 1938, apart from the time that he was recalled to arms to join the British Expeditionary Force at the beginning of the 1914 War.

[20] Letter of 2 September 1913 from Williams to Francis Acland MP. We are indebted to Garry Tregidga for this reference. There was no truth in the rumour about the man and the girl.

[21] The Gasworkers' Union provided strike pay to 946 men during the ball-clay strike, although only a quarter had actually joined the union.

[22] *Report on Strikes*, HMSO, 1914, pp.ii–xiii.

[23] However, the local Labour Councillor, J.M. MacTavish, opposed Mann's militancy, being more interested in collective bargaining. Peter Galliver, 'Trade Unionism in Portsmouth Dockyard', in Kenneth Lunn and Ann Day (eds), *History of Work and Labour Relations in the Royal Dockyards*, London, 1999, p.119.

[24] Donal Nevin, *James Connelly*, Dublin, 2006, pp.418, 473.

[25] Letter from Francis Acland MP to J.F. Williams of 22 October 1913, courtesy of Garry Tregidga.

[26] A.L. Dennis, *Cornwall County Council*, Truro, 1989, p.109.

[27] *BCW*, November 1913, p.238.

[28] Penderill-Church, n.d., pp.11–26; *BCW*, October, 1913, p.211.

[29] *Cornish Guardian* 7 August 1914. We are indebted to Catherine Lorigan for this reference.

[30] Stuart Dalley, 'The Response in Cornwall to the Outbreak of the First World War', in Philip Payton (ed.), *Cornish Studies Eleven*, Exeter, 2003, pp.85–109.

[31] *CCTR*, June 1919, pp.25–26.

[32] Garry Tregidga and Lucy Ellis, 'Talking Identity', in Philip Payton (ed.), *Cornish Studies Twelve*, Exeter, 2004, p.98.

Chapter Eighteen
Adding Local Value to Clay

As we saw in Chapter Nine, from the beginning of commercial extraction of clay and stone in the West Country, the possibility of increasing local income and employment by developing activities that used these materials was recognised. We have already described early examples of crucible production in Cornwall in Chapter Two, the processing of alum in the Bournemouth and Poole area in Chapter Six and the establishment of potteries at Plymouth and North and South Devon in Chapters Nine and Eleven.

Most of these ventures were small scale, but in the second half of the nineteenth century a new urgency in the quest for additional employment was generated by the decline of the dominant metal-mining sector. In Cornwall alone, mining gave work to some 36,000 men, women and children, over a third of the total working population, but over the next half century this number dwindled to 7,000–8,000. Patriotic landowners, merchants and industrialists such as the Williams, the Bolithos, the Foxes of Falmouth, Hain of St Ives, Bain of Portreath and Smith of Camborne united in attempts to diversify the economy.

They helped to finance, inter alia, a branch railway line to open up the Lizard peninsula, a bacon factory at Redruth, dairy processing plants in other places and explosives factories along the dunes of Hayle and Perranporth.[1] The Foxes and the Royal Cornwall Polytechnic Society, of which they were prime movers, were particularly concerned to promote new activities. 'Now mining, one of the staples of the industry, has failed,' Howard Fox told the Society, 'it behoves us to look to other staples.'[2]

The Revd Edward Collins' pottery proposals

Most of these initiatives focused on West Cornwall, but adding value to china-clay production in Mid Cornwall was another obvious possibility. In the summer of 1867, just a year after the great 'copper crash', the Sheriff of Cornwall invited County Magistrates to consider ways of dealing with the crisis. One of those who responded was the Revd C.M. Edward Collins of Blisland. Having previously lived at Chudleigh in Devon, he was familiar with ball clayworks and the Bovey Pottery in that area, and since moving to Cornwall he had noted the discovery of china clay nearby on Bodmin Moor, particularly at Temple. In a paper of 1868 presented to the Royal Cornwall Polytechnic Society, he advocated establishing a Cornish pottery, suggesting Fowey as a suitable site. As he pointed out, this port was already close to sources of clay at St Stephens and Roche, and when the Lostwithiel to Fowey railway line was opened it would have easy access to clay from Temple. Ball clays from Dorset and Devon could also be brought to Fowey, if a pottery was built there, much more cheaply than to potteries in Staffordshire, Bristol or Liverpool.[3]

Edward Collins was fully aware that the relative cost of fuel for firing kilns was a crucial factor in determining the location of a pottery: a Bovey pottery had to pay £1 a ton for coal, whereas Staffordshire potters only paid 7s. (£0.35). However, he believed that 'good pottery coal' might be shipped into Fowey at 14s. (£0.70) per ton, and while this doubled the cost compared with Staffordshire, he was confident that other economies might offset this disadvantage. Freight costs for china clay, ball clay and flint would be less, and he optimistically asserted that Cornish engineers who had raised the efficiency of steam engines to unparalleled levels, could apply the same ingenuity to improvements in pottery furnaces.

In addition, Edward Collins was familiar with the popularity of Parian Ware, described in Chapter Twelve, which used china clay in the manufacture of monuments, tables, chimney pieces and mosaic floors. Interestingly, he also knew of the growing threat of a takeover of the local china-clay industry by Americans and other outsiders, to be discussed in Chapter Nineteen. If Cornishmen did not move soon to set up an integrated china-clay and manufacturing sector, he warned, others would 'pick our pockets in front of our face'. Finally, he advocated profit-sharing schemes between employers and workers which, he optimistically claimed, 'worked wonders' in improving industrial relations in the North of England.

Edward Collins had clearly given a good deal of thought to his scheme but his ideas fell upon deaf ears. Practical businessmen showed no enthusiasm for them, yet he persisted in his efforts and, seven years later, he addressed the Polytechnic's members again. By this time clayworking had become established at Temple on Bodmin Moor and a rail connection to St Austell and Fowey had been completed. Potteries, including terracotta works, had been successfully set up in South Devon (to be discussed later in this chapter), and Edward Collins now suggested that similar art potteries, as well as general ceramics works, should be opened in Cornwall. Reproductions of Wedgwood pieces made in South Devon would, he asserted, 'provide pleasure, at an affordable price, to every cottage in the land'. He also saw a future for decorated ceramic 'frescoes', illustrating local Cornish scenes.[4]

Again he received no encouragement, but nevertheless he tried his luck a year later before another influential local body, the Royal Institution of Cornwall, of which he was a member. On this occasion he offered Wadebridge as a possible alternative to Fowey as a site for a Cornish pottery, and he was supported by no less an authority than J.H. Collins, FGS, the County Analyst and an expert geologist. Opposition still came from members who argued that coal transport costs made the venture unprofitable, although one member, R. Symons, was sufficiently impressed to write to the *Mining Journal*, to which he was a regular contributor, asking readers to comment on the viability of Edward Collins' proposals.[5]

Experiments with tin waste in West Cornwall

Once more all these initiatives met with indifference, but in the same year of 1876 R.N. Worth, Curator for the Polytechnic Society at Falmouth, gave a talk on 'Modern Pottery' and again asked why a works could not be set up in Cornwall. Accepting that fuel cost was a key factor, he nevertheless argued that this would not be so critical if high-value items were made, instancing the success of the South Devon potteries. 'Judicious patronage', he claimed, would make such an undertaking a success.[6] In 1878 the subject was raised yet again at a polytechnic meeting, this time by its President, Richard Taylor. The son of the great industrialist and mining magnate John Taylor, he was also President of the Miners' Association of Cornwall and Devon, and had added to the polytechnic's membership, at a time when it was losing mining supporters, by enrolling half a dozen managers of his mines in France, Spain and elsewhere. Taylor proposed sites for potteries in West Cornwall at Gwennap, Feock or Kea, at the epicentre of mining collapse. Commenting upon 'refuse' that he had seen being mined at Heathfield in Devon by the Candy Co. (to be discussed later) to make 'beautiful pottery', he suggested that the clays extracted at two or three brickworks between Falmouth and Truro could be used.[7]

A Cornish mining specialist, Edward Borlase had been experimenting with the manufacture of ceramic products, but from finely ground tin waste (or tailings) rather than china clay. Born in St Austell in 1820, Borlase had been employed in the slate industry at Delabole before working for mining companies in the USA at Connecticut.[8] At the 1879 Exhibition of the Polytechnic Society at Falmouth he showed examples of articles made from mine waste that had been awarded a Special Premium by the *Mining Journal*, including bricks, tiles, drainpipes and small garden ornaments. He had, he claimed, been experimenting on these items for six years in his own kitchen, and a ceramicist had also used his materials to make pots.

The tin waste and sand with which he worked had come from the Red River on the north coast, near Camborne, although he had also sampled slimes from the east and west of Cornwall. In a presidential address to the society, Canon Rogers remarked that it was 'a great pity that Cornwall does not start a manufactory on a larger scale for the promotion of this industry'.[9] A couple of years later the proprietor of the *West Briton*, E.G. Heard, well versed in industrial affairs, joined in the campaign. 'Is there no patriotic Cornishman,' he demanded, 'who will find the capital to build potteries as in Devon at Newton Abbot?'.[10]

Twenty years had passed in which calls to the local patriotism of these influential men had all been met with apathy. In 1887 R.N. Worth, in a comprehensive account of clays of all types in Cornwall and Devon, noted that 'many remember the zeal with which Mr E. Borlase from time to time brought before the Polytechnic Society the importance of sundry of the neglected clays of Cornwall', and that John Phillips of the Aller Vale Pottery, near Newton Abbot, 'believes that excellent ware can be made out of the tin slimes, though as he has failed so far to get any patriotic Cornishman to send him some to try, he contents himself with excellent materials close to hand'.[11]

Variations in coal consumption for different types of pottery production

Outsiders began to join in the demand for adding value. In 1891 Thomas Fenwick, a London speculator, in an unsuccessful attempt to buy up the entire china-clay industry of Cornwall and Devon, proposed to build a large pottery in Truro. In 1895 the Lord Mayor of Liverpool was invited by the Cornish architect-adventurer Silvanus Trevail to be the principal speaker at the latter's mayoral banquet in Truro. Taking as his theme the way to boost the local economy, he posed the same question: 'What about potteries using china clay?'[12] The objection was always the same. 'Fuel,' Heard had reported in 1881, 'is said to be the main problem,' while the *Royal Cornwall Gazette* in 1895 commented that 'practical men doubted that potteries would pay' because of the cost of importing coal.[13]

These arguments were not without foundation. Coal was the only important industrial mineral that the West Country lacked, which handicapped it in maximising the benefits from exploiting all the other valuable minerals that it possessed. The share of employment in engineering and metal manufacturing lagged well behind that of other industrial regions that had coal. Nevertheless, in the debate on adding value to clay, we have noted several references to apparently viable potteries in areas which, like Cornwall, had no coal. Why were they successful when, according to the conventional wisdom in Cornwall, they were bound to fail?

First of all, wide variations in the amount of coal used to fire pottery occurred, depending upon the relative price of fuel and the type of product fired. In the Staffordshire Potteries, since fuel was readily available from nearby mines and some producers such as Adams and Wedgwood had interests in coal mines, fuel efficiency was perhaps not of paramount concern and potters elsewhere were able to compete by paying more attention to their firing methods. Secondly, the placing of chinaware in containers called saggars increased the amount of heat required, leading to considerable variations in kiln temperatures for different products. For some, five tons of coal per ton of clayware were needed, for others up to 12 tons.[14] By concentrating upon ceramics with lower coal requirements, potters in coal-importing regions were more able to compete. Thirdly, for some low-grade items such as drainpipes, the expense of carriage to the customer was high relative to production costs, and so producers located near the main markets could compete even if they had to import coal.

Claymen and potters in Dorset

The development of some potteries will now be examined to see to what extent the factors just outlined contributed to their success. Potteries in the heathlands of Verwood in East Dorset date back to medieval times and probably peaked in the early 1800s, when 13 firms employed over 300 hands. They produced a whole variety of products, but competition from Staffordshire and elsewhere halved their numbers and reduced their range to utilitarian items. Industrialisation was also affecting food production and as people began to buy bread, butter, cheese and bacon from shops, the demand for earthenware utensils to make, bake and store these products declined.

However, this was offset by an increase in demand for art pottery. As the newly developed railway network extended the market, new firms sprang up. In 1855 the Architectural Pottery works in Hamworthy, Dorset, was set up by a partnership of Staffordshire and local ceramicists using the fine white-firing white ball clays from the locality, together with Devon and Cornwall china clays. The firm won

Advertisement for Carter & Co., April 1896.

international awards in London, Dublin and Paris. In 1861 James Walker, Chief Technician at the works, established another pottery nearby but became bankrupt in 1873, when the business was taken over by Jesse Carter, a Surrey-based builders' merchant and ironmonger. His firm sold a wide range of products, including cement, lime, bricks and terracotta gardenware, and he challenged the Staffordshire potters by the quality of his glazed faience and decorative tiling for public houses, shop fronts, mosaic flooring, tiled advertising panels and fire surrounds.

In 1895 Carter acquired the Hamworthy works for £2,000 and became associated with William de Morgan and *art nouveau* stylists.[15] Apart from these artistic products, which became collectors' pieces, tiles were increasingly used industrially, commercially and in a domestic setting for porch entrances, hallways and garden paths, and by late Victorian times over 500 tile works existed in England.[16] Tiles became an intrinsic feature of buildings in the so-called Queen Anne style, used in the works of such architects as Norman Shaw and in the Arts & Crafts movement.

A failed attempt to set up potteries and clayworks on Brownsea Island, near Wareham, in the 1850s will be described in Chapter Nineteen. Near Wareham, the Sandford Pottery began in 1860 and was more successful. Potters were brought in from Staffordshire, and at the height of production over 70 hands were employed, housed in substantial cottages built of the local pale-yellow brick. The intention was to make high-quality tableware, but in practice their sales came mostly from 'bread and butter' lines like bricks, chimneypots, drainage pipes, garden path edges and kitchen sinks. However, the pottery continued in the hands of the Shaw family for many years from 1895.[17] Also near Wareham, Lady Ann Baker, who had already set up a pottery in the New Forest, took over a brickyard at Keysworth in 1910 to produce terracotta ware. Another works, Crown Dorset Art Pottery, was established in Poole in 1905 by Charles Collard, in two houses with a small staff, to make decorative ware from local clay and built up a useful export market.[18]

South Devon potters

In South Devon, what had been called the Folly Pottery described in Chapter Nine continued to flourish.[19] The owner of the pottery, Divett, built five water-wheels and a number of kilns and employed up to 250 hands in his works. In the 1840s, when the South Devon Railway was approaching the area, he called a town meeting to raise support for a branch line to carry chinaware and also local ball clay to Teignmouth. The railway extended to Plymouth and Torquay and the Great Western Railway workshops expanded at Newton Abbot, where the population rose from 2,000 to 12,000 by 1900. Apart from selling further afield, potters thus had a growing local market of residents as well as tourists who bought holiday souvenirs. After Divett's death in 1885 his sister and cousin carried on the Bovey pottery trade, and tried to burn lignite (or 'Bovey coal', as it was called) as fuel, but met with no more success than in earlier times. The pottery nearly went out of business in the 1890s but was rescued by new owners who used local clay from Bluewaters, at Bovey Tracey.[20]

Meanwhile, in the early 1870s, Frank Candy founded 'The Great Western Potteries, Brick, Tile & Clay Works' on about 20 acres of workable clay land near Newton Abbot. Candy did not employ the finest clays himself, but sold these elsewhere and used coarser clays to make bricks, salt-glazed wares and garden path edging which won a Gold Medal at the Paris Exposition of 1878. Perhaps he was specialising in items which did not require a high ratio of coal to clay. As the business expanded, Candy took on Jeffry William Ludlam, proprietor of the Marland Brick & Clay Works south of Bideford in North Devon, who became production manager and majority shareholder in a new firm, called 'Candy & Co.'.[21]

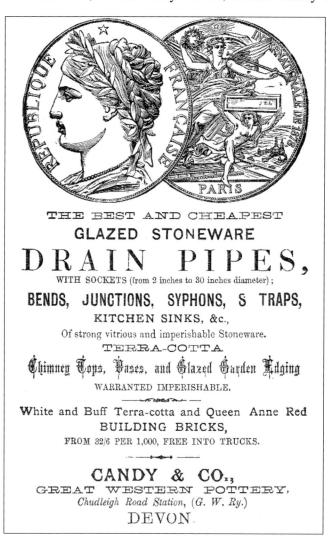

THE BEST AND CHEAPEST

GLAZED STONEWARE

DRAIN PIPES,

WITH SOCKETS (from 2 inches to 30 inches diameter);

BENDS, JUNCTIONS, SYPHONS, S TRAPS,

KITCHEN SINKS, &c.,

Of strong vitrious and imperishable Stoneware.

TERRA-COTTA

Chimney Tops, Vases, and Glazed Garden Edging

WARRANTED IMPERISHABLE.

White and Buff Terra-cotta and Queen Anne Red
BUILDING BRICKS,
FROM 32/6 PER 1,000, FREE INTO TRUCKS.

CANDY & CO.,

GREAT WESTERN POTTERY,

Chudleigh Road Station, (G. W. Ry.)

DEVON.

Candy advertisement, from Cock's Treatise, 1880.

The company's aims were to 'dig, work and search for China Clay and China Stone and potters' or pipe clay' and to 'make, prepare and burn bricks, pipes, tiles or other ware'. With new partners, the capital rose from £40,000 to nearly £70,000. In 1885, however, the partners did not have sufficient resources to fight the case against the Great Western Railway, referred to in Chapter Two. A year later they sold their works to the solicitors and bankers Fowler, Fox & Co. By

1900 it was a major supplier of ball clay to the Staffordshire Potteries as well as a leading manufacturer of bricks, drainage pipes and tiles, terracotta garden statuary, balustrades and chimneypots.[22] Fowler and Fox then purchased the Devon & Courtenay Clay Co., which had been extracting ball clay from a site called Decoy, near Newton Abbot,[23] and in 1916 they brought in the brothers E.A. and H.O. Jones from Staffordshire to manage both Candy and Devon & Courtenay.[24]

In another part of South Devon, the discovery in 1865 of fine red clay during building work in the grounds of Watcombe House on the outskirts of Torquay led to the formation of the Watcombe Terracotta Clay Co. This red clay was mixed with Dorset clay, flint and other materials to make a range of products.[25] The works produced busts, statues and classical urns and vases, and to ensure a high quality the owner, Allen, brought in Charles Brock from Staffordshire as manager and William Higginbottom as art director and master turner. In 1875 a rival firm, the Torquay Terracotta Co., set up nearby at Hele Cross, and Higginbottom soon joined it, winning national awards for his products.[26]

John Phillips at Aller Vale

Close to Newton Abbot, at Aller Vale, John Phillips set up another important pottery in 1881. In Chapter Eleven we described how, in 1862, he had been dismissed as manager of Lee Moor china-clay works after the death of his father, who had done so much to develop it. According to Penderill-Church, the new lease-holder, William Martin, felt that John Phillips had been shabbily treated and gave him a job as china-clay sales representative for the Martin brothers to cover the area west of Bristol,[27] but Phillips had possibly learned something of the potters' art from his father and made high-quality art ware, as well as sanitary ware, drainpipes and terracotta. By the 1890s he employed some 70 hands. In 1883 a fourth pottery began in Torquay, the Longpark China & Terracotta Works, although at first it only produced common chinaware, and in 1891 the Exeter Art Pottery opened, making products similar to those of Aller Vale under the direction of the principal, William Hart, recruited from Phillips' company.

By this time potters were turning to the Arts & Crafts style. Queen Victoria's sixth child, Louise, who took an interest in ceramics, opened an Arts & Crafts Exhibition at Torquay in 1890, and the Aller Vale workshop sold 'Princess Louise Ware'. After her visit to their works in 1894 they produced, at her suggestion, designs incorporating Dartmoor pixies. Phillips also ran arts and crafts evening classes in several colleges, covering wood, copper, brass and ironwork, as well as pottery design and manufacture, which were well supported by local communities.[28]

In 1896 the Exeter Art Pottery closed and William Hart and the brothers, Alfred and Joseph Moist, turner and decorator respectively for the company, started up in business as the Hart & Moist Devon Art Pottery. A year later Phillips, owner of Aller Vale Pottery, died, and his works were acquired by Hexter Humpherson & Co. of Kingsteignton, mentioned in Chapter Eleven, who also took over the Watcombe Art Pottery when it became bankrupt in 1901.

These potteries turned out a vast range of popular wares such as inkwells, tobacco jars, dressing-table trays and sets, shaving mugs and ladies' hatpins. They commemorated every event: the Jameson Raid in South Africa and Boer War battles, royal weddings, jubilees, coronations and funerals, 'Votes for Women', the arrival of the New Zealand 'All Blacks' rugby team and the opening of a local waterworks. They made mementoes for the tourist trade in Scotland, Ireland, Holland, Canada, Australia, the USA, South Africa and Hong Kong. This was the golden age of the South Devon potteries, although they did not come to an end until much later, in the 1960s.

Interestingly, while Candy & Co combined clay working and pottery production from the start, and some ball-clay merchants bought South Devon potteries, the largest operators, Watts, Blake, Bearne & Co., did not venture into local pottery manufacture. This was certainly not due to any lack of initiative. The partners traded in related commodities such as ochre and 'shining ore', a micaceous type of iron ore found locally and used in the manufacture of paints. They also dealt in granulite, obtaining this from the Meldon Stone Quarries at Okehampton, owned by Charles Green, mayor of that town and a lifelong friend of C.D. Blake.[29] The latter became consulting engineer to the Okehampton firm and also bought the Clokie & Co. pottery of Castleford in Yorkshire. Watts, Blake, Bearne & Co. added to their interests by trading in china clay, buying it from Cornish producers and selling it to their pottery customers. At one time they acted as middle-men between John Lovering in St Austell and H.D. Pochin in the North of England, but when they encountered difficulties in getting china clay they acquired their own works on Dartmoor at Headon, Cornwood and later at Shaugh.[30]

North Devon brick makers and potters
The progress of the Marland pottery in North Devon was discussed in Chapter Eleven, where we saw that it was an example of the integrated enterprise that Edward Collins and others wished to encourage in Cornwall. Not only did its owner, W.A.B. Wren, extract ball clay and use it on the spot to make bricks, tiles, terracotta and chinaware, he ran the light railway that brought in his coal and other supplies, carried commodities for local landowners and sent out his own products. Yet, while he appeared to make money from the export of ball clay, the brick and tile works, although turning out large quantities, never seemed to make a profit, even though it was leased by several different groups of local and external adventurers. Possibly the two-way cost of bringing materials to a relatively remote spot

and then sending products back again accounted for this lack of profitability.

Art potteries in North Devon, although not so numerous or so prolific in output as those in the south, nevertheless produced some fine wares. One of them was run by Charles Hubert Brannam, son of a potter, who had worked for the potteries owned by the Rendell family of Barnstaple. Brannam took over his father's works from 1879 and soon built up a high reputation for his 'Barum Wares', typically with designs incised, in the North Devon tradition, through a blue and chocolate-coloured slip. In 1886 he opened premises and showrooms which were visited in 1905 by Princess Christian Helena, another daughter of Queen Victoria. Most of the products were functional objects such as pots, pitchers and flower vases, but commemorative pieces and caricatures of political figures were also made. Brannam and his designers drew inspiration from many sources: Egyptian, Greek, Indian, Italian and, from the turn of the century, *Art Nouveau*, which sold well in exclusive London stores.[31]

Brannam had been joined in 1884 by William Leonard Baron, born in Sidmouth, South Devon, and trained at the Royal Doulton works in London. In 1893 he left Brannam and soon set up on his own to produce 'Barum Ware' in a small family business in Barnstaple involving a dozen or so staff, which continued until his death in 1937. Like the other potters, he made a variety of wares, including commemorative items and products for the tourist trade. Another Barnstaple potter who produced on a smaller scale was Alexander Lauder, who was trained as an architect, designing Wesleyan chapels and schools locally and in London, as well as shops in Barnstaple. He also advised on garden layouts, using ornamental bricks and figures from his pottery. He had given art classes at the Barnstaple Literary and Scientific Institute, his pupils including Charles Brannam. As a potter Lauder designed mugs and vases until 1914.[32]

At Fremington, a few miles to the east of Barnstaple, Edwin Beer Fishley produced glazed terracotta ware from 1860 onwards, with traditional designs incised through glazes made from local clays. Later his work was influenced by the art pottery style of Brannam, Baron and the patronage of John Phillips of Aller Vale. Edwin Fishley's grandson, William Fishley Holland, learnt the trade from him but sold the business on Edwin's death in 1912 to a Staffordshire potter, Ed Sadler, who sold it in turn to Brannam. William Fishley Holland, meanwhile, moved to a newly established pottery at Braunton, some miles north of Fremington, to exploit the blue clay recently discovered at Knowle, but when this did not fire well he switched back to Fremington clay, making mostly utilitarian wares.[33]

The Belleek Pottery in Northern Ireland

One of the most remarkable episodes in the history of Victorian ceramics was the continued existence of a pottery in a small village in a remote and depressed corner of north-west Ireland that won a world-wide reputation for its Parian Ware, especially in the form of latticed baskets, that has survived to the present day. It began in the 1850s, when a patriotic Irish landowner, John Caldwell Bloomfield, set out to create local employment by developing crafts including basketry and lace-making, as well as tourism. A geological survey revealed deposits of some china clay and feldspar-rich rock on his land near Belleek, a village of some 200 people on a bend of the River Erne, a few miles upstream of the port of Ballyshannon. The feldspar content, at 80 per cent, was higher than that of Cornish china stone at 60 per cent. Through the sale of these minerals on a small scale to potters in Worcester and Staffordshire, he encountered Robert William Armstrong, an Irish architect who had designed potteries in those places, had a flair for chemistry and was keen to establish a pottery in Ireland.[34]

Since neither Bloomfield nor Armstrong had much capital, they persuaded David McBirney, a Dublin merchant and railway developer, to finance a works at Belleek, where a water wheel could power all the machinery and where local peat, together with a little imported coal, could furnish the fuel. Armstrong designed a palatial factory, resembling an English country mansion, at a cost approaching £40,000, capable of holding 500 workers, although no more than 180 or so were actually employed. Armstrong ran the works and soon built up a profitable trade in high-quality domestic and sanitaryware and floor tiles, as well as producing some bricks and cement. McBirney chaired a company, in which he sank another £20,000, to build a railway to the coast that passed through Belleek.

All seemed set fair for the future of a thriving earthenware business, but the partners had much loftier ambitions. As patriotic and proud Irishmen, they wanted to put Belleek on the map and make it the rival of the great European centres of ceramic production. Armstrong enticed skilled engravers, designers, modellers and enamellers from the Goss works in Staffordshire by paying them high wages and building them superior housing. The Parian Ware they produced was indeed of the highest quality and won awards at expositions in Dublin, Melbourne in Australia and later in Paris, and sales also expanded to the USA and Canada.

Mixed fortunes

Their success came at a high price, though. Armstrong, a perfectionist, was said to have dumped wagonloads of rejects in the River Erne. Mineral deposits proved to be insufficient to meet the expanding demands, and china clay had to be imported from Cornwall, ball clay from Devon and feldspar from Norway. In addition, Parian Ware needed more coal to fire kilns to the high temperatures it required, and for this they had to pay £0.90–£1.05 a ton.

All these factors increased the cost of production, and manufacture of Parian Ware swallowed up all the profits made in earthenware. The firm was only kept going through continued subsidies from McBirney. To make matters worse, the railway, as was so often the case, cost more to build than was estimated, while revenues were lower than expected. Bloomfield lost the £5,000 he had invested in it and this, together with the failure of three local iron pyrites mines he had set up, drove him to insolvency. McBirney had to buy his estates, but then, in the early 1880s, McBirney and Armstrong died in quick succession, and, with no benefactor to subsidise production, the pottery closed. It was revived by a group of Ballyshannon merchants, who cut down on production of high-cost, unprofitable Parian Ware and concentrated upon common earthenware and sanitary chinaware. In this way they made a modest 5 per cent profit, which they estimated they could have doubled if they had been able to import coal at the same price as Staffordshire potters.[35] After some financial difficulties over the years, Belleek continues working to the present day.

Value added in Cornwall

In contrast to the considerable numbers of Devon and Dorset potteries, we have only found evidence of one attempt to use local clay in a Cornish pottery. In 1899 a site on the banks of the River Fal at Ardevora Veor, near Philleigh, was leased for 21 years to Herbert Moore of Teignmouth, James Calvert of London and William Daw of Sussex, who formed the Fal River Clay Co. to wash and dry clay.[36] This was mica clay, recovered some 20 miles downstream from the western side of the St Austell china-clay district. It was probably the residue from mica refining channels, a silt consisting of coarse kaolin, mineralogical mica, fine quartz sand and unaltered feldspar. The kaolin in the mixture would have been ground by attrition during its journey downstream, making reworking a proposition worth considering.

In the same year, a geological survey refers to a sample of clay taken from a ford close to Ardevora Veor, described as 'of a whitish and pale yellow-grey colour', which was 'not only used for the manufacture of bricks, but works have also been established for the production of the coarser varieties of earthenware'.[37] There were brickworks found nearby at Trelonk, but there is no sign of a pottery kiln. No other references to this activity have been found, and the lease was surrendered in 1906, so the operation, if any, was short-lived. The ruins of a clay-drying kiln remaining on the site do not reveal whether pottery production actually took place.

Another minor attempt by Cornish clay merchants to add value to their business was to sell ground flint. As early as 1685 Staffordshire potters had used local quartz sand to whiten their pottery, and at the beginning of the eighteenth century they also used ground flint for the same purpose.[38] British earthenware and china manufacturers preferred ground flint, which came mainly from the East and South Coast of England, and later from the coasts of Belgium and France. On the other

Advertisement from Kelly's Directory, 1873.

hand, European ceramicists used quartz sand, rather than flint, for the same purpose.[39] In the 1870s two Cornish china-clay and stone companies were offering flint for sale. John Lovering advertised ground flint and ground china stone and Thomas Olver of Trevear, in the Tregargus Valley near St Stephen, was listed as a china-clay merchant and flint grinder.[40]

The flints were calcined in kilns rather like lime kilns, to make them friable and amenable to grinding in pans, similar to those used for china stone. At this time flint was being imported to Charlestown as ballast and dumped on the beach. Selected flints could have been calcined in one of the Charlestown lime kilns and ground in a china-stone mill at Trevear, and it may be that Lovering employed the Olvers as contractors. Because both ground flint and quartz sand fulfilled a similar function, the term 'flint' was often used by potters to refer to both forms of silica. Even so, potters were reluctant to change their well-tried and tested recipes involving ground flint, and perhaps Lovering was trying to build up a market for ground silica sand. His clay pit at Hendra was known for its clean white quartz rock, often utilised locally for the tops of walls, rockeries and garden edgings.

A missed opportunity for Cornwall?
A historian of the Belleek enterprise concluded that 'despite enormous supplies of china clay in Cornwall, a pottery industry never arose there... the clay had always gone to the coal and not the coal to the clay... it required bravery verging on the foolhardy, allied to outright financial recklessness' to start a pottery at Belleek.[41] Had the Cornish clay merchants acted wisely in steering clear of ceramics? Or had they missed an opportunity to add value? Our discussion of ventures in Dorset and

Devon has shown that some potters had a larger local market, but they also exported their wares worldwide, even although coal cost them twice as much as in the Staffordshire potteries. Whereas the china clay masters steered clear of pottery production, Candy & Co and Hexter Humpherson ran local potteries with success.

Some of the ceramic ventures in Devon and Dorset failed, but then so did some Staffordshire firms. What seems clear is that the higher fuel costs of firing Parian Ware or porcelain made these products uncompetitive in locations where coal was dear, but when potters confined themselves to decorated earthenware and china, and brought in experienced craftsmen and managers from Staffordshire with a flair for design, they were successful. The Cornish clay merchants did not follow their example, although they certainly possessed the necessary organisational ability, business acumen and marketing expertise to do the same.

Apart from pottery, could Cornish adventurers have added value to china clay by making other products from it? By the early 1900s three-quarters of china clay was exported, mainly to paper makers. Earlier, several paper mills existed in Cornwall: could the clay merchants have built up their trade? By the 1880s, wood pulp had largely replaced rags as the basic material, and in 1904 over 10,000 tons of wood pulp was being shipped to paper mills near Teignmouth in Devon from Norway and Sweden. The steamships used were too large for the size of consignments of ball clay sent to Scandinavia, and called at the coal ports of England and Wales to fill up with cargoes.[42] China-clay shipments might have been larger and provided a useful two-way trade if Cornish paper mills had been in operation, but none of the commentators who mentioned the success of South Devon potters ever referred to the possibility of operating paper mills. Nor did the Cornish clay merchants seem to consider it.

To sum up, the conventional view that clay producers did not engage in ceramic manufacture because it was cheaper to send clay to places near coalfields than take the coal to the clay is not borne out by experience in the South West. However, the involvement of West Country clay merchants in adding value through ceramic production varied greatly from place to place. In Dorset the main clay producers, the Pikes and the Fayles, did not diversify into pottery manufacuture, even though the original Fayle had been a London potter. In North Devon, in complete contrast, Wren not only owned clayworks, brick and tile works and potteries but ran a private railway as well. In South Devon, Candy & Co set out to be both clay producers and potters, while the clay merchants Hexter and Humpherson acquired local potteries.

On the other hand, Watts, Blake, Bearne & Co.did not manufacture pottery locally, although C.D. Blake bought a Yorkshire pottery, and the partners diversified into trade in related chemical and mineral products. The Cornish and West Devon china claymen made bricks and tiles, particularly on Dartmoor, but did not engage in pottery. One reason for their lack of involvement may have been that china clay and stone was used in the production of porcelain, which required higher firing temperatures and the use of saggars, increasing fuel consumption per ton of pottery produced. Ball clay merchants, on the other hand, were making coarser products using less coal, and selling some of them to large markets nearby. Another explanation for the success of Dorset and Devon potters was their entrepreneurial, design and marketing skills, which enabled them to create a worldwide demand for products which were different from the traditional ware of Staffordshire and other pottery centres. Finally, it may simply be that the china clay merchants felt no need to diversify because they were fully occupied meeting the rapidly expanding demands for their basic product by papermakers.

As we shall see in Chapter Twenty, the Cornish clay families kept closer to their core activity, in both their business and their public life.

The Blue Pool, 1911. Oil on panel by Augustus John RA. (Courtesy of Aberdeen Art Gallery and Museum, Scotland.) The pool at Furzebrook, near Wareham in Dorset, was an area worked for ball clay for many years, from the seventeenth century to the beginning of the nineteenth century. Over the years it filled and water, and in 1935 it became a tourist attraction, with attractive walks and facilities for visitors. A feature of the pool is the changing colour of the water, caused by diffraction due to the weather and to presence of minerals in the water.

References
[1] Ronald Perry, 'The Making of Modern Cornwall', in Philip Payton (ed.), *Cornish Studies: Ten*, Exeter, 2002, pp.166–89.
[2] Alan Pearson, *A Study of the Royal Cornwall Polytechnic Society*, Thesis for Exeter University, 1973, p.197.
[3] *RRCPS*, 1868, p.31.
[4] *RRCPS*, 1875, pp.59 et seq.
[5] R. Symons, letter to *MJ*, 27 May 1876, p.592. Edward Collins seemed to base his calculations upon the local sale of chi-

naware, which would avoid the double burden of carrying clay to Staffordshire and then bringing pottery back to Cornwall, but he did not appear to have taken note of the small size of the local market compared with that of South Devon.

[6] R.N. Worth, 'Modern Pottery', *RRCPS*, 1877, p.46 et seq.

[7] *RRCPS*, 1878, p.19.

[8] He was granted Patents in 1858 (no. 360), 1859 (2586) and 1873 (269) for separating metallic ores from other substances with revolving round slime tables.

[9] *RRCPS*, 1879.

[10] *WB* 12 May 1881.

[11] R.N. Worth, *The Clays and Fictile Manufacturers of Cornwall and Devon*, Falmouth, 1888, pp.5, 8. Reprinted from *RRCPS*, 1887.

[12] *RCG* 29 October 1895.

[13] *WB* 12 May 1881, *RCG* 29 October 1895.

[14] Lorna Weatherill, *The Pottery Trade*, New York, 1970, p.30.

[15] Leslie Haywood, *Poole Pottery*, Somerset, 1998.

[16] Kenneth Clark, *The Tile*, Marlborough, 2002.

[17] D. Young, 'Brickmaking in Dorset', *PDNHAS*, 93, 1971, pp.213–42.

[18] P. Copland-Griffiths, *Discover Dorset: Pottery*, Wimborne, 1998, pp.59–61; Jo Draper, with Penny Copland-Griffiths, *Dorset Country Potter*, Marlborough, 2002, pp.43–69; Peter Stanier, *Dorset in the Age of Steam*, Dorset, 2002, pp.53–54.

[19] A map of the 1830s shows the Folly Pottery. DeRO, 1508M/E/MPA/Bovey Tracey/ Plans A1 of 1837.

[20] German lignite producers later opened a small (100 hp) power station, worked by a water wheel, to produce electricity to extract 'Montan' wax from lignite, but withdrew in the 1914 War. 'Montan' or 'Ester' wax was especially suitable for the 'lost wax' metal casting process. Haytor Granite Railway Co. also used lignite to make gas. Crawley and King, 'The extraction of Ester Waxes', in *British Lignite and Peat*, London, 1946, p.3. Lance Tregonning, *Bovey Tracey, An Ancient Town*, Exeter, 1983, pp.55–62; Roger Jones, *A Book of Newton Abbot*, Bradford on Avon, 3rd ed., 1986, pp.49–54.

[21] Ian Turner, *Candy Art Pottery*, Melbourne, Derbyshire, 2000, pp.8–12.

[22] In the 1930s it employed 320 hands, together with 250 at a subsidiary, Devon & Courtenay Clay Co., and by the 1950s it had over 600 hands in the main works alone. Art pottery was produced from the 1920s to utilise spare capacity in new kilns but discontinued in the 1950s when this capacity was needed for tile production. Turner, 2000, pp.13–15, 29–30, 45.

[23] DeRO, D1508M/1, 1806C/EAA 78, 1/SS/6/4 of 1904.

[24] L.T.C. Rolt, *The Potters' Field*, Newton Abbot, 1974, pp.55–56.

[25] J.W. Perkins, *Geology Explained in South and East Devon*, Newton Abbot, 1977.

[26] Torquay Pottery Collectors Society, *Torquay Pottery*, Torquay, 2001.

[27] John Penderill-Church, *China Clay on Dartmoor*, 2000, Wheal Martyn Archives, p.11.

[28] John Somers Cock, *Abbotskerswell*, 1995, pp.27–28.

[29] Granulite had been used in the manufacture of glass bottles by Siemens in Dresden, and was claimed to be rich in potash feldspar and of great value in making porcelain, chinaware, tiles and other ceramic products. H. Lindsay-Bucknall, 'An Unused Product of Devonshire', *Notes and Gleanings*, II, 1890, pp.136–38.

[30] Rolt, 1974, pp.46–47.

[31] Audrey Edgeler and John Edgeler, *North Devon Art Pottery*, Barnstaple, n.d. (c.2000), pp.6–24.

[32] Edgeler and Edgeler, n.d., pp.26–41.

[33] Edgeler and Edgeler, n.d., pp.42–48.

[34] John B. Cunningham, *The Story of Belleek*, Belleek, 1992; Marion Langham, *Belleek Irish Porcelain*, London, 1993.

[35] Cunningham, 1992, p.17.

[36] Letter from Lord Falmouth to Charles Thurlow, 1996.

[37] Hill and MacAlister, *Geological Memoir of Falmouth and Truro*, HMSO, 1906, p.111.

[38] R. Copeland, *A Short History of Pottery Raw Materials*, Cheddleton Trust, 1983, p.2.

[39] E. Rosenthal, *Pottery and Ceramics*, Harmondsworth, 1949, p.74.

[40] *Kelly's Directory of Cornwall*, 1873, advertisements for Lovering and Olver; *Harrods Directory of Cornwall*, 1878, advertisement for Lovering.

[41] Cunningham, 1992, p.17.

[42] H.J. Trump, *Westcountry Harbour*, Teignmouth, 1976, p.137.

Chapter Nineteen
Outsiders, Takeovers and Mergers

In the first part of this book we saw how much the development of the clay industry owed to outsiders who moved into the South West or shifted from one part of the region to another. China clay pits were opened in Cornwall by Staffordshire potters and Plymouth dealers and on Dartmoor by the Sunderland manufacturer Phillips. The Dorset merchant Crawford provided an initial stimulus to the South Devon ball-clay trade, while the South Devon Pike family moved in the opposite direction to Dorset, where they became one of the pillars of its ball-clay industry, alongside the descendants of the London merchant Fayle.

Colonel Waugh and Brownsea Island

From the mid-1800s to the 1914 War outsiders continued to play a significant role, mostly with beneficial results, but we begin with one of the most over-ambitious and colourful schemes devised by an external investor – the creation of an entire community of clayworkers, potters and brickmakers on an island near Poole, in Dorset.[1]

In 1852 Colonel William Petrie Waugh, late of the East Devonshire Regiment and the Indian Army, bought Branksea Island, now known as Brownsea Island, for £13,000. This seemed a high price at the time, but the reason for the purchase was that his wife, an amateur geologist, had detected what she thought was 'a most valuable bed of the finest clay', worth 'at least £100,000 an acre' according to a professional geologist whom she consulted. Waugh, convinced that he would make a fortune by producing porcelain, borrowed £237,000 (or half a million pounds in some versions of the story) from the London & Eastern Banking Corporation, of which he was a director. He then embarked upon a vast project that transformed the entire island.

On the north coast he dug a number of vertical clay shafts up to 75 feet deep and built a brickworks to the east of these. On the west of the island he constructed a large pottery building, 190 feet long and three storeys high, with seven down-draught kilns, for the production of fine china. A short distance away another pottery was erected to make earthenware and terracotta. A coal pen held 300 tons and water was supplied by a reservoir built to hold water from a spring. To transport clay, pottery, fuel and supplies he built a tramway and a pier into deep water, enabling three vessels of up to 200 tons each to load or unload at the same time.

This was not all. North of the pier he erected a crescent of cottages for his workers, named Maryland Village after his wife, as well as a public house and a school, surrounded by lakes converted from streams and parklands planted with shrubs and trees. For his own accommodation he lavishly embellished a castle on the south-east of the island with gothic turrets and towers, and at a cost of £10,000, built St Mary's Church nearby.

A guidebook of 1857 announced that the island possessed 'every variety of clay, from the coarsest brick earth to the finest blue porcelain, ochre for colouring and sharpest sand'. Unfortunately, in reality the island's clay was only suitable for terracotta and bricks and the profit from these failed to meet the interest payable on the money that Waugh had borrowed from his bank. By this time, however, the colonel and his wife had fled to Spain, a country which had no

Fragments of glazed drainpipes on the beach at Brownsea.

treaty of extradition for fraudulent bankrupts. The London bank crashed and in 1870 the island was sold for £30,000 to the Rt Hon. G.A.F. Cavendish Bentinck, MP for Whitehaven. Three years later he floated a company with a share capital of £20,000, which in 1881 won a prize for the quality of its drainpipes, supplied to local authorities and merchants in Poole, Basingstoke, Southsea and elsewhere. At one time approaching 300 workers were employed, but by 1887 this number had dwindled to 100 and the works closed.

Developments in Devon

As the colonel was unsuccessfully developing Brownsea Island in Dorset, the Blakes, discussed in Chapter Ten, were moving from Stoke on Trent to South Devon, with C.D. Blake emerging as a dynamic force in the growth of the largest ball-clay partnership, Watts, Blake, Bearne & Co. Around the same time, in about 1870, the Martins of St Austell acquired further china-clay works, including Whitehill Yeo and Cholwichtown on Dartmoor, where they became the foremost producer, joined later by other Cornish families such as the Olvers. Watts, Blake, Bearne & Co. were already into china-clay extraction on Dartmoor and in addition helped to finance and market the ball-clay trade of North Devon. Their precursors had been the Greenings of Gloucester, followed by the Wrens of Bideford.

An age of monopolies and mergers

In the late-Victorian era, the idea of takeovers, mergers and combinations of firms was in the air. Intensified international competition was leading coal and steel, metal processing and chemical producers to band together in trusts in America, cartels in Germany and syndicates in Britain. The cotton textile industry, a major customer for china clay, was the most heavily involved: out of 895 British firms affected by mergers between 1887 and 1900, textile companies accounted for 330.[2] Such combinations included the Calico Printers' Association of 1889, with no fewer than 84 directors and eight managing directors, controlling 85 per cent of the sector's output, and the Bleachers' Association of 1900, comprising 71 of the 94 firms in that industry. Paper producers had their own Amalgamated Society of Paper Makers, although it appeared to have little power, but the Wallpaper Manufacturers' Trust established a virtual monopoly of that trade. Other mergers occurred in activities similar to or connected with china clay, like the Salt Union of 1888, the largest British company of its day, an amalgam of 64 firms that refined and dried salt, and the United Alkali Company, formed in 1891, comprising 58 companies making bleaches and chemicals.[3]

Their aim was defensive rather than aggressive, to combat increasing competition from foreign firms by eliminating inefficient units and preventing British newcomers from entering the trade and forcing prices down. For this reason, perhaps, they did not at first seem to show signs of trying to acquire firms in the ball-clay or china-clay industry. However, the china-clay industry seemed fated for a merger or takeover. Enjoying a worldwide reputation for the quality of clay, it was geologically fairly compact and therefore apparently easy to control. Moreover, its products were, in economists' terminology, price-inelastic. That is to say, an increase in its price had a negligible effect upon the final selling price of the products in which it was used. A rise in the price of clay of about 30 per cent could, according to one clay merchant in the 1890s, have boosted the profits of the industry by £100,000 a year. Yet, he argued, it would only increase the final cost of chinaware by one per cent and of paper by under one per cent. The industry had a strong hand to play, he asserted, if only its producers got together to fix prices, buy up potential clay-bearing land and thereby stop a continuous influx of small china-clay producers, who caused supply to outstrip demand. This played into the hands of combinations of pottery and paper manufacturers who forced clay prices down.[4]

The solution was not as simple as this, however. On the demand side, competition from other whiteners and fillers was intense, as we saw in Chapter Thirteen, and it was probably only because the price of clay kept falling that West Country china clay merchants were able to increase their sales. On the supply side, the china-clay families never seemed to be able to get together for long enough to fix prices or ration output. In the mid-1800s Rebecca Martin, Stocker, Treffry and others had tried to limit total output with agreed quotas and fixed prices, but had failed because some leading players refused to join in. Every time a glut of clay appeared, fresh cries arose for a clay syndicate to control the industry: in 1881 Alderman E.G. Heard, editor of the *West Briton* and an expert in industrial affairs, was among those who urged clay merchants to form a syndicate to eliminate what he called the 'cut-throat competition' that was 'undermining the profitability of the trade'.[5]

External investors: Wright, Baker, Orange

Among the smaller entrepreneurs who snapped up china-clay pits in the later decades of the century were the South Wales merchants Marcus Moxham and John Dowle Jones, who acquired the leases of St Austell and Bodmin China Clay Co. at Carluddon and Bodmin Moor,[6] the Rhyl ironmonger Francis Wright who bought the lease for Kerrow Moor China Clayworks[7] and Richard Hoyle, a Manchester textile manufacturer, who leased a small group of pits, the most important of which was Ruddle, near Trethowel.[8] William King Baker, from an Essex farming family, who had made money abroad, leased a sett near Penzance at Georgia, Towednack. By this time, however, the leading china-clay families already controlled the lion's share of 'best clays', and outsiders often had to be content with 'medium' and 'common' clays. When Wright became bankrupt in 1891 he blamed his downfall on the major producers of 'best' clays, who, in a depression, sold their 'common' clays at or below cost price to release storage space for their higher-grade output.[9]

Another lesser-known clay merchant from up-country was George Flint Orange. Born in Nottingham in 1827, son of a clergyman who was involved with a 'Land Savings Bank' in that city, George assisted in this enterprise for some years. Both father and son spent time in

Orange advertisement from Harrod's Directory, 1878.

Cornwall and in 1866 George married Mary Merrifield from Madron, near Penzance. By 1873 he was advertising in *Kelly's Directory* as a dealer in china clay and other minerals, in 1875 he was recorded as using aniline dye to improve the whiteness of clay at a time when clay producers were usually secretive about such a practice. Three years later he advertised a North American Land and Investment Office at St Austell as well as trading in china clay and china stone. In 1879 he took out a patent (No. 4986) for a mechanised form of drying china clay which does not, however, appear to be practicable.

The 1881 Census records him as living at 27 Trevarrick, St Austell, next door to Edward Stocker, and by 1883 he was running Carwen and Torr, two small pits near Blisland on Bodmin Moor. Ten years later, together with partners including his wife and father-in-law, he signed an agreement to work Slip china stone quarry near Treviscoe. In 1897 this business was sold, and the Bodmin Moor pits are said to have closed in 1901, by which time George Orange would have been 74 years old. He had also advertised himself as a dealer in 'Haematite Ore, Metallic Oxides and Minerals'. In the mid-1870s many mining operations on the edge of kaolinised granite areas in the china-clay district of St Austell produced small quantities of often blood-red iron oxide known as haematite. This occurred as iron nodules and veins and is thought to have arisen from the deferruginisation of granite, part of the kaolinisation process, examples being found on the edges of clayworks such as Wheal Edith near Retew and in separate mines including the Ruby Iron Lode near Trethurgy. In later years the number of such operations declined.[10]

Sessions, North, Rose and Beale

In their quest for china clays, newcomers other than Orange explored the area north-east of St Austell on Bodmin Moor. Clay had been recorded there in a geological survey of 1839, but transport costs to the clay ports were high and it was only when existing producers had monopolised the best setts around St Austell that the Bodmin area was exploited. Local adventurers who developed the district included the Truscotts and Parkyn and Peters,[11] but in 1908 the North Cornwall China Clay Co. took over a firm leasing the Stannon clay sett near St Breward and offered shares in what it proclaimed as 'the largest deposit of high-quality china clay in the West of England'. It covered 440 acres of land owned by Sir W.R. Onslow, W.R. Nicholls, F.J. Gaved and others, and the value put upon the venture in the prospectus was just under £100,000. Of the five directors, only Nicholls, a director in the previous company, was a local man, and none of the other four seemed to take an active part.

The most important outsider was Walter Sessions, who joined the North Cornwall company as manager in 1911. A member of a Quaker family of timber, slate and marble merchants in Gloucester and Cardiff, and a gifted organiser and salesman, Sessions became one of the prime movers of the clay industry in the Edwardian period. Between 1911 and 1914 the company more than doubled its output from under 20,000 to over 54,000 tons a year. Even more remarkably, it much more than doubled its revenue from £25,000 to £66,000, selling clay at £1.20 a ton when average clay prices were only about half that level.[12] Sessions later represented the Bodmin district as one of the three managing directors who ran the English China Clays consortium, which was formed in 1919.

Earlier, William Nicholls had been a co-adventurer in a sett on Meledor Moor with Thomas Olver, and after the latter's death his interest was acquired by William John North for £1,500. North had been trading from Wolverhampton in iron, minerals, feldspar, china clay and stone and acting as agent for another Wolverhampton-based firm, the West of England Iron Ore Co.[13] Two years later North's cousin, John F. Rose, formerly of Surrey, joined him, and in 1897 Rose's brother also arrived as a third partner in what became the important clay firm of North & Rose. As we saw in Chapter Eighteen, they were to be moderating influences in the 1913 clay strike.[14] About this time R.C. Dering Beale, an engineering graduate of King's College, London, acquired a number of leases, including West Goonbarrow and Littlejohns clayworks, as well as Trelavour Clay & Stone Works, Chytane Clay, Brick & Tile Works (later abandoned) and Carbis Brick & Tile Works.[15] Another pre-war outsider was a mysterious clay firm, the Standardised China Clay Co., rumoured to be financed from Germany, with a German manager who experimented with advanced techniques, some of which were adopted by local firms many years later.[16] Their works were at Belowda, to the north of the china-clay area.

Investment by British industrial tycoons

Even in the years leading up to the 1914 War, when china-clay output was approaching a million tons a year, the cost of setting up a modest sized works was small compared with starting up such industrial operations as coal mines or iron and steel works. 'A few thousand pounds well applied,' suggested J.H. Collins as late as 1911, 'will suffice for the instalment of works on a considerable scale.' An outlay of £4,000–£10,000, he claimed, could finance a firm capable of producing 4,000–15,000 tons a year.[17] However ,ball clay and china clay were beginning to interest big businessmen, and among those who appeared upon the china-clay scene were Henry Davis Pochin, Sir Charles Cottier, Sir Charles Hanson and Sir Harry Mallaby Deeley.

Henry Pochin, industrial chemist

Pochin, son of a Leicester farmer, was apprenticed to a Northampton apothecary and, at the age of 25, after gaining experience and professional qualifications in

Portrait of H.D. Pochin working on the clarification of soap, which he also patented.

Manchester and Edinburgh, he borrowed money from his father to partner a Manchester industrial chemist. Marrying the partner's daughter, he took control of the business and around 1855 patented a method of making aluminium sulphate by roasting china clay and then reacting it with sulphuric acid (described in Chapter Twelve).[18] The resulting powder was a substitute for alum, widely used in the rapidly growing paper-manufacturing industry as an important ingredient in the storage 'chests', which held the component materials before they were fed into continuous paper-making machinery.

During the slump of 1879, taking advantage of the low prices for china-clay leases, he acquired pits which were eminently suitable for paper applications at Gothers, near St Dennis, from the Browne brothers. He died in 1895, but by 1914 his family firm had added a pit eight miles north-east of Liskeard and two more at Balleswidden and Leswidden, six miles from Penzance. With these Pochin was able not only to meet his own manufacturing needs but also to supply clay to customers in Manchester and Liverpool. Pochin was an industrial tycoon with interests in 'paper alum' factories in Bristol and Liverpool, as well as large coal, iron and steel concerns in Staveley and Sheffield. At his death he was worth £7 million pounds and could easily have taken over the entire clay and stone industry, if the clay families had been willing to sell. Indeed, as we shall see later, he took shares in an unsuccessful attempt to do just that around 1890. His son-in-law, the barrister and MP the first Lord Aberconway, became one of the dominant figures of the china-clay industry in the early-twentieth century.

Cottier, Hanson and Deeley

Charles (later Sir Charles) Cottier was another ambitious man with wide commercial interests. Born in Plymouth the son of an Irish mariner, he was articled to a firm of local solicitors and later became the senior partner. He soon started to speculate in property, acquiring an interest in a Plymouth theatre and hotel and in 1904, in his early 30s, he set out to develop a china-clay sett on Duchy of Cornwall land in a remote part of Dartmoor, at Redlake and Leftlake, north of Ivybridge. The Martins of St Austell had been the largest clay producer on Dartmoor since they acquired the Lee Moor Works in the 1860s, but Watts, Blake, Bearne & Co., who became Britain's leading ball-clay producers, had opened Wigford Downs China Clay Works nearby at about this time. A decade later they started another pit at Headon, closing down the Wigford Downs operation in 1898.[19]

These works were small, though, compared with Cottier's, whose sett was initially predicted to yield 45,000 tons of 'best clay' a year, with reserves of 50 years at this rate of extraction. The clay slurry was to be conveyed to the dries by a nine-mile glazed stoneware pipeline, while a narrow-gauge tramway eight miles long, built by a workforce of 450 men, brought up supplies, as well as 120 clayworkers.[20] It took years, however, for Cottier to bring his scheme to fruition. First, as mentioned in Chapter Fifteen, he had to overcome objections, including legal action, from local interests, who claimed that pollution from the clayworks would ruin their businesses.

It was not until 1910 that he was able to float his China Clay Corporation with a nominal capital of £400,000.[21] The main shareholders, apart from himself, were Charles (later Sir Charles) Augustin Hanson and Harry (later Sir Harry) Mallaby Deeley. A Yorkshire wool manufacturer, Hanson had built a large mansion overlooking the town of Fowey and served as High Sheriff of Cornwall and later as MP for Bodmin and Lord Mayor of London. Deeley, who acquired several titles of lord of the manor, had made a fortune as founder of a large chain of clothiers, 'The Fifty Shilling Tailors', and was later twice elected MP for London constituencies.[22]

An investment of £400,000 to produce 45,000 tons a year seemed large, compared with the £100,000 capital of the North Cornwall venture, mentioned earlier, which raised 55,000 tons, but by 1913 Hanson, as chairman of the corporation, was claiming that it would 'easily' be able to produce 60,000 tons of clay a year of 'unsurpassed quality'.[23] However, unexpected technical difficulties caused further delays: the pipeline, which the contractors were supposed to complete within four months by August 1911, was only ready to achieve full capacity in 1914, when a collapse in demand due to the war soon brought production to a halt. Hanson handed over to Deeley, who eventually bought up the moribund operation after the war for only £47,000.[24] In the meantime Cottier had diversified into other fields, becoming chairman of the Aerated Bread Co. by 1920, but he had not lost his interest in china clay. After acquiring the shipping agents Balfour Williamson, he was on the point of using this company to acquire John Lovering's clay interests when he unexpectedly died at the age of 51 in 1928. Had he lived to a ripe old age, his drive and ambition might have led the Cornish china-clay industry on a more expansionist path.

'The China Clay Swindlegate'

Given the interest of big businessmen in china-clay extraction, it was nothing new when, in 1889, a syndicate announced that it was raising a capital of £800,000, later increased to £1,500,000, to buy 50 china-clay and china-stone companies. What was surprising was that the initiative did not come from insiders, as was the case with the other industries we have mentioned, but from a group of complete outsiders.[25] Quite by chance, its launch coincided with the appearance of a new weekly paper in the clay district, the *St Austell Star*. The editor, F.R. Ray, proved to be a vociferous opponent of 'the age of syndicates', as he called it, and of the China Clay Syndicate in particular.

As the voice of Free Trade Liberalism in Mid Cornwall, a county councillor and headmaster of a local private academy, Ray was an influential opinion-former. He warned that the clay syndicate would immediately raise prices, cut production and cast hundreds of Cornish working families into starvation and misery. Moreover, he predicted, if it succeeded in creating a monopoly it would be a prime target for nationalisation, in an epoch when radical politicians and trade unionists were calling for public ownership of coal and other industries, including the West Country metalliferous mining sector.

After probing into the affairs of the syndicate with all the zeal of a Fleet Street news-hound, Ray came to a remarkable conclusion: it was 'little more than a gigantic swindling machine that will inevitably end in widespread disaster and ruin' for investors as well as claymen. Fraudulent dealings, both by insiders and outsiders, were by no means unknown in the West Country and were not only widely reported in the press but were also described, in thinly disguised forms, in popular novels. In 1868 R.M. Ballantyne, famous Scottish author of boys' adventure stories, described a plausible and charming newcomer who persuaded the simple people of St Just in Penwith to invest in a bogus company.[26]

Twenty years later a book about the same community caused a scandal, and copies were destroyed, because the local people portrayed in it, far from being gullible, were dishonest mining agents and crooked traders who supplied the mines at exorbitant prices at the expense of outside shareholders.[27] In another story, again based upon real events, H.D. Lowry, a Camborne man, described how an unscrupulous local share broker colluded with a mine captain to 'pick the eyes' out' of a nearby mine, working only the richest lodes, thus temporarily inflating profits and share prices.[28]

A notorious 'company monger'

In the 'Swindlegate' case, the scheme seemed to originate from London, although some local people may have been involved. Ray had discovered that the prime mover, Thomas Fenwick, was a notorious 'company monger' who floated companies and then made off with the proceeds without carrying out any of the promised developments. A year earlier Fenwick had formed the Great Northern Salt & Chemical Works, and before that the Electro-Metal Extracting, Refining & Plating Co. with a capital of £150,000. Both companies, together with Fenwick, disappeared, leaving behind a trail of creditors.

On publishing his findings in July 1890, Ray received a sharp letter from a distinguished London firm of solicitors who claimed to have defended no less a person than the Irish leader Charles Stuart Parnell. They warned Ray that unless he immediately retracted his statements, Fenwick would sue him for libel. Far from retracting, however, Ray instead publicly dared Fenwick to go ahead, and from time to time repeated his accusations, taunting Fenwick to take action. Fenwick never did, even when Ray prefaced his remarks about the syndicate with the headline 'The China Clay Swindlegate'.

Fenwick, an expert financial spin-doctor, floated a raft of aristocratic and apparently worthy names before the public as chairmen, directors, solicitors and stock brokers for his companies, the China Clay & China Stone Syndicate and the China Clay Union. They appeared to have overlapping directorates and officials, and their main purpose seemed to be to put up a smokescreen for Fenwick and his cronies to siphon off commissions by buying clayworks for the syndicate and selling them on at a higher price to the union, not the first time he had used this device.[29] The union, moreover, although widely advertised as a limited holding company, had never been registered at Somerset House, as Ray discovered.

An early runner as putative chairman in May 1889 was Sir Charles Mark Palmer, shipbuilder, coal owner, iron master and landowner of Jarrow, but he gave way by May 1890 to Viscount Bury, whose name was A.A.C. Keppel. Another alleged participant in the scheme was the Hon. D.W.G. Keppel, and these were respectively the eldest and the second son of the Earl of Albemarle and nephews of the soon-to-be-famous beauty Alice Keppel, mistress to King Edward VII from 1898 until his death in 1910.

The Godolphin connection

Viscount Bury did not stay long, however, handing the chairmanship over to his deputy, Lord Frederick Godolphin Osborne of Windsor. Here at least was a

Cornish connection, as his middle name implied, for he was brother to the Duke of Leeds, of Godolphin, whose ancestor gave his name to the Cornish village of Leedstown, and who had leased out china-clay setts near Helston. Lord Osborne appeared to add gravitas to the board, until a London financial paper, *The Oracle*, revealed his association with a defunct Johannesburg hotel company. This link, the paper claimed, was 'sufficient to damn the whole enterprise' in which Lord Osborne was engaged.

To embellish this noble window-dressing, the Earl of Morley, owner of the Dartmoor land in which Lee Moor clay pits operated, was also trialed as a director. Perhaps Fenwick intended to imply that Devon clayworks were joining the syndicate, but the Earl soon disowned any connection with it. Two smaller-scale clay producers, Tremayne of the South Goonvean China Stone & Clay Co. near St Stephen, and William King Baker of Towednack China Clay Works near Penzance, also sat briefly on the board before withdrawing.[30] Adding an aura of respectability was the Revd John Garrett, Doctor of Divinity, Rector of Christ Church, Manchester, and one-time vicar of Paul, Penzance, and shareholder in clayworks, who for a time served as 'provisional chairman'.

Perhaps the most bizarre of the rapidly changing boards, however, was one of August 1890, composed entirely of hoteliers. The chairman was Frederick Thorne, who also presided over the New Explosives Co. and the Submarine & Torpedo Explosives Co. and was director of the Swindon Junction Hotel Co., 'none of which has gratifying results', according to the *West Briton*. Other directors of the Clay Union included George Allen, director of Long's Hotel and the Queen's Hotel in Cheltenham and the Queen's Hotel in Harrogate; John Belton, chairman of Long's Hotel and the Queen's Hotel in Cheltenham and director of the Queen's Hotel in Harrogate; and John Chater, another director of the Queen's Hotel in Harrogate.

Fifty clay merchants join the syndicate

Despite the fleeting acquaintance of most of these names with the syndicate (and at least four successive solicitors and two financial agents came and went), Fenwick, in one of his prospectuses, was able to announce that, acting on behalf of the syndicate, he had signed a contract with 50 china-clay vendors to sell their properties to Sydney Hawgood Lee, trustee of the China Clay Union. He was said to offer the smaller producers £6,000–£8,000 for clayworks that would normally have sold for much less, and it was rumoured that Coode Shilson, St Austell bankers and attorneys, had been offered £10,000 for their clayworks that had recently changed hands for only £800.[31]

The vendors may have believed they would receive cash, but Fenwick paid them in debentures issued by the syndicate. However, they had first claim in the event of a winding up of the company so that, even if it failed, they would at least get their pits back, whereas the ordinary shareholders might lose their entire investment. On the other hand, if the company prospered, they would receive a satisfactory five per cent on their debentures, plus a promise of high dividends if they took up their allocation of privileged founders' shares. These £10 shares, Fenwick assured them, would be worth at least £250 each 'in the not too distant future'.

Nevertheless, despite these inducements and Fenwick's optimistic statements, only a handful of clay producers committed themselves fully to the scheme. They included the Revd Dr Garrett, Tremayne and Baker, already mentioned, and John Nicholls of Balleswidden Clayworks, near St Just in Penwith. Nicholls enjoyed the distinction of being the sole participant who actually succeeded in getting cash from Fenwick, and he only achieved this by catching the latter in his London office and demanding £100, possibly in return for his public support for the scheme, or as a commission on sales of shares. Others, such as Francis Wright, an ironmonger from Rhyl, to whom Fenwick offered cash for his clayworks at Kerrow Moor, only received debentures.[32]

With the major exception of Pochin, Thomas Fenwick only succeeded in enticing the smaller fry into his venture. The leading clay families, including the Martins, Loverings and Stockers, did not join. It was rumoured that they had been offered £100,000 apiece, but as usual they maintained a wall of silence over the whole affair. They already held the lion's share of the higher quality 'best clay', and some small producers, resenting their domination, may have been attracted to Fenwick's scheme because it might have stopped the big firms from unloading 'common clays' at or below cost to avoid the expense of storing it and drive firms that only produced common clays out of business. Francis Wright and the Nicholls family blamed their bankruptcies on having to sell below cost.[33]

Fenwick presses on

Fenwick must have been astute enough to realise that once the major families refused to join in it, the venture would fail. Yet he continued to promote shares with unabated vigour, suggesting that he knew that he had no real hope of success but was simply pocketing as much cash as he could. To help him, Dr Garrett came down from Manchester to the White Hart Hotel in St Austell to sell shares which, he claimed, Mancunians were 'falling over themselves' to acquire. When the editor of the *West Briton* challenged him to state how many had been sold, he replied that shares to the value of £260,000 had been 'bespoken'.

An announcement then appeared in August 1890 in the *Western Morning News* that the shares were trading at a premium, having been subscribed seven times over. However, when Ray of St Austell paid a further visit to the union office in London, the trustee, S.H. Lee, told him that 'hardly any' had been sold, and that he had no idea who had informed the press that the share issue had been oversub-scribed. Yet after Ray published this news, Fenwick immediately despatched telegrams to all the interested newspapers confirming that all the shares had been 'allocated'.

In October 1890 a sparsely attended extraordinary general meeting of the China Clay Union in London was presided over by Lord Osborne. Insisting that he was no longer the chairman, he regretted the 'highly eulogising notices' that had appeared in the press. He also informed shareholders that the stockbrokers were demanding £30,000 commission for launching the share issue, but that all the directors and the former chairman had been disqualified from office for not taking up their allocation of shares. Nonetheless, he asserted that he intended to make the union a success, with clayworks extending 'from Penzance to beyond Plymouth' which would offer such a variety of clays that no other company could compete. The loyal Dr Garrett then proposed a vote of thanks.

Pochin's change of plan

One of the most intriguing aspects of the whole affair was the behaviour of H.D. Pochin. The only backers of any significance were T.E. Spalding of Leeds, with 3,000 shares, Pochin with 2,000, Baker with 333, John Nicholls with 350, Fenwick himself with 64 and S.H. Lee with just one. At the beginning of the affair, in April 1889, Pochin visited St Austell and met Fenwick, who told him that the Stockers, Loverings and Martins were not joining the syndicate, and Pochin also formed the impression that the syndicate was paying well over the odds for the clayworks that it was acquiring. All this would surely have aroused the suspicions of a shrewd self-made business tycoon like Pochin and yet, apparently satisfied with Fenwick's integrity, he offered to sell his Halviggan works for £60,000.

Fenwick countered with an offer of £40,000, and in March 1890 a deal was struck for £40,000 cash plus £20,000 in syndicate shares. Pochin may have thought he had made a very good bargain because he had been trying to sell his Halviggan works for around £12,000 to another clay merchant. He also retained the right to continue to produce his own clay up to 12,000 tons a year from two other works he leased at Gothers and Higher Gothers. Had he decided to move out of the clay trade except for supplying his own needs for his chemical plants? Or did he see Fenwick's scheme as a way of increasing his grip on the china-clay industry by acquiring more shares in it in the future?

Whatever his motives, Pochin then encountered the same problem that had faced many other imprudent investors in the past; how to get money out of Fenwick. Payment of £40,000 had been promised within six weeks, but a year later, in May 1891, after a long discussion with his barrister son-in-law, Pochin decided that his only hope of getting some money was to make Fenwick bankrupt.

In June he instructed his solicitors to dissolve his agreement with Fenwick and the following January he paid the solicitors' bill, but there is no record that he ever received anything from Fenwick. Pochin then changed direction and increased his interest in china-clay production, as discussed earlier in this chapter.[34]

Pochin's action against Fenwick marked the virtual end of the Swindlegate affair, except for some casualties left in its wake. The bankruptcies of Wright and the Nicholls family have already been mentioned and, just before Christmas 1890, the unfortunate Dr Garrett also had a notice for bankruptcy served on him for the debts of the union, Fenwick and the others having disappeared. Having bitterly attacked F.R. Ray for propagating 'bouncing lies' and 'barefaced falsehoods' about him, Garrett now had to swallow his pride and write to the *St Austell Star* begging its readers for help.

Ray replied that he was 'not one to kick a man when he was down', but offered no personal assistance apart from wishing him a Happy New Year. John Nicholls, the only one to get Fenwick to part with money, then amazingly found himself sued for the £100 by Fenwick. The latter claimed that it was only a loan, which he had given out of the goodness of his heart when Nicholls came begging for it. Luckily for Nicholls, he had earlier stood up for Fenwick and written a letter to the *Star* stating that he had received a part payment in cash for his clayworks. He was able to produce a letter from Fenwick thanking him for his support, and Fenwick lost his case.

Cornering the West Country extractive market

Although the Swindlegate bubble burst, the publicity surrounding its fraudulent dealings increased public awareness of the importance of the china-clay industry. However, in the years up to the 1914 War, West Country clay firms tended to be private companies, offering little opportunity to the public to invest in them.[35] It was only through large-scale schemes to buy up the entire industry that the ordinary shareholder could become involved. Throughout the period rumours of such projects persisted, especially after tin prices trebled in the early years of the twentieth century, and the idea of buying up the entire West Country mining industry began to find favour. A group of London speculators was rumoured to be buying up 600 abandoned mines in Cornwall and all the Devon mines.[36]

When this fell through, Collins, international authority on metalliferous mining as well as clay, in a series of speeches, articles and circulars, urged everyone from the Duke of Cornwall, aristocracy and gentry to the small trader and artisan to do their patriotic duty and invest in a Cornish mining syndicate to buy all the mines. After his appeal fell on deaf ears he turned to his American connections to finance his scheme but they too failed to show any interest in Cornish tin mines. However, they were certainly keen to lease clay pits. In 1903 local newspapers quoted a 'sensational rumour' from the national press, including the *Daily Express* and the *Daily Mail*, that 'our American cousins will shortly buy up our china clay industry'.[37]

A group of seven of the largest American clay importers had first visited Newton Abbot and Kingsteignton, where ten or 12 firms were selling ball clay mainly to Staffordshire potters, but with some going to the USA. The arrival of the Americans led to rumours that they were offering millions to buy up all the ball-clay firms of Dorset and Devon and close them down, transferring production to the USA.[38] However, the ball-clay producers refused their offer. The ball-clay families maintained their independence until the 1914 War and, as we saw in Chapter Eighteen, several of them branched out into pottery production as well as trade in other minerals. Some worked closely together, however: Watts, Blake, Bearne & Co. advanced money to the Marland Brick & Clay Works in North Devon, which had been revitalised by William Wren, a descendant of the original proprietors, the Greenings, and acted as sales agents for the Marland Co.[39]

After the American investors were rebuffed by the ball-clay companies, they turned to china-clay firms at St Austell and Lee Moor. A St Austell man, exiled in Kentucky, wrote home to warn clay merchants to shun the syndicate. They would 'make one man do the work of six', he said, then drive out competition, raise prices and lower wages. 'If the American syndicate got hold of the clayworks, in a few years' time the clay industry in Cornwall would be a thing of the past,' he predicted.[40]

He need not have worried. Cornish merchants received the would-be American purchasers with a 'decisive and unanimous refusal'. As Martin of Lee Moor told the *Daily Express* reporter. 'what is good enough for an American trust to buy is good enough for us to keep'.[41] Public interest was aroused in what the *Daily Express* called 'cornering clay', and led the *Western Daily Mercury* to commission a report of 'this large but little known industry' for their investing readership. Written by R. Hansford Worth, a well-known Plymouth engineer,[42] it gave details of clayworking methods at Lee Moor and St Austell.[43] However, animosity towards involvement by outsiders remained strong. A.L. Rowse's short story, 'The Curse upon the Clavertons', set in the years before the 1914 War, reflects these suspicions. Claverton, an ambitious man from 'up the country' who was determined to gain control of the entire industry, cheats an honest but untutored Cornish clay merchant out of his lease. Claverton and his family, however, meet their just deserts and fade into obscurity.[44]

Abortive china-stone and china-clay combines

While there were difficulties in securing unity of effort across the china-clay industry, china stone seemed to pose fewer obstacles. The market was mostly oriented towards the potteries and the number of quarries was small, 20 or so, owned by a dozen producers. Nevertheless, it still experienced the same problems as the clay trade. Supply outstripped demand, keeping selling prices and profit levels down. Some 20 years earlier, when a combine had controlled production, the price of the best stone, called 'purple', had been set at £1.20 a ton, but by 1903 it had dropped to between £0.60 and £0.70. Stone millers in the potteries, who held the whip hand over prices, not only kept them low but also demanded long periods of credit before payment, up to 18 months, with an average of 13 months.[45]

To resolve these problems, china-stone masters in 1904 agreed to reinstate an earlier abortive China Stone Association, raise prices by £0.15 a ton on average and restrict the credit given to the stone millers to three months. The members included Stocker's West of England Co., John Lovering, William Luke, Goonvean and Bloomdale. Each producer received a sales quota and had to pass on orders received above this level to other members. If total demand exceeded all quotas, they were increased proportionally. To ensure fair play, a central office was established, with P.M. Coode, from the respected local solicitors Shilson & Coode (who also owned clay leases) to act as administrators and keep accounts.[46] However, a ring of Staffordshire potters seemed to nip the project in the bud, for no further action by the stone producers was reported.[47]

Nevertheless, as both clay and stone supply continued to outstrip demand, rumours of mergers and associations between the leading producers persisted. In 1907 Martin Bros compensated for a rise in wages that it had granted by increasing the price of clay 'without consulting any of its partners', suggesting that it was normal to discuss price changes. In 1908 it raised prices again and in the same year a China Clay Association was formed by producers of best clays.[48] However, an agreement of that year to raise the price of 'best' and medium clays by £0.25 a ton was broken almost as soon as it was made.[49] Two years later an amalgamation between Stocker's West of England Co. and John Lovering & Co. was said to be imminent, in order to make joint use of a new bleaching process to whiten clay developed by an 'outside party',[50] but nothing came of it. On the eve of the 1913 china-clay strike, the tense situation was further disturbed by news of the acquisition of 'a considerable interest' in Carpalla China Clay Co. by one of its main customers, Spicer Bros, the papermakers, and rumours spread of 'frequent visits by big American buyers' seeking to acquire the entire industry.[51]

The 1913 clay strike demonstrated that, when it came to resisting demands for recognition of a workers' union, clay employers could stand firm and present a united front, but when they tried to combine to fix prices and production they always failed. Of course they were not alone in lacking solidarity. Throughout British industry pressures to combine were strong, but the desire of traditional family firms to retain their own freedom of action proved even stronger, and the result was a series of loose agreements to fix prices and production quotas that were broken as soon as they were made by the offer of secret discounts. Only where a vast capital was needed to enter an industry, as in long-distance ocean transport, could an association like the Shipping Conference maintain their monopoly of prices and keep newcomers out. Many of the unwieldy associations referred to earlier in this chapter fell apart because of the indiscipline of their members and competition from new entrants to the industry.

It took the collapse of overseas demand caused by the 1914 War to force the major producers of china clay to work together, and even then the necessity to appoint three managing directors representing different companies suggested the continuing power of family loyalties. As late as 1927 J.M. Coon summed it up well: 'From time to time Associations have been formed to regulate the price of china clay and during their continuance have been beneficial, but mutual distrust and jealousy usually broke them up... most recognise their desirability, a few are so selfish that they will not cooperate, so most suffer.'[52]

References

[1] The following section is based upon Lawrence Popplewell, 'Brownsea's Lost Railway', *The Dorset Year Book for 1972–3*, Society of Dorset Men, pp.89–92; R.W. Kidner, *The Railways of Purbeck*, Usk, 2000, p.84; Rodney Legg, *Brownsea*, Dorset, 1986, pp.23–46; Jack Battrick, *Brownsea Islander*, Poole, repr. 2000, pp.19–27.

[2] John E, Wilson, *British Business History 1720–1994*, Manchester, 1995, pp.98–105.

[3] P.L. Payne, *British Entrepreneurship in the Nineteenth Century*, London, 1988; C.J. Schmitz, *The Growth of Big Business in the United States and Western Europe*, Cambridge, 1993.

[4] *RCG* 22 October 1903.

[5] *WB* 5 May 1881.

[6] *RCG* 18 May 1877; John Tonkin, 'East Carluddon', *CCHSN* 5, 2003.

[7] *SAS* 12 June 1891.

[8] *RCPS* 1879.

[9] *St Austell Weekly News* 12 April 1890; *Financial News* 3 June 1890; Coode & French, Solicitors, *Francis Wright bankruptcy papers, Kerrow China Clay Co. 1890–91*.

[10] For some of the information in this section we are indebted to John Tonkin. Other sources include Boase and Courtney, *Collectanea Cornubiensia*; Kelly's Directories of 1873 and 1878; CRO, CF/1/3975/1-28 on Slip Quarry; A.W. Brooks, *Iron Ore Mining in Cornwall*, St Austell, forthcoming.

[11] Peter Joseph, 'Durfold Clayworks, Bodmin', *JTS*, 28, 2001, pp.40–41.

[12] Derek Giles, 'North Cornwall China Clay Company Ltd', *CCHSN*, 13, 2006, pp.5–7.

[13] He advertised these interests in David Cock's *Treatise on China Stone and Clay*, published in London and Wolverhampton, 1880.

[14] John Penderill-Church, *The Claypits of Cornwall and Devon*, Wheal Martyn Archives, n.d., p.19; R.M. Barton, *A History of the Cornish China Clay Industry*, Truro, 1966, p.141.

[15] *BCW*, May 1910, p. 41. His business seemed to founder during the 1914 War.

[16] John Tonkin, 'The Belowda Clayworks', *JTS*, 21, 1994, pp.2–21.

[17] J.H. Collins, 'The china clay industry of Cornwall', *MM*, 1911, pp.454–55.

[18] H.T. Milliken, *The Road to Bodnant*, Ilkley, 1975; Eric Mensforth, *Family Engineering*, London, 1981, pp.18–24.

[19] E.A. Wade, *The Redlake Tramway and China Clay Works*, Truro, 1982, pp.13, 19.

[20] *WMN* 24 November 1911.

[21] *BCW*, XVIII, February 1910, p.294.

[22] *SAS*, 20, 27 August, 10 September 1903; Wade, 1982, pp.119–36.

[23] *BCW*, XXII, October 1913, p.211.

[24] *China Clay Trade Review* II, 17, October 1920, p.563.

[25] Except when otherwise stated, the sources of information for the section on 'Swindlegate' are: *WB* 20 April, 5 and 19 June, 3 and 28 July, 4 and 21 August, 16 October 1890, 8 January, 17 December 1891; *RCG* 16 May 1889, 19 and 26 June 1890, 25 May 1893; *WMN* 7 August 1890; *SAS* 5 and 12 April, 10, 24 and 31 May, 28 June, 1 November 1889, 10 and 17 January, 2 May, 25 July, 1, 7, 15 and 22 August, 5 and 26 September, 10 and 17 October, 26 December 1890, 19 March, 26 May 1893; *Financial News*, London, 5 and 26 September 1890; *London Star* 5 September 1890; *Investors' Chronicle*, London, 31 May 1889; *Financial Observer*, London, 26 April 1890. Biographical details come from the *Dictionary of National Biography*, various editions of *Who Was Who? Fulford's County Families*, London, 1916, and Charles Carlton, *Royal Mistresses*, London, 1990.

[26] R.M. Ballantyne, *Deep Down: a Tale of the Cornish Mines*, London, 1868.

[27] Edward Bosanketh, *Tin*, London, 1888. 'Bosanketh' was the pen name of a man whose father had fled the district leav-

ing a mountain of debts. See Justin Brooke, 'A Key to Tin' in the reprint of the book in 1988.

[28] H.D. Lowry, *Wheal Darkness*, London, 1927. The uncompleted work was finished by another writer, his cousin, Catherine Dawson-Scott.

[29] He had formed the Electro-Metal Companies with almost identical titles.

[30] As mentioned in Chapter Eleven, Baker was a man of some substance. From a farming background in Essex, he had worked as an engineer in India before settling near Penzance, where he was of sufficient local standing to be elected to Cornwall's first County Council in 1889.

[31] Coode and French, Solicitors, letter from Thomas Olver, St Stephens China Stone merchant, to Thomas Fenwick, 1890.

[32] Coode and French, Solicitors, *Bankruptcy papers of Francis Wright, Kernow China Clay Company, 1890–1*, available via the internet.

[33] However, evidence at the Bankruptcy Court suggested that incompetence was a primary source of failure. For Wright see *SAS* 12 June, 24 July 1891, *WB* 26 January, 2 February 1891; William Nicholls, William Walter Robins Nicholls and Frederick Hart Nicholls traded as William Nicholls, William Nicholls & Co. and as the Lower Lansalson China Clay Co., see *SAS* 6 July 1894; *BC-W*, June, July, September, November 1894.

[34] H.D. Pochin & Co, *Board of Directors Minute Book*, Box 19, China Clay History Society Archives, 20 April 1889 to 19 January 1892, pp.120–165. We are indebted to Derek Giles for these references.

[35] *NE* 31 January 1913.

[36] Peter Joseph, 'Electricity and Tin: the Rise and Fall of Cornish Consolidated Tin Mines Limited', *JTS*, 31, 2004, pp.30–42.

[37] *Daily Express* and *Daily Mail*, quoted in *RCG*, 10 September, 1 October 1903.

[38] *BCW*, XII, September 1903, p.194, October 1903, p.244.

[39] Rolt, 1974, pp.46–47; Ball Clay Heritage Society, *The Ball Clays of Devon and Dorset*, 2003, pp.48–49.

[40] *BCW*, XII, December 1903, pp.343–44.

[41] *RCG*, 1 October 1903.

[42] Worth later prospected Dartmoor for clay for Cottier and constructed his narrow-gauge railway.

[43] *WDM*, 18 September 1903.

[44] A.L. Rowse, *Cornish Short Stories*, London, 1967, pp.57–75.

[45] *RCG*, 20 October 1904.

[46] CRO, DDCF 3996/3997, document of 23 September 1904.

[47] *RCG*, 27 October 1904.

[48] Derek Giles, 'Some less known china clay families: Martin Brothers', *CCHSN*, 14, 2006, p.4.

[49] Kenneth Hudson, *The History of English China Clays*, Newton Abbot, n.d., p.49.

[50] *SAS*, 28 July 1910.

[51] *NE* 9 May, 4 July 1913.

[52] Jos M. Coon, 'The China Clay Industry', *RRCPS*, 1927, p.664.

Chapter Twenty
Clay in Fact and Fiction

In October 1909 400 china clay and farming tenants of Admiral Sir Charles Graves-Sawle were reported to have gathered at his home, Penrice, to celebrate the coming-of-age of his only son, Lieutenant R.C. Graves-Sawle of the Coldstream Guards.[1] To mark the occasion a committee of tenants, chaired by John Wheeler Higman and with Frederick Augustus Coon as Honorary Secretary, collected the considerable sum of 1,000 guineas. In making the presentation, William T. Lovering expressed the hope that the lieutenant would one day become colonel of his regiment, just as his uncle, Sir Francis Aylmer-Sawle, had done some years earlier. Lovering himself was later to become a lieutenant-colonel in the Territorial Army.[2] Then J. Harris, estate carpenter, presented another handsome gift on behalf of the servants, after which Lovering called for three hearty cheers for the young squire, and Coon demanded three more for the gallant Admiral and Lady Graves-Sawle.

Here we have a cameo of a well-ordered and deferential social structure, with each rank knowing its place in the Mid Cornwall hierarchy. The Cornish clay-country writer Ken Phillipps remembered that the high spots of his mother's social life occurred when Sir John Molesworth of St Aubyn came over to shake her hand.[3] Graves-Sawle was another of the 'claylords' who owned much of the clay-bearing land and who dominated the political life of Mid Cornwall as MPs, High Sheriffs and Deputy Lieutenants, limiting the clay merchants' scope for social advancement via acquisition of land.

History and mythology in the clay industry
How did the merchants and men of the Devon ball-clay industry compare in their contribution to the economic and social fabric of their communities? Before considering accounts of activitities reported in newspapers and journals of the time, it is of interest to consider the way in which the clay industry was portrayed by writers of fiction. For, as Alan M. Kent, leading authority on china-clay literature, has maintained, far from dismissing the writings of romantic novels, we should recognise the way that their interpretations have become embedded in local identity and culture.[4]

In this respect, a vast difference exists between the fictional treatment of the ball-clay and china-clay industries. Ball clay production was, of course, smaller and more scattered, and neither Thomas Hardy in Dorset nor Eden Phillpots in

Devon used the industry as a background to their works, although they must have been aware of its presence. China-clay working, in contrast, has provided the setting for a number of novels and stories by authors, most of them born or bred in and around the clay district. Here we distinguish between two groups, the first consisting of Quiller-Couch ,Joseph Hocking, Anne Treneer and A.L. Rowse (to which we can add Phillpots), the second and later set including Rowena Summers, Mary Lide, E.V. Thompson and Susan Penhaligon.

Quiller-Couch and the romantic novelists

Nowadays, the reader may not appreciate the weight that the writers we have mentioned carried in forming public opinion. Sir Arthur Quiller-Couch, or 'Q' as he called himself, was not only the leading Cornish man of letters in Edwardian times, but was Cambridge Professor of English Literature and an important political figure. His attitude to industrial Cornwall was summed up by Andrew C. Symons in a discussion of one of Q's stories set in the Hayle and St Ives area. Q 'ignored the industrial aspects' and focused on 'a traditional world of ancient crafts, superstition and the Bible'.[5] From his vantage point overlooking the Fowey River, Q, as Harbour Master, commented on the growing number of clay ships, but as far as the clay industry was concerned, his public received an impression of

Painting of Carclaze pit, by Elliot of Newquay in 1892. In the foreground is an area worked for tin and in the background is the site of china-clay production.
(Courtesy of Wheal Martyn China Clay Museum)

dullness, dishonesty and inhumanity. Claywork 'deadened the touch for the finer things in life', unlike copper or tin mining, where 'fire's in the veins and you follow it like a lover'.[6] As we saw in Chapters Thirteen and Fifteen, Q's clay merchants, if they were not consigning men to a watery grave by overloading and over-insuring vessels, were causing their deaths by neglecting to remove overhanging debris in the pits, ruining the environment and corrupting miners' morals. The clay industry simply had no place in Q's genteel and romantic view of a 'Delectable Duchy'.

Silas, Joseph and Salome Hocking had fallen almost into oblivion before Kent rescued them.[7] Yet the over 200 books they wrote were immensely popular in their day throughout Britain, and Silas was the first British novelist to produce a million-copy best seller. The Hockings were brought up in the clay parish of St Stephen, although the brothers soon left, as ministers, to find fame and fortune as authors elsewhere, and their sister followed later. However, they drew upon Cornish sources for some of their writings. Silas' first novel was a seafaring yarn based upon an uncle's experiences, while Salome retold tales she heard from her father about mining and farming, and Joseph included romances set in an earlier period of Cornish history. Clayworking played a very minor part in their writing. As discussed in Chapter Fifteen, the only references that Silas and Joseph made to it were criticisms of its impact upon the landscape. Perhaps, as a later clayland writer Jack Clemo commented, they regarded clay working as a 'messy industry with which they had no concern'.[8] Only Anne Treneer, of the writers with experience of the claylands before the 1914 War, romanticised the physical impact of its operations, in common with a few artists such as Laura Knight and Augustus John. Yet, by focusing upon pools in the worked-out pits and upon small clay works using archaic methods, they presented a misleading picture of entrepreneurial sluggishness and technological backwardness in what was, in reality, a dynamic industry.

An alternative perspective: Eden Phillpotts

Elsewhere we have compared their largely hostile treatment of a major local activity with the attitude of other industrial novelists of the time, such as R.M. Ballantyne, Edward Bosanketh, H.D. Lowry, Arnold Bennett and Eden Phillpotts.[9] The Dartmoor-based Phillpotts presented a well-rounded picture, rather than a caricature, of West Country entrepreneurial performance. His first important work was set in the West Cornwall fishing village of Newlyn,[10] but later tales included backgrounds of slate quarrying at Delabole in North Cornwall, paper manufacture near Totnes, a pottery at Torquay and a china clayworks at Redlake on Dartmoor.[11] To him china clay was a superior and useful commodity, not just another form of mud, but although he lived close to the ball-clay industry it did not seem to interest him. In his books he showed a knowledge of the industrial processes involved and highlighted the pride that employers and workers took in transforming banal materials into useful, and sometimes beautiful, products that enjoyed worldwide sales.

Phillpotts also skilfully wove into his stories the tensions that were developing as smaller paternalistic family firms were taken over by more cost-conscious and profit-maximising entrepreneurs. Neither Q nor the Hockings showed such an understanding in their Cornish novels, although A.L. Rowse later based a short story upon the way in which an unfeeling clerk from 'up the country' connived to gain control of a pit run by a traditional good-hearted Cornish clay merchant.[12] Whereas Q's industrial accidents, discussed in Chapter Thirteen, merely illustrated the negligence or worse of clay merchants, Phillpotts used them in the manner of other novelists of the period to resolve his plots by displaying the nobility of his heroes and the evil intentions of his villains. In the vast slate quarry at Delabole, for instance, the workmen attributed an accident not to the manager's indifference, but to the carelessness of an employee, and the quarry master enhanced his reputation within the workforce by his decisive action in dealing with a disaster caused by a massive fall of rock.[13]

The silence and social mobility of the clay merchants

Rowse was apt to make fun of the way that the clay merchants kept their cards close to their chests, joking that they had inherited the secrecy of the ancient Chinese who had given their name to the clay they worked, but their reticence was both sensible and necessary to protect their own and their customers' interests because of public suspicion of the uses to which china clay was put. Phillpotts certainly recognised this, and in one of his novels an opportunist who wheedled the secret of a ceramic recipe out of a master potter's daughter was punished for this act of trickery by the author.[14] Joseph Hocking, with perhaps a better grasp of business matters than Rowse or Q, also appreciated the need for discretion, in this case about the potential value of uranium in a Cornish tin mine.[15]

The clayland literati also had a good deal to say, again most of it unflattering, about the upwardly mobile clay families. Although Q wrote amusingly, if condescendingly, about the Troytown (Fowey) bourgeoisie and their need to be 'cumeelfo' (comme il faut),[16] he did not apply his comic invention to the clay merchants. Joseph Hocking and A.L. Rowse, on the other hand, had pronounced views on the place of the clay producers in the social hierarchy. In one of Hocking's stories, based on a mine at Terras, near St Stephen, once worked by his father – he related how a farmer had saved enough to send his son to a public school 'so that he might marry the squire's daughter'.

Painting of South Meledor clay pit, based on a photo in a North & Rose brochure of 1910.

To make even more money he invested in clayworks which 'turned out well and gave a kind of standing among monied men'. Not enough, however, to impress the squire, and it was only a bonanza from the radium by-product of a tin mine that enabled him to win the daughter's hand.[17] Reworking his plot in a later novel, Hocking's hero was a farm labourer who made a fortune by striking a rich lode of tin. On this occasion, however, it was not enough to satisfy the squire, and it was only when the hero was found to be the long-lost heir to one of Cornwall's noblest families that he was accepted.[18] Melodramatic and sentimental these tales might be, but they described a society with clear-cut gradations: money made from clay-working was socially less acceptable than riches from metal mining, which in turn was of less consequence than hereditary wealth.

The controversial A.L. Rowse

The ever-contradictory Rowse faced both ways on this use of social mobility. He was only a boy in the years leading up to the 1914 War, but he has left a vivid account of clay-country life in that period, although we must remember that he wrote it 30 years later, at a time when he had famously turned his back on Cornwall and the Cornish.[19] This was an era when British entrepreneurs who had worked their way up from the lower strata of society were increasingly criticised for deserting their industrial milieu and gentrifying their sons through public school and university education to fit the life of a City financier, a barrister or a country squire.[20] Rowse both ridiculed those clay merchants who tried to rise in the social hierarchy and criticised the lack of ambition of those who did not make the effort. 'For all their money,' he said of the upwardly mobile, 'they only built enlarged suburban houses.'[21]

As a social climber,, Rowse was ideally placed to observe the social nuances of clayland life. Brought up with, on one side, the respectable new suburbs of the 'clayopolis', as St Austell was sometimes dubbed, and on the other side the rough working-class 'Higher Quarter', he was, as he said, 'betwixt and between'. His father was a clay labourer, while his mother worked for (and was emotionally attached to) the gentry, frequented the Anglican Church and the Tory Primrose League and ran a small shop. He witnessed both the independence of claymen who boasted they would never call any man 'Sir', and the deference of a newly rich clay family who occupied church pews near the front but not as prominent as those of the gentry, 'whose precedence was acknowledged by all'.[22] In another tale he made fun of the wife of 'The Squire of Reluggas', a clay merchant who added 'a hideous glass verandah', and 'an eyesore of a pink roof' to their 'gallant old house'.[23]

Another clay producer 'The Recluse of Rescorla', built a 'vulgar mansion' with a lodge house. 'For all their pretensions,' said Rowse of the clay merchants, 'they had no taste.'[24] He referred to yet another property as 'an appallingly ugly house built recently... by one of the china clay people, the usual middle-class bad taste.[25] In contrast, in a different story, the sons of 'Old Cap'n William Slade', a clay merchant who could barely read or write, followed the same pastimes as ordinary clay labourers. They were only interested in whippets, spaniels, ferrets and women, differing merely in their order of preference for them.[26]

Changing interpretations by late-twentieth-century writers

These impressions of entrepreneurial lethargy and low social status were moderated by a *volte face* on the part of Rowse, and even more dramatically by the popular romantic novelists Rowena Summers, E.V. Thompson, Mary Lide and Susan Penhaligon. At the same time as he was publishing his unflattering tales about apparently fictitious clay merchants, A.L. Rowse, paradoxical as ever,was heaping praise upon some of the best-known producers. 'The two John Loverings, Senior and Junior, must have been men of worth,' he wrote, 'old Rebecca Martin a strong personality in her own right and T. Medland Stocker a big-minded man.'[27] The late-twentieth-century authors depicted clay merchants in a different light, as thrusting adventurers in a dynamic industry.[28] In the typical manner of the genre their picture was black and white: their heroes were decent and honourable clay merchants, contrasted with villainous and ruthlessly competitive rivals. When

Thompson's clay merchant negotiated an agreement with fellow producers to raise clay prices in order to pay workers a higher wage, the others secretly undercut prices before the ink was dry on the contract so as to steal orders of clay from him, nearly driving him to ruin. The father of Summers' heroine, a loyal and trustworthy clay captain, mediated fairly between his employers and his workers, but was undermined by a fellow clay captain, a 'strike monger' who goaded his men to down tools.

Summers continued the tradition, discussed earlier, of using accidents as opportunities for heroes and villains to reveal their true colours. In one novel a drunken striker steals a wagon laden with clay and crashes into a bakery with fatal results. In another a train carrying clay workers on a feast day excursion comes off the rails. In both cases, her hero, a clay merchant, compensates the victims although he was is no way to blame for these events. In complete contrast, Lide's villain, a blunt north-country mill owner who has aquired a clay works, argues that the only way to deal with discontented workers is to lock them out, which he does. Some of Penhaligon's clay merchants hire thugs who murder a clay worker and set fire to an employer's house and then put the blame on workers.

Literary views of the workforce

As discussed in previous chapters, mining journalists and some historians of the nineteenth century china-clay industry undervalued clay workers as well as clay merchants, dismissing the men as 'common labourers', mere beasts of burden, 'great men for shifting sand' and 'makers of mud pies'. This condescending view was shared by the earlier group of writers, and A.L. Rowse treated them with his usual contempt for the 'swinishness of the masses'. As Philip Payton has noted, he privately despised the 'sheer stupid ignorance' of clay workers, who loyally supported him as parliamentary candidate.[29]

By setting their narratives in periods of civil strife, the late-twentieth-century novelists enhanced an impression of worker militancy. Summers used the food riots of the mid-1800s as a background for some of her novels, and Penhaligon set her story in 1880 but featured events from the 1876 strike. Thompson's plot unfolded in the troubled years leading up to the great dispute of 1913, while the strike in Lide's novel, although set in the 1930s, closely resembled the turbulence of an earlier period. One of Lide's characters derides his fellow workers as 'mule stubborn' and 'too stupid and proud to stick up for their rights', whereas both Thompson and Penhaligon introduced the theme of relatively content Cornishmen being incited to violence by union organisers from up country. When one of Lide's clay workers displays superior qualities, he turns out to be the secret offspring of a land-owning family, echoing a device, mentioned earlier, that was used by Joseph Hocking for one of his clay merchants.

In general this group of writers constructed an identity, both for merchants and men, that was much more forceful, albeit hardly more favourable, of ruthlessly competitive clay merchants and a strike-prone and unruly mob of workers. However, both sets of novelists had one thing in common. They failed to recognise the important improvements in extraction, refining and materials handling tha were transforming the clay industry. The reasons for this have become clear in previous chapters. Clay merchants kept their affairs a closely guarded secret because of public suspicions about the uses to which clay was put; the pace of change was evolutionary rather than revolutionary; and overlapping technology meant that small clay pits followed the time-honoured traditions that romantic novelists and sentimental artists loved to portray.

Clay merchants in reality

How do these literary interpretations of clay merchants compare with reality? In terms of social status, the nature of the clay industry allowed no chances of discovering rich lodes of ore that had propelled some metal mining adventurers in the past into higher levels of society. Nor were china clay producers able to acquire territory and rise into the ranks of the landed gentry. However, one clay merchant who broke through the property barrier was Frank Parkyn, the son of a prosperous local merchant. Parkyn acquired the status of country squire and lord of the manor near Lerryn, on the southern fringe of the clay country.[30] A long-standing habitué of the Café Royal in London, he enjoyed a reputation for high living.[31] Other prominent clay families lived in some style in substantial villas with a retinue of servants and a drive leading up to them, the hallmark of successful men. Elias Martyn had built a residence at Carthew big enough to justify two lodge-houses,[32] and 'Eddie's mansion', as claymen called the home of Edward Stocker, was another large edifice on the edge of St Austell town, from which he rode to church or to his works in his horse-drawn carriage. William Lovering of 'The Grove' at Charlestown, and the self-made Samuel Dyer drove around the country lanes in their expensive automobiles, causing a great deal of resentment among the marching columns of men during the 1913 clay strike. Some left substantial sums of money: John Lovering left £36,000 in 1900, William Martin £28,000 in 1905 and Thomas Martin £63,000 in 1913, which would make them millionaires at present-day values, at a time when an ordinary clay labourer counted himself lucky to earn more than £50 a year.

Politics and religion

The clay country was a Methodist-Liberal stronghold where religious and political differences were deeply felt, although they seldom led to violence or unrest except where religious education was concerned. Some clay merchants and most clay captains and pit managers attended the same Nonconformist chapels, preached as laymen to the congregation and supported the same sporting and social events as the clayworkers. Whether Non-conformist or Anglican, Radical or Tory, the clay leaders performed the customary duties of the well-to-do, patronising Sunday-school tea treats and the parish saints' days, laying foundation stones and contributing to the cost of chapels, churches and workers' institutes. John Gaved, a bible-class leader for 50 years, and John Lovering, both gave generously to the construction of a new Wesleyan Chapel at Trewoon, and John Varcoe of Hendra was a substantial supporter of the Methodist Free Church of St Dennis. The 'clay landlords' often donated the land for such buildings.[33]

As regular church- or chapel-goers, most clay producers made sure that all who worked for them attended religious services. John Lovering senr, while insisting that his workers went to church on Sundays, left them free to choose their place of worship. Some members of the Stocker family were members of St Austell Parish Church, but Thomas Stocker presided over a 'tea meeting' that followed the opening of Mevagissey's imposing Congregational Chapel, and Thomas Medland Stocker was a prominent Baptist.[34] Up on Dartmoor, Christopher Selleck, who developed the Wotter Pit, became the first leader of an inter-denominational organisation, the Pentecostal League of Prayer.[35]

The self-made clay producer Samuel Dyer, never one to mince his words, lectured his audiences as a Wesleyan lay preacher and Sunday-school superintendent on the evils of drinking and gambling, and annoyed some claymen by insisting that they and their families could manage very well on the wages they received if they did not fritter them away. Sometimes the claymen were able to turn the tension between the clay producers' religious beliefs and commercial attitudes to their own

Thriscutt & Bale entry in Kelly's Directory, 1902.

advantage. According to a popular tale in Bugle, the clay merchant Henry Adolphus Bale was a very pious man. So, when he refused to give his workers a rise in pay in line with that of some other employers, they gathered outside his house after evening chapel and bowed down before him, chanting, 'Oh Bael, hear us, oh Bael!' Reminded of his Christian duty, Bale gave them the pay rise.[36]

Of the clay leaders who played a prominent part in the Graves-Sawle celebrations described at the beginning of this chapter, William T. Lovering and John Wheeler Higman were pillars of the local Tory party, whereas Frederick Augustus Coon was one of the prime movers in the formation of the Cornwall Liberal Federation and a frequent speaker and writer on economic and social policies on the radical wing of that party.[37] Only on a few occasions, however, did religion and politics mix with unfortunate results. During the 1876 clay strike a stone was thrown through a church window where Thomas Stocker was preaching and, as we saw in Chapter Seventeen, in the 1913 strike Samuel Dyer caused an uproar when he used a foundation-stone laying ceremony for a Wesleyan church to denounce a trade-union leader.

The clay captain: betwixt and between

How did the china-clay captains, who usually managed the pits, hiring and firing men without much interference from the clay merchants, fit into the social hierarchy? According to A.L. Rowse, 'There was no ill-feeling, let alone class hostility, between the workers and captains, they were all still too close together socially for that; they came from the same people, they attended the same chapels, they shared the same tastes in football and brass bands – they were all Liberals! Many maintained their independence by having a second string to their bow, a smallholding, a shop.' No one, of course, was better placed to observe such nuances of social status than Rowse, whose father was a clay labourer, his mother a small shopkeeper, and certainly some clay captains enjoyed great respect among the workforce, like Tom Yelland, doyen of the West of England Co. With close upon 60 years of loyal service, he greatly impressed important visitors, who he was entrusted to show around the works, with his great knowledge of the industry and the keen interest he took in the welfare of juvenile workers.[38] However, the 1875–76 and 1913 strikes interrupted the idyllic relationship that Rowse portrayed. While the captains kept close to the men in social affairs, they were expected to side with the employers on economic matters, which they did.

For their loyalty the captains received little financial recompense. Samuel Dyer, admittedly a hard taskmaster, paid his own sons a mere £4.50 a month to manage his pits, on the grounds that only in the very largest works could a captain earn as much as £5.[39] This meant that they took home little more than the lowest paid labourer and less than a hard worker on piece rates. Their reward came mainly from the authority that they enjoyed. Their wives or widows were sometimes addressed with their husband's titles, such as 'Mrs Cap'n Tremayne'. According to the mother of the clayland author Ken Phillipps, 'This was Roche etiquette', on the grounds that 'many are called "Mrs" but few are chosen clay cap'ns wives'.[40] It is not surprising if, as R.S. Best put it, 'some of them were conscious of their own importance' and that they were sometimes accused of acting like 'little kings'. But having worked their way up, like Marshel Arthur, from a kettle boy at 6d. (2$^1/_2$p) a day, they saw the growth of union power as a threat to their status. One of them was said to have called out to a union official, 'If I had my way I would cut you down with a stick.'

Municipal service to the community

As we have seen in earlier chapters, some ball-clay dynasties were well established when the china-clay industry was still in its infancy, and members of ball-clay families were prominent in local affairs from the early 1800s. Successive generations of Nicholas Watts were Justices of the Peace in South Devon, and the Pikes successfully unseated a long-serving and influential Member of Parliament in Dorset, and later were prime movers in improving Wareham's water supply, employing their own equipment to bore test holes.

China-clay families, in addition to civic pride and a desire to serve their local community, had a particularly strong financial incentive to engage in local politics. Acts of Parliament created democratically elected local councils and empowered them to carry out reforms in education, health and local government, but imposed the financial burden upon the local ratepayer. Unlike mining adventurers, granite quarrymen and farmers, the clay merchants had to pay rates on the value of their output of clay, as well as upon their private property, and they naturally wished to influence decisions on how much money was raised and how it was spent.

It is not surprising, therefore, that the leading china-clay families had a finger in every municipal pie.[41] Already in the 1840s the St Austell Market House Commissioners included such claymen as William Browne, Elias Martyn, Edward Stocker, William Varcoe and members of the Quaker Veale family. When St Austell Local Board, the governing body of the town, was formed in 1865, its members included William Langdon Martin, Henry Wheeler Higman, Edward Stocker and Charles Truscott. A quarter of a century later, clay family names on the board were Henry Adolphus Bale, Frederick Augustus Coon, John William Higman and F.E. Stocker.

By this time, membership of leading local government institutions read like a roll call of the principal china-clay families. Control of elementary education was a bone of contention, setting Tories and Anglicans against Liberals and Nonconformists, and competition for places on school boards was intense. In the

1890s Coon, Bale and W.T. Lovering made up half of St Austell Board, with Lovering in the chair. When his father, John Lovering junr, died in 1901, he resigned because of pressure of work, but another John Lovering, his brother, took over the chairmanship. This John Lovering was made a magistrate in 1908, at the same time as T. Medland Stocker, who was only 32 years old, although Stocker played a less active role in municipal affairs, resigning from a position as governor of the new St Austell County School because of the demands of work.

When a new tier of local government was set up in 1894, including Urban and Rural District Councils, the clay families took a firm grip upon it. John William Higman served on the Rural District Council for two decades, and W.T. Lovering was another member, both of them occupying the chair from time to time. Henry Wheeler Higman, a solicitor who represented the clay companies in court cases, was not only a member of the Rural District Council but also clerk to the Urban District Council for many years, and F.A. Coon was chairman of the Urban District Council.

The right-wing 'Sammy' Dyer, a newcomer to the employers' ranks, was at first unsuccessful in elections to the Rural District Council, but was later elected and gave years of service to it, as well as to St Austell Parish Council. Another staunch Conservative, Thomas Martin, managing director of the Lee Moor works in Devon for many years, and chairman of Charlestown Foundry and Ironworks, was also a member of Plympton Rural District Council. As discussed earlier, it was he who called for magistrates to be dismissed because they refused to bring in troops to confront the strikers in 1876. It may be remembered that C.R.H. Selleck, mentioned earlier, fell out with this council by ruining some of their roads. Having made his peace with the council by transporting thousands of tons of stone to repair the roads, 'Cap'n Chrissy' was elected a member of the council, as were his two sons, John and Arthur.[42]

The St Austell Board of Guardians, which administered workhouses for the poor, was another institution in which clay interests were represented by Dyer, W.T. Lovering, J.W. Higman , John Varcoe and W.G. Truscott, and before his early death at the age of 39, G.E. Fortescue Martyn was a member of the board, as well as of St Stephens School Board and St Austell Rural District Council. In contrast, Alfred Luke, clay merchant, featured in the local press in a less favourable light when he was fined £5 for assaulting the St Austell stationmaster in a dispute about alleged non-payment of a ticket.

Big fish in a small pool
Although the china clay merchants held the whip hand in local affairs, they did not, as mentioned earlier, extend their influence outside the clay country as MPs, High Sheriffs, Deputy Lieutenants or County Councillors. In this they differed from the metal mining magnates of West Cornwall, and also from some ball-clay leaders of

South Devon. Two of the largest ball-clay firms at the time were Whiteway & Co. and Watts, Blake, Bearne. John Hayman Whiteway was chairman of Teignmouth harbour for many years, while William John Watts served as High Sheriff of Devon and was a partner in a banking firm that advanced money to ship owners.[43] However, these metal-mining and ball-clay dynasties had been established for a century or more, whereas the principal china-clay families had only been prominent for half a century or so, and this was perhaps one reason why they had not reached the higher ranks.

The St Austell constituency was represented in Parliament for over 20 years by a rich Australian, W.A. McArthur, a Wesleyan chairman of the *Methodist Times*, with many industrial and commercial interests, including directorships of the Bank of Australia, British Thomson-Houston, Gulfline and Union Marine Insurance. In the early 1900s J.S. Lovering was groomed as a Liberal Unionist and Tory candidate, but McArthur was succeeded in 1908 by the popular 'Tommy' Robartes (the Hon. T.C.R. Agar-Robartes), whose father, Lord Robartes of Lanhydrock, was formerly MP for East Cornwall and one of Cornwall's largest landowners. Tommy, said to be the most reckless horseman in Cornwall and the best-dressed man in the House of Commons, gained the respect of the clayworkers during the 1913 clay strike, as we have seen, when they lost confidence in such other Liberals as Hodge, sometime chairman of the Urban District Council, and Coon.

Out of 66 councillors and 22 aldermen on Cornwall's first County Council of 1889, the only clay connections were Sir C. Brune Graves-Sawle, Robert Varcoe and William King Baker. Graves-Sawle, a clay landowner, former MP, High Sheriff and Deputy Lieutenant for Cornwall, was elected County Alderman for the first term of the new council and was immediately chosen as chairman of influential county committees. Varcoe, on the other hand, was defeated when he stood for election as county councillor, but was chosen as an alderman by the elected councillors specifically to represent clay interests. As for Baker, far from being the typical Cornish clay merchant, he came from an agricultural background in Essex and had worked abroad as an engineer before settling near Penzance, where he

---

St. Austell School, Cornwall.

HEAD MASTER - F. R. RAY.

London University, Life Member of the College of Preceptors, formerly Cornwall County Councillor, Vice-Chairman of the St. Austell School Board, Cardinal Exhibitioner of Dedham School, &c., &c.

The object of the School is to give a good education at a very moderate cost, and to combine the individual care and attention of private teaching with the thoroughness and comprehensiveness of the education of a Public School.

The domestic arrangements are admirably calculated to supply the want so frequently felt by parents, of a school in which boarders enjoy the watchful care of home.

Part of an entry in Kelly's Directory for 1902. Ray was also editor of the St Austell Star *for many years.*

acquired a small clayworks and later engaged in farming.[44]

Neither Varcoe nor Baker was elected to chair key County Council committees' affairs although both protested, without success, against their exclusion from the Technical Instruction Committee that was responsible for training clayworkers. Nor was Varcoe re-elected as alderman when his term expired in 1892. The county councillor for St Austell town was Francis Barratt (later Sir Francis Layland Barratt, MP for St Austell), whose wealth derived from the great Hodbarrow iron mine in Cumbria.[45] F.R. Ray, county councillor for St Dennis, was appointed to the Technical Instruction Committee, but for his educational background, as the headmaster of a local private academy, rather than for his knowledge of the china-clay industry. Later, Charles Edwin Davis, managing director of Carpalla United China Clay Co., served as a county councillor and JP and was appointed by the County Council to the governing board of the Camborne School of Mines.[46]

China clay merchants stick to their core activities

As discussed in previous chapters, some ball-clay merchants integrated their extractive activity with other operations, a prime example being Wren of North Devon, who ran clay pits, brick and tile works, potteries and also a local private railway. In Devon, C.D. Blake had a finger in nearly every commercial pie. He partly financed and acted as sales agent for the Marland clayworks in North Devon, traded in chemicals and minerals such as granulite from Okehampton, and acquired china-clay works on Dartmoor and a pottery in Yorkshire. Clay merchants in Dorset constructed and ran their own light railways, piers and jetties, while those in Devon sailed their own barges downstream.

In contrast, few china-clay merchants dispersed their capital or their energies into other business fields, one exception being John Varcoe. Principal shareholder in Goonvean China Clay & Stone Co., he also had large holdings in the Consolidated Bank of Cornwall, the biggest Cornish-owned bank, as well as in South African mines. As already discussed, china-clay merchants, while they had originally diversified into clay production from other trades, did not move out of it into the manufacture of pottery, paper or other products that incorporated clay. In this they differed from some Devon ball-clay merchants who sold ochre and other commodities and set up or acquired potteries.

Tourism might have seemed an obvious investment opportunity because of the clay merchants' commercial background and their location near the Victorian watering places of North and South Devon or in Cornwall between the burgeoning resorts of Fowey to the south and Newquay to the north. However, the china-clay families' ventures into the holiday trade were both rare and unsuccessful. When the Cornish architect-developer Silvanus Trevail floated shares in the King Arthur's Castle Hotel Co. at Tintagel, only a few members of the Lovering family took a handful – a wise decision, since it proved a financial disaster.[47] A.E. Gaved, company secretary of Charlestown Foundry and secretary of the St Austell Mercantile Association, was a prime mover of the United Cornish Association, an abortive attempt to promote local tourism.[48] A member of the Lovering family participated in the Newquay Baths & Sanitary Steam Laundry, formed to help local tourist operators, but in the first three years it made a loss of over £500 on a capital of £3,000.[49]

An intriguing, if again financially unrewarding, diversification was that of Lionel Martin, son of Edward. After Eton and Oxford he became a shareholder in the family firm of Martin Bros in 1902, but his passion was for racing bicycles and tricycles and at one time he held the record for the fastest cycle run from London to Lands End, in a little over 22 hours. In 1913 he formed a partnership to modify sports cars and developed the Aston Martin, a name chosen to reflect his success in climbs on Aston Hill between Tring and Wendover. His involvement ended in 1926, after it had cost him over £25,000 (half a million at today's prices).[50]

Why did the china-clay merchants seem less willing or able to diversify into other fields than ball-clay leaders? One possible reason is that a much greater increase in volume of production fully occupied their minds. Whereas ball-clay output nearly doubled in half a century, china-clay production went up 12 times. Another explanation may be that the marketing of china clay required much more attention. While traditional uses in pottery remained the main outlet for ball clay, producers of china clay had to adapt to the differing needs of paper-makers, particularly those in America, as well as a continuing demand from cotton cloth manufacturers, potters and from such industries as alum and linoleum.

The achievements of the clay merchants

How can we sum up the attainments of the Victorian and Edwardian clay families? In the years leading up to the 1914 War, the china-clay merchants were supplying over 70 per cent of world output and completely dominating international trade in it, some of the largest clay-producing countries being among their best customers.

Yet we find little recognition of these great achievements in the works of the clayland literati and artists of the period who failed to acknowledge the impressive technological advances that were made, nor the commercial acumen involved in exporting products to all parts of the globe. By focusing upon small clayworks using archaic methods, they painted a misleading picture of entrepreneurial sluggishness in what was in reality a dynamic industry.

However, in fairness to these authors and painters, and to others who undervalued the achievements of the china clay merchants, we must recognise that people in Cornwall and parts of West Devon still lived under the shadow of the once-great metal-mining industry. A glance at any of the regional newspapers of the time will confirm this. They devoted many columns to metal-mining employment and share prices, contrasting gloomy news of failures at home with upbeat accounts of successes abroad. There seems to be a general lack of appreciation of the worth of non-metallic minerals in the West Country. As we saw in Chapter Eighteen, even when the clay men of South Devon and Mid Cornwall were in the throes of strike action, local newspapers outside the immediate area gave a far greater coverage to a violent dispute involving West Country exiles in South Africa than to events at home. Yet this reflected the interests of many of their readers, for it was often said that more men born in Cornwall and West Devon were working outside their homeland than in it.

Finally, we must give credit to those clay merchants who, while running an activity that was growing faster than any other in the South West, found time to serve the people of their area well as parish, urban and rural district councillors, chairmen of boards of guardians and school boards, and also as wardens and benefactors of churches and chapels. Some observers may have found them dull and conventional, but they were successful in a period when many adventurers were not.

The 1914 War brought the golden age of the independent family firms in the china-clay industry to an end. German U-boats played havoc with international trade and the labour force dispersed, some joining the armed forces, others working in munitions factories up country. Most china-clay family firms, with their china-stone interests, merged to form a giant corporation to face the Great Depression of the interwar years. The ball-clay industry, not so dependent upon foreign customers, was less affected by the maritime blockade, but nevertheless production fell and the workforce declined, although a reduction in domestic demand was partly moderated by increasing sales to potters who made crucibles for the expanding steel industry. Amalgamations among some of the larger family firms only occurred, however, at a later date.

As we have seen throughout this book, the history of the Extraordinary Earths up to the 1914 War has been distorted by folklorists, mining journalists, romantic novelists, sentimental artists and unsympathetic historians. As a result, an impression remains of lethargy, inertia and low labour skills. It is our hope that this book will help to redress this historical imbalance, and give the recognition so richly deserved to the generations of employers and workers who were engaged in the extraction, refining and distribution of china clay, ball clay, china stone and soapstone.

References
[1] *SAS* 21 October 1909.
[2] *JPMBT*, 1 October 1936.
[3] Ken Phillipps, *Catching Cornwall in Flight*, St Austell, 1994, p.49. The author was here referring to a later inter-war period.
[4] Alan M. Kent, *The Literature of Cornwall*, Bristol, 2000; *Pulp Methodism*, St Austell, 2002.

[5] Andrew C. Symons, 'A Study Guide to Ia by Arthur Quiller-Couch', *JRIC*, 2008, p.81.

[6] Sir Arthur Quiller-Couch, 'Step O'One Side', *Q's Mystery Stories*, London, 1937, pp.323–41.

[7] Kent, 2000, 2002.

[8] Jack Clemo, 'The Hocking Brothers', *Cornish Review*, 1969, quoted in Kent, 2002, p.201.

[9] Ronald Perry and Charles Thurlow, 'Literati and Claymasters', *JRIC*, 2006, pp.45–56.

[10] Eden Phillpotts, *Lying Prophets*, London, 1897.

[11] L.A.G. Strong, 'The Dartmoor Novels', in Waveny Girvan, *Eden Phillpotts*, London, 1953, pp.29–48; John Hurst, 'Eden Phillpotts' Cornish World', *JRIC*, 2005, pp.59–70.

[12] A.L. Rowse, 'The Curse upon the Clavertons', *Cornish Stories*, London, 1967.

[13] Eden Phillpotts, *Old Delabole*, London, 1915.

[14] Eden Phillpotts, *Brunel's Tower*, London, 1915; for a similar treatment of a breach of secrecy see his *Storm in a Teacup*, based on paper-making at Tuckenhay, Totnes. New York, 1919.

[15] Joseph Hocking, *What Shall it Profit a Man?* London, n.d.

[16] Sir Arthur Quiller-Couch, *The Astonishing History of Troy Town*, London, 1888, repr. Fowey, 1995.

[17] Joseph Hocking, n.d.

[18] Joseph Hocking, *Caleb's Conquest*, London, 1932.

[19] A.L. Rowse, *A Cornish Childhood*, London, 1944, repr. 1993.

[20] Michael Sanderson, *Education, Economic Change and Society in England*, Cambridge, 1995; David Edgerton, *Science, Technology and the British Industrial 'Decline'*, Cambridge, 1996.

[21] Rowse, 1944, p.123.

[22] Rowse, 1967, pp.96–116. On his background see Philip Payton, *A.L. Rowse and Cornwall*, Exeter, 2005; James Whetter, *Dr A.L. Rowse, Poet, Historian, Lover of Cornwall*, Gorran, 2003; Richard Ollard, *A Man of Contradictions: A Life of A.L. Rowse*, London, 1999.

[23] A.L. Rowse, 'The Squire of Reluggas', *Cornish Stories*, London, 1967.

[24] A.L. Rowse, 'The Recluse of Rescorla', *Cornish Stories*, London, 1967.

[25] A.L. Rowse, 'Peter, the White Cat of Trenarren', , London, 1974, p.22.

[26] A.L. Rowse, 'The Curse upon the Clavertons', *Cornish Stories*, London, 1967.

[27] A.L. Rowse, 'The Cornish China-Clay Industry', *History Today*, July 1967, p.486.

[28] Rowena Summers, *Killigrew Clay*, London, 1986; *Clay Country*, London, 1987; E.V. Thompson, *Ruddlemoor*, London, 1995.

[29] Philip Payton, *A.L. Rowse and Cornwall*, Exeter, 2005, pp.102, 116.

[30] Andrew Foot, *A History of St Veep Church*, St Veep, 1989.

[31] Derek Giles and Edward Best, 'Some less well-known clay firms: Parkyn and Peters', *CCHSN*, 11, 2005, pp.5–7.

[32] Close to Wheal Martyn Museum. The residence has been demolished but both lodges still stand.

[33] We are indebted to John Probert, Derek Giles and John Tonkin for some of this information. See also Roger F.S. Thorne, 'Chapels in Cornwall', in Sarah Foot (ed.), *Methodist Celebration*, Redruth, 1988, p.37.

[34] Richard Pearse, 'One of the Pioneers', *ECC Review*, Christmas 1957; R.S. Best, *Clay Country Remembered*, Redruth, 1986; *West Briton* 20 January 1883.

[35] A.D. Selleck, *The Selleck Family and the China Clay Industry*, 1970, p.5.

[36] We are indebted to John Tonkin for this story.

[37] Coon was also Secretary of the St Austell Technical Education Committee, lecturer on china-clay production and member of the Royal Cornwall Agricultural Association.

[38] John A. Service, 'Potters Clay and how it is obtained', *American Pottery Gazette*, repr. West of England China Stone & Clay Co., c.1910, pp.12–13.

[39] *SAS* 21 November 1900.

[40] Phillipps, 1994, pp.53–54.

[41] Apart from the references quoted, information in this section comes from Fortescue Hitchins and Samuel Drew, *The History of St Austell*, repr. St Austell, 1894; William White, *History, Gazetteer and Directory of the County of Devon*, Sheffield, 1878/9; *SAS* 1889 to 1915; *NE* 1913; *Kelly's Directories* of the period; *CCTR*, Vol I 1919/20; *BCW*, Vols III to XXII, 1814–1914.

[42] Selleck, 1970.

[43] H.J. Trump, *Westcountry Harbour*, Teignmouth, 1976, pp.55–56, 81.

[44] *WB* 20 June 1889, Supplement on Cornwall County Council.

[45] A. Harris, *Cumberland Iron*, Truro, 1970.

[46] He was a principal witness in the dispute between his company and the GWR, described in Chapter Two.

[47] Ronald Perry, 'Silvanus Trevail and the Development of Modern Tourism in Cornwall', *JRIC*, 1999, pp.33–43.

[48] *RCG* 20 May 1909.

[49] *SAS* 31 May 1895, 6 January 1898.

[50] Charles Thurlow, Lionel Martin, 'Founder of the Aston Martin Motor Company', *CCHSN*, 2007. He died in 1945 at the age of 67, following an accident on his tricycle.

Index